Military and Civilian Pyrotechnics

Dr. Herbert Ellern

Military and Civilian Pyrotechnics

by

DR. HERBERT ELLERN

formerly Staff Scientist
UMC Industries Inc.
St. Louis, Mo.

CHEMICAL PUBLISHING COMPANY INC.
New York 1968

Printed in the United States of America

Foreword and Acknowledgments

Modern Pyrotechnics (1961) was the first pyrotechnic primer and collection of up-to-date information in the English language—written mainly in order to help the newcomer to the field of military pyrotechnics. Judging from the many comments I have received, it served this purpose well.

This book retains the basic approach and framework of my first book. The subject matter has been greatly enlarged as indicated by the increase in the number of formulas from 91 to 201, the addition of numerous tables, and a near-doubling of the references—from 354 to about 700. The subjects treated in the first book are presented in greater detail and, of course, with addition of new information from the years 1961 to 1966. Some of the subjects that formerly were under security restrictions—now lifted or relaxed—could be added, though not all in full detail. A special effort was made to enlarge on the *commercial*, or as I would rather call it, *civilian*, aspects of modern pyrotechnics.

These concern the subject of matches, fireworks, the thermite process, and certain minor devices. An additional feature is a special chapter on spontaneous ignition of common materials and one on the dispersion of chemical agents, such as tear gas and insecticides. It has also been possible to add, with discretion, some of the matter on clandestine activities from and since World War II—a subject that has been part of my and my company's effort.

Ever since I became engaged in the match industry (1937) and concerned with the chemistry of matches—specifically the safety book match—I have considered this seemingly commonplace prime ignition source as the most fascinating phenomenon of pyrotechnics. And since this subject is not found in the generally accessible literature, except under the historical aspect, and the chemistry of matches is rather obscure even to chemists, the chapter on matches has been greatly enlarged. As to fireworks for pleasure, I became convinced that a book on pyrotechnics, without some details on this subject, might disappoint many readers. Since here my specific knowledge is extremely limited,

I had to ask for help. Chapter 17 on fireworks was written by the Reverend Ronald Lancaster, amateur pyrotechnician and consultant to a firework manufacturer, Pains-Wessex Ltd., Salisbury, England. Only a few minor alterations and interpolations were added by me. The formulas on the subject contributed by Mr. Lancaster should be of special interest since up-to-date formulas connected with commercial interests are generally hard or impossible to come by.

Otherwise, the profusion of facts and figures that I present in this book are a combination of experience, literature study, and advice from others. In World War II, my company, Universal Match Corporation of St. Louis, Missouri (now UMC Industries, Inc.), became engaged in such activities as comminution of magnesium and manufacture of colored flame signals and other military devices—all of which broadened my pyrochemical knowledge. Since the Korean War, both production and development work added greatly to the diversification of my personal experience when for many years I directed the research and development effort of the then newly-founded Armament Division. A branch of the latter is now the Unidynamics Division, Phoenix, Arizona, of which I was staff scientist and consultant until my retirement in the summer of 1967.

The reader will have little difficulty in discriminating between those subjects of the military pyrotechnic field, in which I speak from personal experience, and those where I had to rely all or in part on information by others. However, a book written by one person, in my opinion, permits a greatly preferable, comprehensive and cohesive approach, even though some unevenness must be tolerated. The other way is to present a subject as "written with a pair of scissors," as German publishers used to call it, i.e. as a collection of perhaps competent but more-or-less disjointed articles, often leaving other areas of the field unmentioned.

A number of people and organizations have helped my effort in a variety of ways. UMC Industries, Inc., with whom I was affiliated for over a quarter century, has supported the preparation of the manuscript to such a generous extent that it could be completed in little more than a year. During this period, the untiring effort of my secretary, Mrs. Ruth Levine, extended far beyond typing and retyping, and included most of the burden of checking on references, patent numbers, and the other tedious by-play of getting out the final typed copy. Mrs. Mae Chaudet, librarian of Unidynamics, St. Louis Division of UMC Industries, Inc., has been most helpful in procuring an unending stream of printed information.

The manuscript was read and scrutinized by my friend and former colleague, Mr. E.R. Lake, now of McDonnell Company, St. Louis, Mo., and by my son, Lt. James B. Ellern, USN, now (1966) USNR, and a graduate student in chemistry at the California Institute of Technology, Pasadena, California. These two helped me to eliminate mistakes, numerous faults of style, and—alas!—many Germanisms, especially long phrases. The reader will have to bear with those faults of "foreign accent" of the written variety that only a superior stylist can overcome when he acquires a new country and a new language at the midpoint of his life.

Several people in industry and government have been specifically helpful. Detailed information that was of value for incorporation in the text was received from the following, to whom my gratitude is herewith extended: J. H. Deppeler, Vice President and General Manager, Thermex Metallurgical, Inc.; B. E. Douda, Research Branch, NAD, Crane; Dr. R. Evans, Denver Research Institute; Ronald G. Hall, Technical Director, Brocks Fireworks, Ltd.; J. F. Sachse, Vice President, Metals Disintegrating Corp.; H. E. Smith, Product Director, Reynolds Metal Co.; Garry Weingarten, Feltman Research Laboratories, Picatinny Arsenal.

Even though I have now retired from industry, I will be happy to receive comments and criticisms at my home address, 25 Shireford Lane, Ferguson, Mo. 63135, in the hope that this will augment the dual purpose of the book—to give information and to stimulate future research and development.

November, 1967 Herbert Ellern

Table of Contents

PART VI—NOISE

PART VII—HEAT PRODUCTION PER SE

PART VIII—CHEMICAL PRODUCTION

PART IX—BASIC BEHAVIOR AND PROPERTIES OF MATERIALS

PART X—SPECIFIC MATERIALS

PART XI—A FORMULARY OF PYROTECHNICS

PART XII—THE LANGUAGE OF PYROTECHNICS

PART XIII—AFTERMATH; AFTERTHOUGHTS

List of Tables

List of Figures

Part I

General Outline

chapter 1

Definition and Scope

> As the Egyptian Phoenix rose resplendent in its red and gold plumage from the ashes of its past, so may a new development in the art rise from the ashes of the present antiquated methods of manufacture.
>
> Faber, 1919[1]

Pyrotechnics, the Fire Art—from the Greek words *pyr* (fire) and *techne* (an art)—is one of three closely related technologies: those of *explosives*, *propellants*, and *pyrotechnics* proper. These three not only have a common physicochemical root, but their functions and purposes overlap. In their typical manifestations, explosives perform at the highest speed of reaction, leaving gaseous products; propellants are gas-formers and of brisk reactivity, but slower than explosives; and pyrotechnic mixtures react mostly at visibly observable rates with formation of solid residues. Numerous exceptions to these definitions may be cited, in which cases the purpose of the action classifies the item.

Chemical reactions, in general, either require heat input throughout the whole process (*endothermic* reactions) or they give off heat (*exothermic* reactions). The heat released in the latter case may be insufficient to cause a self-sustaining reaction, or it may produce flame or glow throughout the substance or substances either at high reaction speed, or gradually, making it suitable for pyrotechnical purposes. Such an exothermic reaction may either be the result of interaction of two or more substances, or stem from the decomposition of a single compound. *Interaction* between solids is the rule in pyrotechnics, *decomposition* in explosives, while there are typical propellants of either kind. Pyrotechnics is nearly exclusively concerned with solid ingredients, while explosives and propellants may be solids or liquids.

A pyrotechnic process differs from ordinary combustion by not (or at least not predominantly and essentially) requiring the presence of the ambient air. Once it is triggered by a small external force, it may take its course in complete isolation from external chemical influences. As we come to descriptions of actual devices, it will become clearer how versatile and relatively simple most pyrotechnic devices are. Except at the point of initiation, they employ no moving parts or delicate components and perform equally well on land, on water (even immersed in it), in the air, and in outer space.

A pyrotechnic process is not normally intended to result in the production of useful chemicals. It is specifically devised for the utilization of the evolved heat, or—expressed in terms of thermodynamics—by the energy the chemical system loses. This energy becomes a useful agent in a number of ways: One is the creation of flame or glow itself from materials at ordinary temperature and with minimal mechanical effort. All primary ignition and initiation for pyrotechnic items, rockets, and explosives is based on this, and in civilian life the ordinary *match* is the most frequently used of all prime igniters. Next, pyrotechnic heat is brought to a main item either directly (*first fire*) or by a gradually advancing flame or glow front (*delay* fuse or delay column). If we concentrate such heat in devices of varying size or design, we can use it as either a destructive force (*incendiaries, document and hardware destructors*) or as a beneficial heat source (*heat cartridges*). Combined with a variety of volatile substances, the heat source becomes a smoke-producer or a dispersant for toxic substances, or the chemical reaction itself creates the smoky material and evaporates it (*pyrotechnical aerosolization*). If the reaction furnishes mainly gases, it can be useful as a source of *kinetic energy*.

All this will appear to the layman as rather far-fetched and subordinate to what he regards as pyrotechnics. *Light* and also *sound effects* are most intimately connected with the word pyrotechnics. *Fireworks*, high intensity *white lights*, and *colored signals* are the best-known and most conspicuous devices, based on the high energy concentration and the allied radiation phenomena in and near the visible part of the spectrum.

In some cases, the typical pyrotechnic system—prime ignition, self-contained reaction, relatively slow completion, "portable"—is able to furnish pure *products* such as oxygen gas, a metal, or a special alloy in pure state. Such systems will also be described. Not included are explosives and propellants, except in a few limited functions such as explosive *sound* producers, light *flashes* and *dust explosions; small-scale rocketry;* and propellants for confined *kinetic action*.

From the dictionary definition of being "the art and science of making fireworks," pyrotechnics has grown, so as to deserve a wider and more precise definition as follows:

> Pyrotechnics is the art and science of creating and utilizing the heat effects and products from exothermically reacting, predominantly solid mixtures or compounds when the reaction is, with some exceptions, nonexplosive and relatively slow, self-sustaining, and self-contained.

The major types of pyrotechnically useful pyrochemical reactions and devices are the following:

1. *Prime ignition sources*—actuated by friction, impact, electricity, mixing or contact with air; used in the form of matches; stab, percussion and electric primers; and self-igniting incendiaries.
2. *Light producers of relatively long duration*—in the form of flare candles, sparklers, stars, streamers, etc., both white and colored, also sometimes utilizing nonvisible radiation for illumination, fireworks, and signals, including infrared (IR) augmentation.
3. *Light producers of short duration and explosive action*—for creation of intense flashes, mainly as illumination for photographic purposes.
4. *Smoke formers by chemical reactions*—for signal and small-scale screening smokes.
5. *Gas-forming mixtures*—with the heat and gas aerosolizing dyestuffs, toxic chemicals, or nuclei for water vapor, and used as colored smoke, tear gas, or cloud seeding generators.
6. *Gas-forming mixtures and compounds*—as energy sources for actuating cutting devices or electric switches, opening and closing valves, and performing involved sequencing of mechanical operations culminaing in catapult initiation for the ejection of pilots from airplanes.
7. *Explosive sound producers*—as warning signals, for pleasure, and for gunfire simulation.
8. *Whistling sound producers*—for pleasure and in training devices.
9. *Heat-transferring powder trains, surface coatings, and discrete subassemblies*—used as fuses for fire transfer, delay timing, first fires, and rocket igniters.
10. *Heat sources of extreme heat concentration*—for welding or destruction of metallic items; also to furnish pure metals and alloys.
11. *Low or high-gassing compositions*—for ignition of structural timber and other combustibles as firestarters and incendiaries.
12. *Fully encapsulated low-gassing heat sources*—for melting, warming of food cans, and other uses of indirect heating.
13. *Gas formers*—for the production of useful chemical entities such as respirable oxygen.

The arrangement of the described phenomena and items in the

following chapters corresponds with minor exceptions to this enumeration. The reader may gauge from the foreword, as well as from the text itself, what is mainly taken from books, manuals, specifications, and reports by others, and what I have done or seen myself.

While following the general arrangement of *Modern Pyrotechnics*, I have abandoned the separation of each chapter into a main text appended by notes containing references and subordinate matter. It caused frequent duplications, though I thought it would help the beginner retain the main thread of narration without interruption by less important matter. As before, all the references are combined, consecutively numbered, in a single chapter as are all the formulas. The glossary, specifically adapted to the content of this book, should not only be helpful to the reader but is also intended for use by the baffled secretary who, no matter how good her general education is, must be stumped by the niceties of some differentiations (fuse, fuze, fuzee) as well as by many unusual words themselves.

chapter 2

The Literature of Pyrotechnics

Much has happened in the field of pyrotechny, as indeed, elsewhere, during the last twenty-five years. . .

Brock, 1949[2]

Generally available information of practical use to the modern formulator or designer of military and civilian pyrotechnics is not abundant. None of the few books written prior to World War II contains more than fragments of applicable information. Faber's *Military Pyrotechnics*[1] (1919) combines historical background, detailed descriptions of the manufacturing techniques of World War I devices and, surprisingly, extensive information on fireworks for pleasure. While its formulations—for military devices at least—are obsolete, some modern military items such as parachute flares released from airplanes and signal lights fired from rifles or pistols are described in their early modifications.

The first book that concerned the state of the art of contemporary military pyrotechnics and became known in the United States came out in 1954 in the Russian language. It was preceded by a first edition (1943) of which only a few copies seem to have reached this country. The 1954 edition of this book by A. A. Shidlovsky,[3] a professor at the University of Moscow, was translated into English in 1964 and this version is available through U. S. Government channels. A third edition, of which I received an autographed copy from the author soon after its publication in 1964, has also been translated into English as a government report.*

My own book, *Modern Pyrotechnics*,[4] of which this is a rewritten and expanded edition, was published in 1961. It filled a severely felt need for a commercially available introduction to the subject, with emphasis on, but not entirely restricted to, military requirements.

Neither Shidlovsky nor I had any idea of each other's existence when

*Specific information quoted from Shidlovsky in later chapters is taken from the third edition.

he wrote his second and I my first edition. Still, both books have a good deal in common. Both use a strictly empirical approach with theories and calculations being proffered in a subordinate manner. On the other hand, both books attempt to impart a subjective understanding of the principles of the underlying phenomena rather than offering mainly a profusion of formulas and detailed "how-to-do" facts. Reluctance to perhaps say too much is obvious in the Russian book, especially in the 1964 edition, where the author quotes many newly added formulas from my 1961 edition rather than reveal too much of his more up-to-date native information.

Shidlovsky's references show that a sizable pyrotechnical literature exists in Russia. Since this is as good as unknown in this country, a selection from the second edition going back no farther than 1936 is referenced 3a—3ee on the basis of interesting titles, though the books and articles themselves are unknown to me. Additional references, especially newer ones from the third edition, will be quoted in the appropriate context.

Pyrotechnic information also appears in *Aerospace Ordnance Handbook*[5] published in 1966.

The British work by Taylor, *Solid Propellent* and Exothermic Compositions*[6] offers a limited amount of pyrotechnical facts and data and contains a sizable amount of valuable information on related subjects such as gas-formers, propellants, and explosives.

The modern fireworks industry can be represented by three books. George Washington Weingart's *Pyrotechnics*[7] is a practitioner's manual full of useful details. It has recently been reprinted (1967). Some of the information on fireworks, both from Weingart and other sources, is found in Tenney Lombard Davis' *The Chemistry of Powder and Explosives*[8], which is somewhat more accessible and also, notwithstanding its age, a mine of useful information on explosives. Izzo's book,[9] written in Italian, is a rather condensed manual of fireworks, with emphasis on design of the more elaborate arrangements.

The fireworks industry is rather conservative and three works dating back to the last century have been cited to me by an expert in the field as still of some practical use. The student who wishes to delve into these sources of information must break the language barrier, since they are written in German,[10] French,[11] and Italian.[12]

A History of Fireworks by Alan St H. Brock[2] is an eminently readable

*Regarding spelling of this word and others, as well as definitions, see Glossary.

work full of interesting details, which also contains some military material, but essentially deals with historical development. For readers interested in the history of fireworks, this book contains about one hundred references, mostly of older works as far back as 1225. One of these, *De La Pirotechnia* by Vannoccio Biringuccio, can be warmly recommended even to those who have no specific interest in the history of chemistry or of pyrotechnics. It is available in an English translation [13] and gives information on the state of metallurgy at the time of the Renaissance, as well as on gunpowder, explosive cannon balls, incendiaries, and fireworks. It closes with a chapter entitled "Concerning the Fire that Consumes Without Leaving Ashes" (Love!). Somewhat in the same vein is the rather dry and scholarly *History of Greek Fire and Gunpowder* by Partington.[14] Once the acolyte of the "art" has become interested, he will want to extend his knowledge of the historical background and—if he is a chemist—of the history of chemistry and alchemy.

The series *Science in World War II* contains occasional references to pyrotechnics, especially in the volume Chemistry by R. Connor et al.,[15] which has some information on smokes, light producers, and explosives.

The German book, *Sprengstoffe und Zündmittel,* by Dr. C. Beyling and Dr. K. Drekopf[16] has a misleading title, since the "Zündmittel" (igniter materials) refers only to initiation of explosives rather than to ignition in general.

Pertinent information is sometimes discovered in unexpected places. Thus, a slender volume by Alfred Stettbacher[17] contains a detailed chapter on sabotage incendiarism and explosive activities, which is linked to pyrotechnics proper.

More recently, Louis Fieser has revealed some interesting details on the otherwise barely accessible field of incendiarism and demolition in clandestine operations during World War II. One wonders if the rather baffling title of his book, *The Scientific Method*[18], was chosen with tongue in cheek.

J. Bebie's *Manual of Explosives, Military Pyrotechnics and Chemical Warfare Agents*[19], notwithstanding its grand title, is actually only an extremely limited glossary of terms.

It is obvious that the manufacturers of the few utilitarian products of civilian pyrotechnics are as reluctant to broadcast their formulas and techniques as are the fireworks people, and information on such things as railroad torpedoes and fusees is scarce. The match industry is more open-minded as evidenced by the splendid series of articles by Crass[20,21] and articles on the subject in the encyclopedias, though reliable formulas are less prominent than mechanical techniques. What the reader will find in Chapter 12 of this book is undoubtedly the most extensive study on the pyrochemical aspects of the subject.

Since articles in the encyclopedias of general knowledge as well as in the technical encylopedias are written by specialists,* one may find authoritative information in those works, though an oblique approach may be needed in order to find some subjects. Thus, Ullmann's *Encyklopädie der technischen Chemie*[22] relates detailed information on the metallurgical and other commercial uses of the thermite reaction under Aluminothermie (Vol. 1) and treats the subject of fireworks rather extensively under Pyrotechnik (Vol. 14). This edition has not as yet (1966) reached its final volume, which will contain the subject of matches under Zündwaren.

The first edition of the *Encyclopedia of Chemical Technology*[23] contains an authoritative article on military pyrotechnics in Volume 11, written by the late Dr. David Hart (d. 1958) of Picatinny Arsenal. The second edition[24] of the same work is remarkably detailed on a subject that in part ties in with pyrotechnics, i.e., Chemical Warfare (Vol. 4).†

The encyclopedias, such as *Americana*,[25] *Britannica*,[26] *Collier's*,[27] *Chamber's*,[28] and the German *Brockhaus*[29] are generally strong on the subject of matches, especially the historical aspect, but sketchy on other branches of pyrotechnics. Nevertheless, they often make informative reading on side issues such as the history of the thermometer scales devised by Celsius, Réaumur, and Fahrenheit, the principles of optics, the phlogiston theory, etc.

The largest amount of pertinent and up-to-date information on pyrotechnics is found in reports on the work being performed in government arsenals and laboratories as well as by private contractors for branches of the Defense Department and other government agencies, such as the National Aeronautics & Space Agency (NASA). Most of these reports refer to a specific development, and rare are the investigations that busy themselves with a systematic, scientific (in the empirical

*It has been said that the last person who could cover the whole scientific knowledge of his time was the German naturalist Alexander Humboldt (1769-1859), as shown in his conception of the physical world in *Kosmos*, written at the midpoint of the last century (Britannica).

†I have completed the articles on Matches and on Pyrotechnics in forthcoming volumes.

and inductive sense) treatment. Unfortunately, some of the most informative reports carry the forbidding label "confidential" or "secret" —sometimes for specious reasons. However, awareness of the need for more generally available information, at least to the pyrotechnic fraternity concerned with military and space applications outside of federal institutions, is evident from those reports quoted in the later chapters, which are basic studies. Prominent are the Research Laboratories at Picatinny Arsenal (PA), Dover, N. J.; Harry Diamond Laboratory (HDL, formerly DOFL), Washington, D. C.; Frankford Arsenal (FA), Philadelphia; Naval Ammunition Depot (NAD), Crane, Indiana; Naval Weapons Laboratory (NWL), Dahlgren, Va., Naval Ordnance Laboratory (NOL), White Oak, Md., and several others.

Government reports generally available are listed in *Government Research Reports* and other lists of titles. Registered defense contractors can receive reports without charge through the Defense Documentation Center (formerly ASTIA) at Alexandria, Virginia. Abstracts of these reports appear in *Technical Abstract Bulletin* (TAB). Purchasable reports—some subject to special restrictions but all of them restricted to the qualified users—are also available through the Clearing House for Federal Scientific and Technical Information, Springfield, Virginia (formerly OTS). The more recent reports carry an "AD number" that identifies them, and this number has been added to cited reports where available.

Foreign material that has been translated into English (or French) is abstracted in the semimonthly *Technical Translations* (U. S. Department of Commerce), which gives instructions on how to order the translated reports.

A comprehensive manual[30] on which the three Services have labored mightily for many years is in the process of being published while this manuscript is being prepared. It should fill a long-felt need. Two accessory parts of it have already appeared.[31,32,32a]

Valuable, though often obsolescent, information is contained in the *Technical Manuals of the Army and the Air Force*[33-42a] and a number of similar publications by the Department of the Navy.[43-44] Essentially, these are catalogs of standardized and proven "hardware" such as parachute flares, smoke grenades, booby traps, simulators for training purposes, ignition devices, and many others. Formulas are rare, dimensions sketchy, performance data scant, but altogether there is much useful information for the pyrotechnician in all of the literature cited.

The same agencies have also issued more specialized and detailed books and manuals, such as the *Ordnance Explosive Train Designers' Handbook*[45] (now declassified) which, however, shows its age in this fast-moving era. Special manuals of gas cartridges[46,47,47a] of recent or recently revised origin exist. The most recent manuals describe items for unconventional warfare,[47b,47c] and one of them is quite detailed on the subject of incendiaries. Much of the same information is also available in loose-leaf form with frequent additions in a handbook issued by Frankford Arsenal.[47d]

A review of World War II foreign ordnance[48] adds to our knowledge of pyrotechnics in Germany at that time. The brilliant comprehensive *Encyclopedia of Explosives*[49] unfortunately has not advanced beyond Volume 3, after the retirement from government service of its guiding spirit, Dr. Basil T. Fedoroff. It is replete with pyrotechnically useful information as well as with data on those explosives that are linked with pyrotechnical devices, and its literature abstracts are prodigious.

A Navy report on *Toxic Hazards Associated with Pyrotechnic Items*[50] is incidentally a large formulary of light and smoke signals, though devoid of materials specifications and other details because of the special purpose of the publication. The hazards of processing and handling pyrotechnic items with special regard to unforeseen or unforeseeable flashing or explosive behavior and the safety precautions applicable to the laboratory or the manufacturing area are described in condensed form in the *General Safety Manual*[51] and in the *Military Pyrotechnics Series Handbook*,[31] in greater detail in a PA Handbook,[52] and in an NOL Manual.[53] Also of interest is the pamphlet *Ordnance Safety Precautions*,[54] an article "Safety in the Chemical Laboratory,"[55] and an article on high energy fuels.[56] Additional references will be found under the specific subjects.

The foundations of pyrotechnics are chemistry and physics. In elementary texts on physics, and even in works on physical chemistry, pertinent knowledge on solid reactions, heat and heat flow gets short shrift. Some books on optics, such as those by Sears[57] and Hardy and Perrin[58] go into the important subject (to us) of radiation, at least enough to furnish some fundamentals, but others confine themselves exclusively to the refractive aspects of optics.

The picture is much brighter as to chemical information, though the modern tendency to stress even on the elementary level highly generalized theoretical knowledge rather than the despised "masses of unconnected facts," makes many modern texts rather frustrating reading to the student of pyrochemical phenomena. Books on chemistry, with a predominantly empirical, factual approach, are by Ephraim[59] and Remy,[60] both in English translation. The author's "bible," however, which has guided him from his student days (second edition, 1919) to the present (eighteenth edition, 1965) is the text by Hofmann-Rüdorff,[61] *Anorganische Chemie*, but available only in German. The factual and theoretical approach has been happily combined in *Inorganic Chemistry* by Therald Moeller.[62]

Data of heats of formation of compounds for the purpose of determining the enthalpies (heat outputs) of pyrochemical reactions, together with other pertinent data (on phase changes, melting points, etc.), are found in *Circular* 500 *NBS*[63] and *Metallurgical Thermochemistry* by Kubaschewski and Evans.[64] *Lange's Handbook*[65] contains reprinted data from the former.*

The giants of chemical information, Gmelin,[66] Mellor,[67] J. Newton Friend,[68] and Beilstein[69] are indispensable to the unearthing of basic information, but caution is advised as to the validity of older "facts" sparse as they are on dry reactions, and more so of figures such as those on heats of formation. After all, these "handbooks" (nothing to hold in your hand—rather to be kept *at* hand for comprehensive reference, implied by the meaning of the original term *Handbuch!*) are simply compilations of data elsewhere reported.

Theoretical approaches to solid reactions are represented by two books: *Chemistry of the Solid State*[70] by W. E. Garner contains two chapters that touch on the subject of pyrotechnics. Chapter 9, The Kinetics of Exothermic Solid Reactions, presents data on substances that decompose exothermically. Chapter 12, Solid—Solid Reactions, by A. J. E. Welch, treats interaction of solids but without reference to pyrochemical reactions.

K. Hauffe's *Reaktionen in und an festen Stoffen*[71] (Reactions In and On Solid Substances) does not contain specific information on pyrotechnic reactions either, but it has considerable information on metathetical reactions of mixtures of solids, the mechanism of reactions of powder mixtures, and other subjects that are, in a wider sense, of interest to the pyrochemist.

In the course of this book, a large amount of additional literature will be cited when a particular book, pamphlet, article, or report is applicable to the specific subject in question. It should be noted that

*I found only one misprint therein (tenth edition): ΔHf^0 for sodium nitrate should be -111.54. This has been rectified in the *revised* 10th editiion.

the author has not always been able to study the original literature, even though its citations may give the impression. Also, when it is obvious that the original, e.g. a Japanese or Turkish reference, might be useful to only a few people in this country, I have let the citation in *Chemical Abstracts* stand in its stead.

Part II

Primary Flame and Glow

chapter 3

General Remarks

Pyrotechnic mixtures, propellants, and explosives need a small external energy source in order to bring them to the point of reactivity. The external energy is applied by impact on a small, sensitive explosive charge or by electrically heating such primary explosive; rarely is it applied by other means such as friction. The resulting "spit" of flame and of glowing particles is, in pyrotechnics, usually only the first link in a chain of ignition sequences leading eventually to the activation of the main charge in order to bring it to a self-sustaining reaction.

Modern everyday fire making by means of matches is based on the same kind of easily induced reaction, though its technique and requirements are different from primer initiation.

The whole intriguing field of primary ignition deals with a large array of diverse chemicals and arrangements. It comprises virtually all the elements; gases, liquids, and solids; inorganic and organic compounds; and numerous fuel-oxidizer mixtures. Some of these ignite entirely spontaneously in air or on contact with certain reagents. While these reactions are more difficult to use as prime ignition systems than primers, they have their place in some unusual designs or deserve discussion because of the hazards they present.

Spontaneous ignition in air is a special case of autoxidation—the process we observe in everyday life in the rusting of iron, the souring of certain foodstuffs, the decay of paper, or in the hardening of paint. The heat evolved in such oxidation is normally imperceptible because of the slowness of the reactions. In spontaneous ignition (also called self-ignition or pyrophoricity), the oxidation in air or the reaction between several substances becomes accelerated to the point where ignition temperature is reached, i.e. enough heat is created to bring adjoining layers of the substance to self-sustained exothermic reaction.

While a few substances are prone to such spontaneous reaction under all conditions (except perhaps at extremely low temperature), others must be converted into the pyrophoric state by increasing the surface area. These circumstances will be discussed in the following chapters. Materials will be treated in approximate order of declining intensity of pyrophoric behavior, followed by the practical aspects of use as primers and matches. The hazards of unwanted self-ignition have been regarded as sufficiently important to be included in a separate chapter within the subject of Primary Flame and Glow.

chapter 4

Self-Igniting Hydrides and Their Uses

Some gaseous or liquid compounds of nonmetals and hydrogen ignite on contact with air and burn with a luminous flame. . These are the *phosphines*, *silanes*, and *boranes*—the hydrides of the elements phosphorus, silicon, and boron, respectively. Of these, the luminously burning phosphines have practical importance because they can be generated by moistening of solid phosphides with plain water according to the (simplified) equations:

$$Ca_3P_2 + 6H_2O \longrightarrow 3Ca(OH)_2 + 2PH_3$$

$$Mg_3P_2 + 6H_2O \longrightarrow 3Mg(OH)_2 + 2PH_3$$

$$AlP + 3H_2O \longrightarrow Al(OH)_3 + PH_3$$

Actually, the chemistry of the phosphines is more complex. The first member of the series (PH_3) of b. p.—87.4°C is spontaneously flammable in air only when completely dry, but the moist gas can become self-ignitible under diminished air pressure. It shares this curious behavior with elemental white phosphorus, whose emission of cool light in prevented by the increase of oxygen pressure (e.g. in pure oxygen), and which is more readily spontaneously ignitible when access in air is diminished. These phenomena point toward a complex reaction mechanism involving intermediary oxidation products.

In practical use, the self-ignition agent is the compound P_2H_4 (diphosphine)—a liquid at room temperature. This compound is the major reaction product from water and technical calcium phosphide. The latter, made from lime and red phosphorus, also contains about 50% of calcium pyrophosphate. On the other hand, when magnesium chips, aluminum powder, and red phosphorus are ignited (a pyrochemical reaction in itself!), a product with the approximate formula $Mg_3P_2 \cdot 2AlP$ results. This product, however, furnishes with water a phosphine gas mixture with only 1% P_2H_4—unable to ignite spontaneously by itself.[72] It is therefore used in combination with 20% of technical calcium phosphide in emergency signals and spotting charges such as the *Drill Mine Signal Mk* 25.[50] A somewhat different scheme is employed in the *Marine Location Marker* (*night*) *Mk* 2[33] [formerly called Depth Charge Marker], in which a large amount of calcium carbide (CaC_2) produces, with infiltrating water, the illuminating gas acetylene (C_2H_2), and a smaller charge of calcium phosphide furnishes the self-ignition material. This signal, after an ignition delay from time of launching onto the surface of the water of 70—90 seconds (more in very cold weather), creates a yellowish flame about 9 inches high of 150 cp (candlepower) for about 50 minutes, visible from 4 to 10 miles, depending on the aspect. It should be noted here that such duration of pyrochemical action is never found in more conventional pyrotechnical devices.

The chemistry and behavior of the silanes is equally involved. According to the work by R. Schwarz, based on the studies of Stock and Somieski (1916),[73] only the lower gaseous members of the homologous series SiH_4, Si_2H_6, Si_3H_8, . . . are stable. They are formed from magnesium silicide (Mg_2Si) and hydrochloric acid in very complex reactions in which much hydrogen is formed.[74] Parallel reactions take place:

$$Mg_2Si + 4HCl + 2H_2O \longrightarrow SiO_2 + 4H_2 + 2MgCl_2$$

$$Mg_2Si + 4HCl \longrightarrow SiH_2 + H_2 + 2MgCl_2$$

Silene (SiH_2), an unsaturated compound, polymerizes and the lower polymers form with water the silanes and a prosiloxan (H_2SiO), which in turn decomposes in water, forming SiO_2 and H_2:

$$(SiH_2)_2 + H_2O \longrightarrow SiH_4 + H_2SiO$$

$$(SiH_2)_3 + H_2O \longrightarrow Si_2H_6 + H_2SiO$$

$$H_2SiO + H_2O \longrightarrow SiO_2 + 2H_2$$

The need for acid rather than water for silane formation from silicides is one of the reasons why these gases have not found practical application as illuminants. Another reason is that the flaming is accompanied by some explosive action. Chemists, who burn magnesium ribbon on ceramic dishes or heat the powder in glass tubes and subsequently attempt to clean the vessels with hydrochloric acid, experience the crackling noises accompanying the scintillating combustion of the silanes. The magnesium metal has in part reacted with the silicon (IV) oxide in the glass or ceramic according to the equation

$$SiO_2 + 4Mg \longrightarrow 2MgO + Mg_2Si$$

A third group of related hydrides, the *boranes*, will be mentioned only in passing. They are of interest for high-energy propellants and as starting materials for other high-energy fuels. All are endothermic compounds, their heat of combustion being the sum of the heats of combustion of the components augmented by the heat of formation of the original compound. The lower members of this group are *diborane* —a gas (B_2H_6); *tetraborane* (B_4H_{10}) of b. p. 17.6°; *pentaborane*-9 (B_5H_9); and *pentaborane*-11 (B_5H_{11})—low boiling liquids. Boranes, with the exception of the lowest member, diborane (B_2H_6), are found in the reaction products of hydrochloric acid and magnesium boride. The latter is made by reducing boron (III) oxide with an excess of magnesium powder:

$$B_2O_3 + 6Mg \longrightarrow 3MgO + Mg_3B_2$$

Out of a welter of contradictory statements about the pyrochemical behavior of the boranes, only two will be quoted: "Diborane and the two pentaboranes ignite in air; the other boranes, if they are pure, are not self-flammable"[61]; and from a description of *Borane Pilot Plants*[75]: "Solid boranes in presence of lower hydrides or in mixture are unpredictable and may ignite from moisture."

While the connection between the structure of the hydrides and their pyrophoric behavior seems to be obscure, it should be pointed out that the compounds of hydrogen fall into three distinctive groups: the above-described phosphines, silanes, and boranes, which have covalent bonds; the salt-like hydrides of the alkali and alkaline earth metals; and the interstitial, nonstoichiometric or "berthollide"—type hydrides of the transition metals—e.g. the rare earths, titanium, and zirconium. Pyrophoric compounds are found in all three groups.

While formerly only the phosphines and silanes were of any interest or even known at all for pyrophoric behavior, the search for new high-energy fuels has brought more and more of the salt-type hydrides, described below, into the laboratory and even made them commercially available.

Sodium hydride (NaH)[76] is stable in dry air, but in moist air, self-ignition may occur; when dispersed in oil, which is then removed by washing with solvents, the exposed hydride may be pyrophoric because of its small particle size.

Potassium hydride "catches fire in air."[61] No information on the pyrophoric behavior of the hydrides (or deuterides) of rubidium and cesium appears in otherwise rather detailed descriptions in Supplement III of Volume II of Mellor[67] or in Volume VI of the Sneed-Brasted series.[77]

Magnesium hydride, normally stable and commercially shipped without special precautions, exists in a pyrophoric form.[78]

Beryllium borohydride $Be(BH_4)_2$ (sublimes at 90°C) flames in air[61] and aluminum borohydride $Al(BH_4)_3$ (m.p. -64.5°C, b.p. +44.5°C) is spontaneously flammable in air, besides having an extraordinarily high heat of combustion of 989 kcal/mole,[60] or 14 kcal/g.

Uranium hydride (UH_3), a dark powder of d=10.92, but very low apparent density, "can ignite spontaneously in air."[79]

Cerium hydride (CeH_3) and the higher thorium hydride (ThH_4 or Th_4H_{15}) are extremely active chemically and ignite spontaneously in air. CeH_3 will also combine with nitrogen at room temperature but treatment with dried carbon dioxide will cause some surface passivation to make the compound more manageable.[79] The amount of hydrogen shown in these two formulas is, or may be, the maximum out of an unbroken series of compounds with varying percentages of hydrogen, depending on temperature and pressure at the time of synthesis. The formulas with whole figures are given for convenience and the same applies to other compounds described later, such as TiH_2 and ZrH_2, which in reality always contain less than two hydrogen atoms.

Finally, some not clearly classifiable, self-ignitible compounds[61] will be mentioned here: the reaction products of Si_2Cl_6 and ammonia—the compound $Si_2N_4H_6$ and silicocyan, $(Si_2N_2)_n$, (R. Schwarz); $(SiH_3)_3N$ (A. Stock, 1919); Siloxen ($Si_6O_3H_6$) (H. Kautsky, 1925); polysilenes $(SiH_2)_x$, (R. Schwarz, 1935), and the corresponding polygermenes $(GeH_2)_x$ (R. Schwarz and P. Royen, 1933).

Amberger and Boeters[80] have described the compound $P(SiH_3)_3$ as a colorless liquid that ignites spontaneously in air.

Stewart and Wilson,[81] besides giving a review of the silanes, report on one case of spontaneous ignition of digermane (Ge_2H_6), a liquid boiling at 29°C. L. Spialter and C. A. MacKenzie[82] describe a family of self-flammable chlorosilicon hydrides for the purpose of producing finely divided silica (SiO_2) and for skywriting.

chapter 5

Self-Flammable Metal-Organic Compounds

Compounds of metals and alkyl or aryl groups are frequently spontaneously ignitible in air, especially the lower members of homologous series, such as the methyl and ethyl compounds. The metals in the most active compounds are the alkali metals and aluminum, zinc, and arsenic. The nonmetals boron and phosphorus also furnish active members, which conventionally are classed with the metal-organic (sometimes called organo-metallic) compounds. It is noteworthy that the alkyls of silicon do not flame in air.

The first discovered members of this class were dimethyl zinc and diethyl zinc (Frankland, 1849). They are relatively easy and safe to handle—for example, blanketed by carbon dioxide gas with which they do not react, whereas, remarkably, beryllium and magnesium alkyls can react under glow with carbon dioxide gas. Many of these compounds will react explosively with water and also with carbon tetrachloride.[83] However, spontaneous ignitibility in air and stability toward various reagents may well occur with the same compound.

Borderline cases of spontaneous ignition are frequent and it should be kept in mind that probably no substance can unequivocally be called "spontaneously ignitible in air." Three parameters must be taken into account: temperature, air pressure, and dryness of the air. Even if spontaneous combustion occurs at extremely low ambient temperature, the reaction may be slow or uncertain enough to make it useless or impractical in a device such as a flame thrower. Experimental addition of the more active alkyls to liquid fuels and pairing the manageable ones with oxidizers as high-energy rocket fuels seem to be the only practical applications found for these metal alkyls.

Table 1 of the better known and characterized metal-organic compounds is based on information from major text books[61,68] and handbooks,[66,69] specialized works,[84] and some separate articles.[85,86]

By strict definition, alkyls of boron and phosphorus are not metal-organic compounds but are included because of the similarity of their properties.

Table 1
Spontaneously Ignitible Metal-Organic Compounds

	mp, °C	*bp, °C*	*Properties**
$Li \cdot CH_3$			white solid, red flame
$Li \cdot C_2H_5$	95		brillant red flame
$Li \cdot C_3H_7$(n-)			liquid!
$Na \cdot CH_3$, $Na \cdot C_2H_5$; $Na \cdot C_6H_5$; $Na \cdot CH_2 \cdot C_6H_5$			white powders; declining vigor with size of molecules
K, Rb, Cs alkyls			increasing flammability with atomic weight of metal
$Be(CH_3)_2$		subl. ca.200	white, volatile, solid, flames also in CO_2
$Be(C_2H_5)_2$	−13 to −11	93—95 at 4 mm Hg	liquid
$Mg(CH_3)_2$			higher alkyls more active, explosive with water, glows in CO_2
$Zn(CH_3)_2$	−40	46	blue flame, clouds of ZnO
$Zn(C_2H_5)_2$	−25	118	
$Zn(C_3H_7)_2$		135-7	flames only if large area is exposed
$Cd(CH_3)_2$			fumes strongly, ignites on paper
$B(CH_3)_3$	−161.5	−20	trimethyl borane
$B(C_2H_5)_3$	−92.9	95	green sooty flame, explosive in pure O_2
$Al(CH_3)_3$	15	122	
$Al(C_2H_5)_3$	−52.5	194	
$Al(C_4H_9)_3$ (iso)	1.0	113.8	fumes strongly or ignites
$Al(C_6H_5)_3$	196-200		decomposes in air, flames with water
$4\ Al(CH_3)_3 \cdot 3C_4H_{10}O$ (ethyl ether)			mostly flames in air
$Ga(CH_3)_3$	−15.8	55.7	trimethyl gallane, ignites at −76°C, explosive at room temp.
$Ga(C_2H_5)_3$	−82.3	142.6	ignites in air with purple flame and brown smoke
$Tl(CH_3)_3$	38.5		
$H_2P \cdot CH_3$		−14	methyl phosphine gas, fumes or flames
$HP(CH_3)_2$		25	colorless liquid
$P(CH_3)_3$	−85.3	37.8	fumes or flames
$P(C_2H_5)_3$		127.5	soaked in paper, explosive in air
$HAs(CH_3)_2$		35.6	dimethyl arsine, colorless liquid
$HAs(CH_3)(C_2H_5)$		71	
$(H_3C)_2As—As(CH_3)_2$	−6	165	"cacody," heavy oil
$(H_5C_2)_2As—As(C_2H_5)_2$		185-187	"ethyl cacodyl"
$(H_5C_6)_2As—As(C_6H_5)_2$			"phenyl cacodyl," white needles
$CH_2{=}CH{-}CH_2 \cdot As(CH_3)_2$		108-110	dimethyl allyl arsine, ignites in air on filter paper

$Sb(CH_3)_3$		80.6	trimethyl stibine, flames in larger quantities
$Sb(C_2H_5)_3$		158.5 [730 mm Hg]	
$Sb(C_3H_7)_3$	80-81		ignites or carbonizes on filter paper
$Bi(CH_3)_3$		110	explodes on heating in air
$Bi(C_2H_5)_3$		107 [79 mm Hg]	fumes or ignites in air
$Te(C_2H_5)_2$		138	red-yellow liquid

* All compounds are described in the literature as flaming readily in air, presumably at "room temperature"; only exceptions are noted.

chapter 6

White Phosphorus

Paracelsus (c. 1490—1541), physician and philosopher on the nature of man and the origin of sickness, seems to have been the first to discover "the element of fire" in the residue of the pyrogenic decomposition of urine. More than a century later (1669), the alchemist Brandt discovered or rediscovered the waxy material that miraculously glowed without heat in the dark and that caught fire on the slightest provocation. Brandt had "distilled" (i.e. decomposed by heat) urine in the absence of air in the course of an experiment that was to lead to the conversion of silver into gold.

After the accidental discovery of phosphorus, better methods of manufacture were found, but the material, at a price of 50 guineas per ounce, remained a mere curiosity until another hundred years had passed. Its use as an ignition material from then on will be traced in Chapter 12.

Pure phosphorus is a completely transparent colorless solid[61] of m.p. 44.1 °C, but the commercial product is waxy white and after short exposure to light, yellowish; hence both the names white and yellow phosphorus for the same material.

Its ignition temperature is variously given between 30 and 60°C, the higher figure being more realistic. The bulk material has the paradoxical property of tending to spontaneous ignition at 15.5°C if access of air is partially restricted by covering the surface with cotton, powdered resin, sulfur, animal charcoal, or lampblack.[67] Because of its ease of ignition, white phosphorus is kept and preferably handled under water. It may ignite otherwise when cut with a knife or rubbed against a wooden board. When scattered by the force of a burster charge, such as the *Igniter AN-M9*,[44] the substance will ignite easily in air and, in turn ignite various fillings of incendiary bombs. Similarly, it will ignite when dispersed from items such as the *WP Smoke Shell M302* for 60 mm mortars,[41] the *M105* for 155 mm howitzers,[41] the *WP Smoke Hand Grenade M15*,[38] or the *Navy Smoke Bomb AN-M47A4* with either

100 lb of white phosphorus (WP) or 74 lb of plasticized white phosphorus (PWP).[44] The tactical advantages and disadvantages of phosphorus as a smoke or incendiary material will be dealt with in the proper places. For the purpose of this chapter, it should only be pointed out that burning phosphorus after being doused with water will often reignite spontaneously after drying. This is also true of red phosphorus because during its burning some of it will volatilize and condense from the vapor in the white modification.

White phosphorus dissolves easily in carbon disulfide (CS_2). If such a solution is soaked into filter paper or cotton cloth and exposed to air, the highly volatile solvent evaporates readily and finely dispersed phosphorus residue bursts spontaneously into flame. Henley[87] describes a "magical act" of igniting paper while breathing on it, which is based on this phenomenon. Addition of gasoline or toluene to the phosphorus solution delays the spontaneous ignition because of slower evaporation rate of the added vehicles.[47c]

The spontaneous flaming of white phosphorus in air can also be made more reliable when a eutectic mixture of 55% of it with 45% of of phosphorus sesquisulfide (P_4S_3) is formed.[3] This liquid of d=1.84 and m.p.=40°C was used in World War II[15] in one-shot flame throwers designed to protect tanks against suicide attacks by foot soldiers (*E1 Antipersonnel Tank Protector*).

Another liquid eutectic (freezing point −7°C) of white phosphorus contains 25% sulfur[3], but nothing seems to have been reported on its ignition characteristics.

chapter 7

Miscellaneous Pyrophoric Substances

This chapter will deal with a few additional spontaneously igniting substances. They act in this manner in what might be called their natural state or, more often, in a special condition of dispersion or preparation.

The alkali metals form a series having increasing tendency to spontaneous flaming with increase in atomic weight: Lithium and sodium will not ignite in air at ambient temperatures and even up to their melting points; potassium may oxidize so rapidly that it melts and ignites, especially when pressure is applied.[67] The latter was the case when a 2-cm^3 piece exploded and caught fire after several cuts had been made with a stainless knife.[88] Rubidium will sometimes ignite in dry oxygen,[61,67] and cesium burns in air as it is removed from an inert covering oil.[67] The reactivity of the alkali metals is further increased when they are melted under an inert liquid hydrocarbon, such as xylene, of boiling point higher than the melting point of the metal, and the mixture is stirred vigorously.[89] This process converts the bulk material into small globules.

That moisture in the air or in the oxygen plays some part in these oxidations is evident according to Holt and Sims[89] from the fact that sodium or potassium can be distilled under oxygen that has been dried with phosphorus pentoxide. Moreover, the burning metals are said to have their flame extinguished when "immersed" in thoroughly dried oxygen.

A more practical way of extinguishing alkali metal fires is to deprive them of oxygen by covering them with dry alkali chloride, soda ash, or graphite (carbon dioxide, carbon tetrachloride, and even sodium bicarbonate are useless or dangerous).[90] Graphite takes a special place with molten potassium, since it is immediately wetted, swells up, and disintegrates the metal. If those mixtures are deprived of excess potassium by heating in vacuum, pyrophoric compounds are found to exist of unusual structure but of definable formulas, C_8K and $C_{16}K$.[91] Similar "insertion" compounds of rubidium and cesium, from metal vapor over graphite, have been described more recently in a French article.[92]

Hauffe[71] discusses the surface oxidation of alkali metal, quoting the theory of Pilling and Bedworth[93] that the smaller volume of the formed oxide is responsible for the progressivity of the reaction. On the other hand, where the oxide volume is greater as on some heavy metals, it forms a protective layer.

However, this theory does not explain the existence of pyrophoric iron or the phenomenon of pyrophoricity at all, as shown in Table 2 by a comparison of the ratio of volume of oxide to volume of element. Also, in analyzing the problem of autoxidation of the alkali metals, one must consider than quite different types of oxides result from burning of these metals in air, e.g. Li_2O, Na_2O_2, KO_2, RbO_2, and CsO_2.

Table 2

Ratio of Volume of Oxide to Volume of Element

(according to Pilling & Bedworth)[93]

Na	0.32	Mg	0.84	Cu	1.70
K	0.51	Al	1.28	Si	2.04
Li	0.60	Pb	1.31	Fe	2.06
Sr	0.69	Th	1.36	Mn	2.07
Ba	0.78	Zr	1.55	Co	2.10
Ca	0.78	Ni	1.68	Cr	3.92

Sodium and potassium form alloys, among them a eutectic of −12.3°C (9.95°F) freezing point with 77.2% potassium.[90] This liquid alloy, perhaps because of its quickly renewable surface area, is spontaneously ignitible, especially if dropped from a few feet onto a wooden surface.

"NaK" (pronounced *nack*), as the eutectic is called, as well as the alkali metals in general, have found new interest as heat dissipating and heat transfer agents in nuclear reactors and in motor valve stems of internal combustion engines. These uses have led to a special series of articles in the literature on the subject of handling these substances and their properties.[90,94,95,96]

The reactivity in air of the alkali metals is more or less preserved in their previously-described hydrides and also in a few little known compounds such as their silicides (W. Klemm 1948).[61] Brauer[97] describes the preparation of these compounds and characterizes them as follows: NaSi, self-flammable as loose powder; KSi puffs and flames in air; with water, all four (i.e. the silicides of Na, K, Rb, Cs) ignite explosively.

We turn now to the elements and compounds that in bulk, or in a more or less coarsely powdered state, are stable or at least are not spontaneously ignitible. In order to convert them into the pyrophoric form, a variety of procedures are employed, all tending to increase the surface area of the substance.

When the azides of strontium or barium are heated in vacuum to 140° and 160°C respectively, nitrogen is released and the alkaline earth metals are formed. These very finely powdered particles of both strontium and barium will burst into flame on admission of air.

This formation of spontaneously ignitible strontium and barium has been reported by Tiede.[98] He also mentions the decomposition of calcium and lithium azide under the same conditions, but remarks only that lithium azide is rather explosive on heating.

The following are a number of more generally applicable methods by which self-ignitible heavy metals and related substances can be produced:

1. Pyrogenic decomposition at "low" temperatures of organic salts such as tartrates. This process yields self-ignitible metal powders of small particle sizes.
2. Reduction at relatively low temperatures of finely powdered oxides by hydrogen.
3. Transformation of an alloy into a "sponge," or simply the alloying process itself, even if the other metal is inactive as in the case of some amalgams.
4. Electrolysis of solutions of certain metal salts, using a mercury cathode. The resulting amalgams or mixtures will yield pyrophoric metals after removal of the mercury by distillation.

Processes 1, 2, and 4 produce metal powders of very small particle size, but the reason for the pyrophoric behavior is not yet fully understood in every instance. Some experiments seem to indicate that adsorbed hydrogen is the cause. In other cases, it is obvious that the metal of greatest purity is most ignitible and that the presence of hydrogen reduces activity. These differences are, however, not irreconcilable since they may be caused by the different ways in which the hydrogen atoms are bonded to the reacting metals.

Both Gmelin[66] and Mellor[67] quote a variety of conditions under which metals are produced as pyrophoric substances. An important observation in cited references is that mixtures of iron oxide with aluminum oxide will tolerate much higher temperatures (red heat) and still yield pyrophoric iron on reduction by hydrogen.

Increase of surface area by mere mechanical comminution may sometimes lead to self-flammable products. Such is the case in the ball-milling of certain stainless steels and of "misch metal" (cerium alloy) under a hydrocarbon. The process leads to formation of minute flakes with the character of bronzing powders. Spontaneous ignition after removal of the liquid in the case of the stainless steels is due to the absence of a passivating protective layer upon the newly created surfaces.

The best known pyrophoric metals (using the term in the sense of spontaneous ignitibility only) are lead, iron, nickel, and cobalt.

Self-flammable lead powder can be made by heating certain organic salts alone if care is taken to exclude air. Michael Faraday (1791—1867), in a delightful little volume of lectures,[99] describes the "lead pyrophorus" made by heating dry tartrate of lead in a glass tube closed at one end and drawn to a fine point at the other and then sealed. On breaking the point and shaking out the powder, he demonstrated to his "juvenile auditory" how the finely divided lead burned with a red flash.

Beside the tartrate, other organic salts—notably the oxalate and the formate—yield the more easily produced pyrophoric metals, i.e. lead, nickel, and cobalt, if the pyrogenic decomposition is performed at "low" temperature. These organic salts or the oxides of nickel, cobalt, and iron, under hydrogen at about 300°C, also furnish the self-flammable metals, but in the case of iron, a pyrophoric ferrous oxide (FeO) may first appear by reduction with hydrogen as well as by reduction under carbon monoxide.

The reactivity of nickel is increased by converting it into a "sponge," also called "Raney" nickel, which is used as a catalyst in the hydrogenation of organic compounds. It is made by alloying nickel with aluminum in equal parts and removing the aluminum by dissolving it with an alkali. The residual spongy nickel must be kept under water or oil since it bursts into flame when dry.[60] As a consequence of its manufacture from the alloy with aluminum, this pyrophoric metal always contains some aluminum and hydrogen. On removal of the latter by prolonged aeration of the water slurry, the material loses its pyrophoricity.[100]

When nickel is deposited electrolytically on a mercury cathode, it appears to form an intermetallic compound containing about 24% nickel, but when the mercury is distilled off, finely divided, strongly pyrophoric nickel is formed. Iron and cobalt are deposited by electrolytic process from their salt solutions as very fine suspensions on mercury cathodes. The iron and cobalt so deposited possess pyrophoric properties.[101]

Chromium, molybdenum, and uranium have also been described as pyrophoric, and Remy[60] mentions that chromiun, obtained in a finely divided state from the amalgam by distilling off the mercury, is pyrophoric in air. A recent (1964) Russian article [102] refers to pyrophoric manganese and manganese nitride, made from manganese amalgam. The metal is obtained after removal of the mercury through vacuum distillation, the nitride when the removal is accomplished under nitrogen.

Evidence for the high reactivity of uranium and some uranium compounds is found in the report by Hartman, Nagy and Jacobson[103] that uranium metal in thin layers ignites at room temperature within a few minutes after exposure, as do uranium hydride (UH_3) and thorium hydride (ThH_2).

Katz and Rabinowitch[104] call the powdered uranium "usually pyrophoric," UO_2 (from uranium and water) "pyrophoric," and UH_3 "often pyrophoric and ignitible on moistening."

Recently, processes have been invented to produce metal powders in "ultrafine" condition, i.e. below 0.1μ particle size. The fuel-type metals such as aluminum, of an average size of 0.03 μ, are pyrophoric and must be kept under a hydrocarbon.[105,106,107]

Cerium, thorium, and zirconium can be made truly pyrophoric by alloying. Cerium amalgams, the alloys of cerium and mercury, become solids if the cerium content exceeds 2%. At 8—10% cerium content, these amalgams are spontaneously flammable in air.[22,67] Thorium and zirconium, when alloyed with silver, copper, or gold, become much more ignitible, bursting into flame on rubbing. Thorium-silver alloys in approximately a 1:1 proportion are actually self-igniting. These alloys of 40—60% thorium with silver, made by a sintering process of powders, are described by Raub and Engels.[108] They also describe the alloys of thorium with copper or with gold as "less pyrophoric." Alloys of zirconium with silver, copper, or gold require a rubbing action in order to exhibit self-ignition, and the alloys of titanium with silver, copper, or gold are stable.

Brief mention will be given to some self-flammable lower oxides of active metals. This curious fact—that even lower oxides of elements can exhibit a strong enough tendency to further oxidation so as to become spontaneously flammable—has been mentioned by Ephraim[59] for indium oxide (InO) made by hydrogen reduction; chromous oxide (CrO) made by oxidation of the amalgam of the metal; and vanadium sesquioxide (V_2O_3). Ferrous oxide (FeO) has been mentioned above.

Fricke and Rihl[109] have described pyrophoric, pure white ferrous hydroxide, $Fe(OH)_2$, forming Fe_2O_3 when it had free access to air, but Fe_3O_4 on slow oxidation. The ferrous hydroxide is not stable even in absence of oxygen.

A pyrophoric iron sulfide made from hydrated iron oxide and hydrogen sulfide under gasoline has been described recently.[110]

The easy flammability of some carbonized organic matter (tinder) is well known and can be increased when organic substances are heated with various additives in the absence of air. Such pyrophoric or semi-pyrophoric carbon had some practical uses in the past. Portable, self-flammable carbon-containing mixtures in sealed glass tubes are mentioned in the literature as late as the second half of the last century under the name of "the pyrophorus," a word that today is as good as forgotten.

Crass,[20] in his description of the precursors of the modern match, goes into some detail on these mixtures. They included "Homburg's pyrophorus," a roasted mixture of alum, flour, and sugar, hermetically sealed, which was essentially an activated charcoal; and "Hare's pyrophorus," similarly made from Prussian blue heated to redness and immediately sealed. In 1865 Roesling patented a similar substance that, "packed in air-tight vessels," was to be sprinkled on the bowl of a pipe and brought to red heat by suction, thus lighting the pipe.

Up to this point, all pyrophoric behavior has been reported as taking place under ambient conditions, meaning the normal atmospheric pressure. When circumstances are such that a metal is exposed to an active gas such as oxygen under greatly increased pressure, some metals may ignite at room temperature even in massive state. This has been reported for "cleaned" zirconium at oxygen pressures of 300 $lb/in.^2$ or higher and for Zircaloy-2 under certain conditions.[111]

chapter 8

Mechanically-Induced Pyrophoricity

If ambiguity is to be avoided, the term "pyrophoric" should be used only when ignition occurs completely spontaneously on exposure to air. However, we have seen in the previous chapters that sometimes a modicum of energy input makes the difference between mere rapid oxidation and flaming or glowing. This energy may do no more than renew the surface of a liquid or it may simply bring fresh, nonvitiated (i.e. of normal oxygen content) air to solid surfaces, intensifying the rate of reaction in both cases.

When we read however, that an alloy of copper, zinc, and barium metal is described as "pyrophoric, stable in air"[112] we realize that a second meaning, that of "producing sparks when ground or abraded,"[113] is attached to the word pyrophoric, causing a need for clarification where the word pyrophoric is used without further amplification. Metals and metal alloys of the spark-producing category—borderline cases both of pyrophoric behavior and of pyrotechnic interest—are the subject of this chapter.

The ancient method of striking together flint and steel or flint and pyrites not only requires considerable effort, but it produces relatively "cool" sparks that ignite tinder or black powder but not liquids such as alcohol, benzene, or gasoline. These liquid fuels are easily ignited, however, by the very hot sparks from alloys of cerium.

Cerium, a rare-earth metal of low ignition temperature (160°C), is a prodigious producer of white-hot sparks when scratched, abraded, or struck with a hammer. Most of its alloys share this property. In practice, the basis for spark-producing "flints" as used in pocket lighters is a mixture of about 50% cerium with 25% lanthanum, 16% neodymium, and smaller amounts of other rare earth metals. It is called misch metal (from the German *Mischmetall* meaning metal mixture); its ignition temperature has been given as 150°C. The complex composition derives from the nature of its raw material, the monazite sand.

Lighter flint alloys contain additions of iron (up to 30%) for greater chemical and physical resistance. The properties of cerium and its alloys are described in some detail in the major inorganic encyclopedias and textbooks.

Misch metal, ground under xylene in small ball mills and incorporated into bridgewire sensitive ignition mixtures and also in some delay formulas, was still in use in Germany during World War II.[48] It seems to be quite unsuitable for items that have to be stored for a considerable time.

Zirconium/lead alloys in a wide range of proportions spark or catch fire on impact and have been proposed for use as impact igniters for incendiary bullets, or as tracers that mark the impact of a projectile by the light effect.[114] Similar alloys mixed with organic fuels are claimed in a U. S. patent as self-igniting incendiaries.[115]

Alloys of either cerium or zirconium with soft metals such as lead or tin have been described as prodigious spark producers.[67,116] Very hot sparks are also obtained by abrading uranium metal.[67]

As a final contribution to the use of sparking metal, it should be mentioned that prior to the invention of Davy's safety lamp in 1815, devices seem to have been used that continuously created sparks to provide some illumination in mines. Mencken[117] quotes: "Life is as the current spark on the miner's wheel of flint; while it spinneth there is light; stop it, all is darkness." However, the only reference this writer could find on the subject, and a meager one, was by Beyling-Drekopf,[16] where the devices are called *Lichtmühlen* ("light-mills"). May one speculate that it was recognized that such sparks were safer than rushlights, candles, or oil lamps because they might have prevented the explosions caused by firedamp? Biringuccio[18] speaks only of oils and resins as illuminants in mines. His famous contemporary, Agricola, did not mention the subject at all in his *De Re Metallica*.[117a]

chapter 9

Spontaneous Ignition of Common Materials

Fires and explosions in factories and shops, and more rarely in homes, may be caused by spontaneous ignition. It is important to know the circumstances under which such events may take place so that safety precautions can be taken. This knowledge will also help in differentiating between an intentionally caused fire or explosion and true accidents. As we shall see, it is also possible to engender a seemingly accidental "spontaneous" fire and, not quite as easily, an "explosion without an explosive."

The oxidation of unsaturated oils such as linseed oil in ambient air is a strongly exothermic (heat evolving) reaction. Rags used in "washing up" of printing presses or in cleaning painting utensils are sometimes left in heaps or bundles under conditions where air has access but cannot circulate freely enough to carry away the heat of autoxidation. Such oil-soaked rags may get hotter and hotter in a few hours, begin to smolder, and when a temperature of about 350°C is reached (or less when other easily ignitible materials are present) open flaming may occur. Commercial oils can be roughly divided as to potential hazard on the basis of the degree of chemical unsaturation in the molecule. A measure of this property is the so-called *iodine number*. A high iodine number of 170 to 190 means high unsaturation and is found in tuna, linseed, Chinawood, perilla, menhaden, bodied linseed, sunflower, corn, herring, salmon, codliver, and walnut oils. Intermediate or low are sardine, olive, peanut, and castor oils; lowest are palm, cottonseed, and cocoanut oils. The hazardousness extends also to substances naturally impregnated with these oils such as fish meal, fish scrap, or peanut meal.[118,119]

A person understanding the principles of this mechanism can deliberately set a fire that not only leaves no incriminating evidence but also has a delayed action feature that can be determined experimentally beforehand with some degree of accuracy and that depends on the ambient temperature; the type, amount, and dispersion of the oil; contact with other combustibles; and the control of air access in a specific arrangement. Such systems are depicted in a recent technical manual.[47c]

Morris and Headlee[120] have described a laboratory experiment that is a model of the above-described phenomenon. The relation of the diameter of spherical piles of various materials to their self-ignition tendencies has been studied.[121] Other authors[122,123] have discussed similar situations of spontaneous combustion of oiled textiles such as freshly made oil cloth.

Linoleic and linolenic acid are the oxygen acceptors in the linseed oil. The heat of oxidation of linolenic acid was given to the author[124] as 172 cal/g, which ties in with the figures of about 53 kcal/g mole for ethyl linoleate or linolenate of 308.51 and 306.49 m. w.[125] The heat of polymerization of linseed oil has been reported as 230 cal/g.[126]

When certain fabricated organic products are stored while still hot from the manufacturing process and without a chance for rapid dissipation of the heat, destructive exothermic decomposition reactions that are self-accelerating may occur. This has been observed for piles of fiber-boards.[127,128] K. N. Smith[129] has studied this self-heating that starts at 325°F and is said to be proportional to the percentage of material extractable by a mixture of alcohol and benzene. This is mainly the natural wood resin in the cellulosic fibers.

When wet hay is piled up, bacterial action takes place with self-heating when a high relative humidity is maintained (95—97%). These biological processes may raise the temperature to 70°C, at which time the bacteria become the victims of their own metabolism and further exothermic bacterial action is stopped. However, chemical reaction may take over from here on and eventually lead to ignition.[130]

Another hardly ever considered source of heat, under circumstances similar to the above-described ones, is hygroscopicity, Cellulosic fibers such as paper or cotton normally contain about 5—12% water in equilibrium with the humidity in the surrounding atmosphere. When such moisture is removed by heating, the fibers absorb water from the air as vigorously as the best chemical drying agents; this absorption is a strongly exothermic process. In the paper industry, it may lead to charring, and since some papers tend to be overdried when leaving the paper machine, the precaution is taken to retain or readmit at least 5% water before winding.[131]

A few years ago, in the suburban community of Ferguson in St. Louis County, Missouri, where the author lives, the fire department was called to a home with smoke pouring out. A basket filled with warm laundry just removed from an automatic clothes dryer had been set on a bed and had started to smolder. On emptying the basket, its contents burst into flame. It was impossible to determine what specifically caused the accumulation and production of additional heat, but similar cases have been reported elsewhere, showing how close to home spontaneous exothermic reactions may occur.

In addition to these ignition reactions in air, there are occasional hazards from fuel/oxidizer contacts that cause fires and explosions. By classification, they would belong in the next chapter, but as hazards in normal civilian life, their description follows here.

The increased use of calcium hypochlorite bleaching powder, $Ca(OCl)_2$, for disinfection of swimming pools has led to unexpected accidents. While ordinary solid bleaching powder consisting of chlorinated lime, CaCl(OCl), with half the oxidizing power can cause spontaneous flaming, the hazards of the newer hypochlorite are much more severe. A mixture of calcium hypochlorite and an organic fungicide has caused a destructive fire and a trailer containing both soap powder and the bleaching powder in bags has been reported to have exploded.[132]

Not so very long ago, the slaking of burnt lime was an impressive little spectacle wherever the old-fashioned building mortar was prepared from hydrated lime and sand. It was known in antiquity that a moistened mixture of quicklime and an easily flammable material such as naphtha or sulfur created enough heat to ignite the fuel-type material. Partington[14] quotes various reports of this process being used or suspected of being used either for incendiary purposes or for pretended magic with religious overtones. We shall meet the heat-producing properties of lime once more in Chapter 26.

Undoubtedly, the intrusion of more and more highly-reactive chemicals into everyday life will further increase the danger of fires and explosions of seemingly obscure origin.

A quite different hazard from spontaneous ignition that is of great concern where finely powdered combustible substances are produced or handled is the destructively explosive combustion of dusts. It concerns the pyrotechnics manufacturer because of his handling of finely powdered metals of great susceptibility to accidental ignition.

In the realm of common dust explosions, we move along the farther borderline of pyrotechnics, but a few remarks are not amiss because the phenomena, except for their industrial safety aspects, fall into a no-man's land. They are not treated in the literature of explosives and do not exactly belong in pyrotechnics.

Dust explosions resemble gas explosions because of the enormo force that both exert on structural parts. Roofs are blown into the a and building walls burst apart. In gas explosions, we have a gaseo hydrocarbon mixed with air and the progression of combustion is th able to proceed with the speed of an explosive deflagration. In du explosions, the conditions are quite similar, each fuel particle being surrounded by air. It is in the nature of such explosions that they exert a considerable pressure for a longer period (long by the standard of explosive action, i.e. of milliseconds) on large surface areas, which explains the force and extent of their destructiveness. In contradistinction, a properly set off high explosive such as dynamite sticks or blocks of plastic explosives detonate with a much more shattering force, but their effects are more local as long as relatively small amounts are involved. Thus, the unfortunate acts of terrorism in civilian life, performed mostly with stolen dynamite, generally do little damage, whereas a gas leak followed by lighting of a match may completely erase a building.

The term "static electricity" is understood to mean a high-voltage low-current electrical discharge in the form of often visible sparks caused by contact and separation of dissimilar nonconductors and also of insulated conductors of electricity. It is nowadays in constant evidence in connection with synthetic "plastic" films and fibers and also with wool and silk used in clothes, carpets, upholstery materials, etc. In the case of dusts, static electricity is created by the movement of the particles, and certain sensitive metal powders can ignite by being poured from a sheet of paper or plastic film. Low relative humidity favors buildup of static electricity.

Normally, the cause of a dust explosion will remain unknown, especially in a factory where sparking electrical contacts, hot surfaces from steam lines, clandestine smoking, and other ignition sources may have existed but no proof or evidence could be found under the rubble.

The *National Fire Protection Handbook*[133] furnishes interesting statistics on the extent and variety of dust explosions in the United States throughout the first fifty nine years of the century. Of 1110 major accidents of this type, the most damaging and most injurious to life have been those affecting grain elevators and processing plants for starches, flour, wood, feed, and cereals. Metals are frequently involved, coal dust much less than might be thought. Cork, sugar, sulfur, plastics, bark, cotton, coffee and spices—all have caused loss of life and property through dust explosions.

Fieldner and Rice[134] and Hartmann and Greenwald[135] have rated numerous fine metal powders and commercial products as to susceptibility to accidental ignition and explosion. Kalish[136] has brought the hazard into a somewhat different focus by relating it to the settling rate of metal powders in air: Particles of 100 μ size settle at about 60 ft/min; of 50 μ at 10 ft/min; and of 10 μ at ½ ft/min. Because of nondirectional turbulent flow of air in a room of about 25 ft/min, fine metal powders will disperse readily and float for prolonged times if given an opportunity to escape from their container during scooping, etc.[137,138]

A recent U. S. Bureau of Mines report also rates the metal powders, with the rather surprising result that aluminum dust is classed with seemingly more hazardous metal powders such as magnesium and thorium in the most susceptible group.[139] Another report discusses the prevention of such ignition by the use of powdered additives with inerting properties.[140]

Accidents involving flammable dusts occur frequently in two stages —an initial "blow" that in turn disperses more material, which then may also explode. Systems have been devised that claim to be able to sense within ten milliseconds the heat and pressure from a dust or gas explosion and counteract it by discharge of water or inert gas. Thus, an incipient accident can be effectively suppressed.[141]

A relatively small explosion with high explosives can be arranged in such a manner that it provokes a sizable dust explosion. This was done during World War II in semi-military sabotage activities. Optimum conditions for dispersion and ignition of materials in ships or warehouses are rather difficult to achieve, and studies on the subject may be found in certain reports and manuals that fortunately are not accessible to the general public.

chapter 10

Self-Igniting Fuel-Oxidizer Systems

The spontaneous or easily induced incandescent reactions involving a fuel and an oxidizer other than air comprise an array of dissimilar pairs of reactants and numerous modes of activation. They will be presented in the following order:

1. Halogens and halogen compounds with fuels;
2. Water-activated systems in which the water is either the oxidant itself or acts indirectly through hydrolytic reaction or as a physical medium;
3. Liquid oxidizers and fuels and the closely related acid-activated fuel-oxidizer mixtures;
4. Spontaneous reactivity of solid oxidizers with solid fuels.

The free halogens fluorine and chlorine gas, liquid bromine, and solid iodine form a series with decreasing reactivity with increase of atomic weight. Fluorine was discovered by Henri Moissan (1886) who tested its reactivity extensively. At ordinary temperatures this gas combines in incandescent reaction with all but the noble metals but only superficially with those that form a protective coating on the solid surface—magnesium, zinc, copper, tin, nickel, bismuth, and lead. The direct combination with carbon under glow is remarkable, as is an incandescent reaction with silica (SiO_2) of large surface area. Many organic substances are ignited on contact with fluorine gas; ordinary protective clothing may catch fire when hit by a stream of the gas. Conversely, some organic substances such as sugar and mannite are curiously unreactive.

Moissan's often-cited assertion that liquid hydrogen and solid fluorine will combine explosively at —253°C has been refuted by Aoyama and Kanda.[142] Other modern investigators (Bodenstein, 1934; Grosse, 1955) have found that the reaction of hydrogen and fluorine at very low temperatures depends on the presence of catalytic agents and impurities and even on the choice of container.

Chlorine gas combines with flame or glow with powdered arsenic, antimony and bismuth, and with copper, magnesium, zinc, and steel wool. White phosphorus explodes with liquid chlorine. The reaction of hydrogen gas with chlorine has found much interest, and aroused speculation during the nineteenth century, because it proceeds in proportion to the amount of absorbed visible or ultraviolet radiation. Thus, if the two gases are mixed in the dark in a transparent container, the mixture will be quite stable but will react explosively when irradiated by burning magnesium. Based on this reaction, Draper invented an actinometer in 1843, later improved by Bunsen and Roscoe.

Bromine reacts with incandescence with white phosphorus, arsenic, and antimony, but is generally less reactive than chlorine. Iodine, in turn, is still less reactive than bromine and some of its flaming reactions require increase of intimate contact by pressure because it is a solid. In the case of aluminum powder and iodine, the presence of water promotes intimacy of surface contacts, causing flaming reaction.[61]

The effects of the halogens on the alkali metals vary with temperature, dryness, and other conditions and consequently the reported phenomena, especially in secondary references, are sometimes contradictory. Coating of the metal surface with at least a temporarily-impervious layer of the halide may stop progression of the reaction.

The interhalogen compounds are highly reactive substances, especially those containing fluorine.[143] Chlorine trifluoride (ClF_3) of b.p. 11°C is in many respects more reactive than fluorine and flames instantaneously not only with many organic substances but even with aluminum oxide and magnesium oxide. Similar in behavior is bromine trifluoride (BrF_3) of m.p. 8.8°C and b.p. 135°C. Other interhalogen compounds, all very reactive, are ClF, BrF, BrF_5, IF_5, and IF_7. More recently discovered was the compound chlorine pentafluoride (ClF_5) of m.p. approximately −103°C and a b.p. of −18.1°C.

Chlorine trifluoride seems to have been considered as an incendiary material in World War II by the Germans, who built a 1000-ton plant for its manufacture. A welding and cutting torch for copper using chlorine trifluoride and hydrogen has been described.[67]

Even more reactive and called one of the strongest oxidizers in existence is the compound ozone difluoride (O_3F_2),[61,144] which, however, already decomposes at −153°C. Other highly active compounds are OF_2 and O_2F_2, and the explosive compounds O_2NOF and O_3ClOF, both derivatives of a hypothetical hypofluoric acid.

Many of the above-named halogen compounds have found use, at least experimentally, as high-energy components of propellant systems;

others are useful for the synthesis of organic compounds.

Because of the high heat of formation of many metal halides, the other halogens, especially chlorine, may play a part in some quite unexpected spontaneous reactions even though the chlorine compound involved is seemingly unreactive and, unfortunately, unsuspected of any possible mischief.

Sodium-potassium alloys have been reported to detonate on contact with silver halides. The word "detonate" seems to be used freely in the general literature to emphasize violently-explosive reactions without specific reference to a measurement of high-order detonation velocity. However, the reaction of the sodium-potassium alloys with organic halogen compounds such as carbon tetrachloride (CCl_4) has been specifically reported as high-order detonation, able to initiate substances that are difficult to detonate. It has been stated that the shock sensitivity of this combination is two hundred times as great as that of mercury fulminate.[145] Similarly, one finds explosive reactions of the single alkali metals with carbon tetrachloride, chloroform ($HCCl_3$), and diiodomethane (CH_2I_2).[8,69]

Since many of our thought processes and ensuing convictions and attitudes (posing as "sound common sense") are based on mere analogies and jumping to ill-founded conclusions, it is understandable that carbon tetrachloride, the popular nonflammable solvent and cleaning fluid, has been wrongly endowed with the property of inertness. As shown above, it is a most hazardous oxidizer in combination with reactive fuels. Triethylaluminum has been reported[146] as exploding violently with this solvent when a cooled mixture reached room temperature.

Lindeijer[147] reports a fatal accident from the explosion of powdered aluminum with carbon tetrachloride, and van Hinte[148] describes a case of spontaneous ignition of clothing, apparently contaminated with aluminum dust, while being degreased with trichloroethylene.

The instability of organic chlorine compounds in the presence of zinc and moisture is of interest in pyrotechnic smokes (see Chapter 18).

Water is the oxidizer in flaming reactions with potassium, rubidium, or cesium metal, which explosively burst into flame on contact. A plum-sized piece of potassium thrown into water "as a joke" was said to have killed a German university student in the early part of the century. Sodium will not burst into flame when thrown onto cold water as long as the piece of metal can skim freely over the surface of the water. The moment it attaches itself to the wall of the vessel or is purposely held in one place, it will burst into flame, ignite the developed hydrogen if accumulated, and often will disintegrate explosively. The same occurs if the water is thickened by some colloid, such as starch paste, if the water temperature exceeds 40°C, or if the metal is porous or finely divided.[66] This increased activity of the alkali metals can be used for ignition by water of gasoline gelled with napalm if as little as 2% of dispersed sodium or potassium (in xylene) are admixed to the gel.

Lithium is the least active of the alkali metals but a dispersion of it will ignite if thrown on water.[149]

That the flaming reactions of the alkali metals and water need no air may be shown by the fact that sodium (and after some delay, lithium) ignites in an atmosphere of steam and argon, and a lithium dispersion will ignite in liquid water under argon.[150]

Even if we disregard the element lithium whose behavior at the top of the vertical column in the periodic system tends to resemble that of the elements of the second column (Ca, Sr, Ba), there is no coordination between spontaneous flaming and heat output from the reaction with water, though there may be a connection with the melting point of the metals when we keep in mind the influence of renewal of the surface on reactivity. Table 3 shows these relations. In calculating enthalpies in pyrochemically unusual cases of this kind, one must clearly state if the water is included as a reactant and also if the considerable heat of solution plays a part. In the cited study,[150] the heats of formation of the hydroxides dissolved in an infinite amount of water were given, rather than the true heats of reaction, which may have been intentional.

Table 3
Reactions of Alkali Metals and Water*

	Li	Na	K	Rb	Cs
Molar Heat of Reaction (kcal) $Me(s)+H_2O(l) \rightarrow MeOH(s)+1/2H_2$	48.13	33.67	33.46	30.6	28.9
same per g Me	6.9	1.5	0.9	0.4	0.2
$Me(s)+H_2O(l) \rightarrow MeOH\ (aq\ \infty)$	53.19	42.92	46.68	45.6	45.9
same per g Me	7.7	1.9	1.2	0.5	0.4
M.p. of metal°C	180.5	97.8	63.6	38.8	29.7
B.p. of metal°C	1370	883	775	680	700

* Figures taken, expanded, and modified from Reference 150.

The hazardous practice of using the reactions of the alkali metals with water to perform a practical joke has been described by H. Allen Smith.[151] Henley[87] mentions a magic trick whereby drinking water is "ignited" when it is poured into a brass bowl onto a small piece of potassium covered with ether. Aside from such unusual practices, the reactivity of alkali metal and water has limited use. Certain Navy incendiary bombs contain an igniter loaded with white phosphorus for ignition on land or filled with sodium for use over water, such as the *Igniter AN-M9* (formerly called *Igniter E2*) or *Igniter M15*.[44] A burster scatters the sodium, which ignites or continues to burn on contact with water, and will ignite the incendiary filler of the bomb. Sodium and methyl nitrate in separate compartments and brought together by breakage under pressure were used in a land mine by the Germans in World War II.[48] Methyl nitrate, called "Myrol" was regarded as an explosive more powerful than nitroglycerine.

In ordnance design in general, highly reactive components of fuel-oxidizer systems are a poor choice as long as a similar effect can be achieved with less-active substances. One must consider not only the difficulties of handling such materials prior to loading but also the consequences of leakage during storage of the finished item. Damage or insufficient sealing may cause at best inerting through influx of air or moisture and at worst a premature reaction and destruction of the unit and its surroundings.

On the other hand, the number of highly reactive chemicals that advance from the stage of laboratory curiosities to commercial items is constantly increasing, Some of these are sodium hydride (NaH),[76] lithium aluminum hydride ($LiAlH_4$),[152] lithium borohydride ($LiBH_4$),[153] aluminum and beryllium borohydride, $Al(BH_4)_3$ and $Be(BH_4)_2$,[154] the sodium salt of nitromethane "sodium methane nitronate" or if one prefers "sodium nitro methanate" (H_2CNO_2Na),[69,155] and barium carbide (BaC_2),[156] all of which can flame on contact with water. Again, it must be stressed that particle size and other conditions of exposure may determine whether there is flaming or merely a violent reaction on exposure to water, air, or both.

In a somewhat different manner, the hydrolytic and heat-forming influence of water prevails in some fuel-oxidizer systems that are reactive in the common manner of pyrotechnical combinations, but will also be brought to an intensely-hot flaming reaction by the addition of small amounts of water. Most of these combinations contain sodium peroxide as the oxidizer. Aluminum and magnesium, mixed with sodium peroxide, have been recommended as igniters for the thermite mixtures described later. The magnesium/sodium peroxide can even be initiated by carbon dioxide. The high reactivity of these combinations makes such ignition mixtures quite perishable, especially when exposed to ambient air, which is never completely devoid of moisture.

Sodium peroxide will also ignite organic matter such as sawdust, paper, and numerous organic compounds, e.g. hexamethylene tetramine, when the mixture is moistened. Since sodium metal burning in air forms sodium peroxide, it is not enough to let accidentally ignited sodium burn itself out. The newly formed peroxide becomes a secondary fire hazard and must be destroyed. Moreover, if a larger pool of sodium burns, a part of the metal tends to stay unburnt, since nodules of the peroxide form that enclose unreacted metal.[157]

Water may cause spontaneous ignition in some fuel-oxidizer systems without taking part in any chemical reaction at all. In the case of mixtures of aluminum powder and iodine mentioned earlier, it appears to be simply a matter of promoting contact. In a different manner, a mixture of magnesium powder and finely crushed silver nitrate will burst into flame with a dazzling flash on moistening. The water causes electrochemical exchange between the magnesium and the silver ion, due to their respective positions in the electromotive series. The heat of reaction of this exchange provokes the pyrochemical effect. This remarkable, often violent reaction seems to have escaped even such diligent compilators as Gmelin and Mellor, perhaps for the good reason that it never has been described in the scientific literature. A somewhat similar but far less spectacular exchange reaction (and, in fact, difficult to produce) takes place between cupric nitrate and genuine tin foil[67] and is said to produce flaming or sparking after intimate contact. Cupric nitrate will also ignite paper spontaneously on prolonged contact.

An Italian patent[158] refers to an "incendiary mixture activated by water drops" of powdered magnesium, anhydrous copper sulfate, ammonium nitrate, and potassium chlorate. The water permits both the exothermic reaction $Mg + Cu^{++} \rightarrow Mg^{++} + Cu$ and the metathetical reaction of the salts, which produces the unstable ammonium chlorate. The latter then reacts with excess magnesium. While it is claimed that "little heat is produced in moist air," this mixture is undoubtedly unstable and hazardous. There is a possibility of spontaneous ignition whenever a chlorate and an ammonium salt are present, though in some formulas they appear to be harmlessly combined.

Self-ignition under the influence of moisture has another and more complex aspect, which often has plagued fireworks manufacturers in the past. In fireworks formulas with chlorate and sulfur, the oxidation of the sulfur in the presence of moisture can lead to acidification of the system. In the absence of neutralizers, chlorine dioxide (ClO_2) or (at slightly elevated temperature) chlorine and oxygen gas are evolved. These, in turn, act as strong oxidizing agents. If the rate of the initial reaction, induced by gradual infiltration of moisture, is slow, nothing worse than decay of the composition will result, but it can readily be seen that the decay can become self-accelerating to the point of self-ignition.

Under ordinary conditions, the alkali chlorates will ignite or explode spontaneously or nearly spontaneously only when in combination with the most reactive fuels, such as white or red phosphorus, powdered arsenic, or selenium. The chlorates of the alkaline earth metals and of other heavy metals are even more reactive and spontaneous ignition can occur quite readily with a number of fuels and in the presence of moisture.

Amiel[159] has studied the reactivity of the alkaline-earth and heavy-metal chlorates with sulfur in the presence of a little water and has found that below 50—60°C, chlorine dioxide is liberated, but that above these temperatures, chlorine and oxygen are evolved. This leads to occasional flaming in the presence of organic matter. Amiel also reports incandescent reactions between selenium and alkali chlorate.

Taradoire*[160] reports that wet sulfur, and barium or lead chlorate, flame spontaneously during drying, but that no flaming occurs with potassium chlorate. However, potassium chlorate has caused deterioration when used commercially in such items as railroad fusees. These examples show the relative instability of chlorate-containing systems in the presence of moisture. However, in matches where chlorate, sulfur, and organic binders are combined and exposed to the atmosphere, decomposition is virtually unknown. This shows that proper formulating with inclusion of acid-neutralizing substances can successfully break the chain of gradually destructive oxidation-reduction reactions.

The strong oxidizing power of the free chloric acid can easily be demonstrated by saturating filter paper or linen cloth with the 40% acid, whereupon the organic substances will catch fire with some delay.

Fuming nitric acid has the ability to oxidize many organic compounds vigorously and instantaneously under incandescent reaction. This characteristic is of great practical importance. The reactions not only furnish heat and flame, but produce only gaseous reaction pro-

*_Not_ Ta_b_adoire as misspelled in C. A. **28** (1934).

ducts; therefore they have found use as self-starting, so-called hypergolic rocket propellant combinations. The system of nitric acid and aniline with nitromethane added is an example of such a reaction system. Other oxidizers in hypergolic systems are nitrogen tetroxide and tetranitromethane, or concentrated (80%) hydrogen peroxide, in combination with hydrazine, water, and methanol. Highly active oxidizers such as chlorine trifluoride or fuels such as triethyl aluminum have been considered in hypergolic propellant or propellant-ignition systems but the field is really outside strictly pyrotechnical aspects if we discount the possibility of using such technically rather complex systems as "liquid flares" or incendiaries.

Several acid derivatives have very strong oxidizing power and some of them have been tried as igniters for hydrocarbons in flame throwers. Such compounds are chromium trioxide (CrO_3) and chromyl chloride (CrO_2Cl_2)—the anhydride and chloride respectively of chromic acid, and manganese heptoxide (Mn_2O_7)—the anhydride of permanganic acid ($HMnO_4$).[161]

Other extremely reactive low-boiling substances, all discovered fairly recently, are: chromyl fluoride (CrO_2F_2);[162,163] chromyl nitrate, $CrO_2(NO_3)_2$;[164] permanganyl fluoride (MnO_3F);[165] chloryl fluoride (ClO_2F); perchloryl fluoride (ClO_3F); and a number of active nitrogen-fluorine compounds[166] such as NF_3, N_2F_2, N_3F, HNF_2, H_2NF, $ClNF_2$, NOF, NO_2F, and NO_3F. To these might be added the well-known and stable compound, nitrosyl perchlorate.[167]

In contradistinction to these substances that are difficult to handle and for pyrotechnic purposes overly reactive, there are combinations that are noncorrosive and relatively harmless in themselves. They can be reliably brought to flaming by the addition of the corrosive, but otherwise easily manageable, concentrated sulfuric acid that liberates the active acid from its salt. Most often used is potassium chlorate with sugar or nitrobenzene. Dry mixtures will burst into flame on the addition of a few drops of sulfuric acid, or will explode if mercury fulminate is present. They have been frequently employed as incendiaries for sabotage and in bombs. Their advantage as incendiaries lies in the fact that the reaction can be delayed for hours, days, or weeks by making use of the corrosive action of the acid to work its way through some barrier.

In another setup, the acid gains in volume by absorbing moisture from the air. This leads eventually to a siphoning action into the active material. The diluted acid will react with the chlorate though more slowly than it does in the concentrated state. Similar contraptions utilize tilting and spilling or breakage to bring acid, a fuel such as gasoline, and chlorate together.

The use of the chlorate and sugar or nitrobenzene mixtures for purposes of political assassination and sabotage as in "partisan mixtures," and their modifications to produce delayed reaction mixtures such as "Molotov Cocktails" are described by Stettbacher.[17] He attributes the death of Tsar Alexander II of Russia in 1881 and the Irish baggage explosions in London in 1938 to bombs containing such ingredients. The author also quotes in this book articles by himself about chemical ignition.[168,169]

For more genteel uses, the reaction between chlorate and sulfuric acid was used early in the nineteenth century for a type of match that will be described in Chapter 12.

Acid-activated prime ignition fails with perchlorates because of the greater stability of the perchloric acid. However, the free, *anhydrous* perchloric acid reacts explosively with combustible substances such as charcoal, paper, wood, ether, and even with phosphorus pentoxide. In the latter case, the effect is due to the formation of the explosive and unstable chlorine heptoxide (Cl_2O_7), a volatile oily liquid.

A severely destructive explosion in Los Angeles in 1947 was caused by a mixture of 68—72% perchloric acid with acetic anhydride used for electropolishing of aluminum plates.[170] In general, the perchloric acid as usually encountered, in concentrations of 70% or less in water, is rather harmless aside from the fact that it is a very strong acid. However, if it should be soaked up accidentally in combustible material, it would greatly aggravate an existing fire.

A Safety Data Sheet[171] on perchloric acid gives excellent advice about the handling of the acid at different concentrations. This is particularly difficult in a case where the hazard is conditional rather than straightforward, as is the case with the 70% perchloric acid. It is clearly stated that spillage of this acid on organic matter may become hazardous under the following conditions: additional heat, an existing fire, or presence of strong dehydrating agents.

In contradistinction to the perchlorates, the oxidative power of the permanganates surpasses that of the chlorates. Glycerine brought together with dry crystals of potassium permanganate will burst into flame within 8—10 seconds after contact. Because of the high viscosity of the glycerine and the consequently slow wetting of the permanganate, this reaction is improved by a slight dilution of the glycerine with water.

On the other hand, a more reactive and thinner organic fluid such as dimethyl sulfoxide, $(CH_3)_2SO$, and finely powdered potassium permanganate will flash instantaneously.

The reaction of permanganate and glycerine is mentioned by Stettbacher.[17] Rathsburg and Gawlich[172] describe analogous reactions in which the glycerine is replaced by substances such as ethylene glycol, acetaldehyde, or benzaldehyde.

The stability of the salts of acids containing easily reacting oxygen decreases with increase in the valence of the cation. Consequently, calcium permanganate is even more reactive than the potassium permanganate. It can ignite alcohol, paper, and cotton.

When a permanganate and concentrated sulfuric acid are combined, the highly reactive manganese heptoxide (Mn_2O_7) and some ozone (by decomposition of the anhydride) are formed—both powerful oxidizers that easily cause flaming reaction with many fuel-type substances. The manganese heptoxide itself tends to explosive decomposition at somewhat elevated temperatures. Therefore, a potentially hazardous condition may exist when potassium permanganate is placed in a desiccator where sulfuric acid is the drying agent.

All the previously described mixtures contain at least one gaseous or liquid component, or liquid is added in order to increase intimacy of contact. Otherwise, as in the case of iodine, pressure has to be exerted with some solid fuels to produce flaming.

One combination of two solids exists in which a flaming or even explosive reaction may take place on merely pushing the powders toward each other or on exertion of very light pressure. This reaction occurs when the powdered components are completely dry and the fuel is not superficially oxidized. The two materials are red phosphorus and potassium chlorate and a demonstration of their reactivity should be performed only with a few milligrams of each component. When the phosphorus has been kept for some time in an ordinary reagent bottle, the spontaneity of the reaction may not be so obvious, but the final effect may be just as disastrous, as has been shown many times when high school students have appropriated and mixed together the two chemicals.

This reaction is undoubtedly the most fascinating, and perhaps theoretically the most interesting, solid reaction. It has been ingeniously tamed in the modern safety match, which will be treated extensively in the twelfth and final chapter on primary ignition.

Red phosphorus and chlorate can be mixed in comparative safety in the presence of a liquid vehicle, provided both reactants are thoroughly moistened by the vehicle before they come into contact. Using an aqueous binder solution, small dabs of such a mixture form the explosive ingredients of toy caps. If alcohol or ether is used as the vehicle, it is possible to load larger amounts of the mixture in specially designed containers, so constructed as to allow the liquid to evaporate. The capsule is then converted into a formidable antipersonnel mine, because slight pressure will explode the device.*[17]

The above-described combinations of red phosphorus, potassium chlorate, and a binder are called *Armstrong's mixture.* Its preparation and application has been described by Davis.[8]

Toy caps are called *amorces* in Germany and sometimes in England from the identical French word meaning a percussion primer or priming mixture. While the word is neither in Webster's Dictionary[173] nor in the Oxford English Dictionary,[174] it appears without comment in the title and text of a World War II British Intelligence Objectives Sub-Committee (BIOS) report.[175] The Germans used toy caps in World War II as hand-grenade primers because of a shortage of mercury fulminate. The two formulas given in the report as well as the formula from Davis' book (Formulas 1, 2, and 3) will be found, with all the following, in Chapter 47.

The phosphorus/chlorate/binder combinations are at the borderline between spontaneous reaction and manageable, easily initiated, but stable systems of reactive fuels and oxidizers. Those that have practical application will be treated in the following chapter on primers. At this juncture, mention will be made merely of some highly sensitive solid-solid combinations of little or no practical value, though of course odd mixtures may some day be useful agents for the fulfillment of new requirements.

Red phosphorus forms mixtures that burst into flame or explode on mild friction or impact, not only with chlorates but also with permanganates, lead dioxide (PbO_2), perchlorates, and other active oxidizers such as silver nitrate. Chlorates form friction- and impact-sensitive mixtures with most pyrotechnic fuels, in particular with sulfur and thiocyanates. Chlorates are not only sensitive with arsenic but it has been reported[66] that arsenic (III) oxide, As_2O_3, is self-flammable with sodium chlorate. Mellor[67] quotes an old German reference[176]

*A variation of this item using small amounts of the mixture has been proposed as an anti-infiltration device, causing flash and sound but no injury.

according to which sublimed arsenic* when triturated with an excess of silver nitrate "takes fire immediately when shaken out on paper."

*In German, the word *Arsenik* refers exclusively to the compound As_2O_3, while the element bears the name *Arsen*. It is possible that the identity of "arsenic" might in some places get lost in the translation.

chapter 11

Primers

Spontaneous ignition, as previously described, has only a few legitimate uses other than in hypergolic propellant systems. The exposure to air or the bringing together of liquids and solids are actions that are rarely suitable for quick, reliable, and easily performed ignition—not to speak of the design, loading, surveillance, and safety aspects of such systems. On the other hand, if we provide a certain defined small energy input acting on an often minute quantity of a solid, stable, but sensitive mixture or compound, we can produce instantaneous and reliable flash and flame. In certain cases, as will be shown later in this chapter in discussing electric primers, even relatively insensitive mixtures can be employed as prime ignition charges.

Primers fall into two categories according to the mode of activation —by impact or by electricity. Impact-initiated primers can in turn be divided into stab primers (detonators) and percussion primers. Electrical initiation is generally performed by means of a heated resistor that in most cases is a short piece of very fine resistance wire. The items are called electric matches, electric primers, squibs, actuators, etc.

Since priming devices are merely subordinate components of pyrochemical systems, such as flares, guns, rocket motors, and blasting caps, an extensive and fully detailed treatment is not attempted in this book.

The salient points have been taken from the *Ordnance Explosive Train Designers' Handbook*,[45] recently declassified, and some other sources, such as *NavOrd Report* 6061[177] and the AMC Pamphlet *Explosive Trains*.[178] Commercial catalogs,[179,180] which have also been consulted, mostly let the customer guess as to actual compositions, though the major ingredients may be indicated. Percussion primers are always standardized and commercially available mass-produced units that change little except by adaptation of formulas to modern requirements for greater stability and heat resistance; in contradistinction, the field of electrical initiation is in constant flux, with energy input requirements and the choice of active materials changing from year to year.

Stab igniters are the most sensitive mechanically-actuated primers used where the available mechanical energy is small. Because of this sensitivity requirement, the priming mixture contains a large percentage of prime explosive such as lead azide or, in older formulas, mercury fulminate. Consequently, the effect is one of high brisance (shattering). Stab primers (mostly called stab detonators) are used mainly for initiating detonations, though ignition and mechanical action (secondary actuation of a percussion primer or breaking of a glass vial) can be performed by some mixtures. The firing pin is a sharply pointed cone or pyramid and numerous fine details of geometry, shape, and material of the striker must be observed, since stab primers are merely tubular, explosive-filled cylinders closed with flat disks on the ends and mostly do not contain an "anvil."

It may be mentioned in passing that strike-anywhere type ("kitchen") match tips and formulations based on these (see Chapter 12) can be initiated by stab primer pins. Again, the difference between proper functioning and mere splitting of the match tip may depend on the profile and roughness of finish of the pins, as was experienced in World War II in connection with the *M1 Delay Firing Device* ("delay pencil")[18] about which more will be said later.

The basis of stab primer mixtures is the primary explosive lead azide. Potassium chlorate with either black antimony sulfide or lead thiocyanate (synonym sulfocyanate, also called—rarely in the United States—rhodanate or rhodanide) furnishes additional explosive potential and hot gases and particles. Deviating somewhat from this scheme is the formulation NOL-130, which uses barium nitrate in lieu of the chlorate. The addition of tetracene controls sensitivity. Typical mixtures—PA-100, NOL-130, and the older No. 74—are shown in Formulas 4, 5, and 6. They are used in a number of different stab primers (detonators) such as the *M26* (Army) that can be used as an igniter, while the *M41* that also contains an intermediate charge of lead azide and a base charge of tetryl qualifies definitely as a detonating device; the *M45* with PA-100 mix and ignition powders as intermediary and base charges; and the tiny *Navy Primer Mk 102 Mod 1*, which uses the NOL-130 mix, thus replacing the No. 74 mix of the *Mk 102 Mod 0*. Mercury fulminate mixtures are disappearing even from civilian uses.

Percussion primers are small metal cups that are dented but not perforated by a blunt firing pin. This prevents escape of the gaseous products of reaction, a containment of great importance in the now prevailing obturated (confined) devices. The confinement of the gases not only increases efficiency of fire transfer but is indispensable for reproducible functioning of pyrotechnical delay-timing columns independent of ambient conditions, be it at ground level or under vacuum at high altitude. (It is, of course, necessary than the unit as a whole be leak-proof, e.g. that the percussion primer is sealed into the larger unit in such a way that no gases can escape even under considerable pressure buildup.) The primer cup contains a curved metal insert called an anvil, which promotes the exertion of a crushing force between cup and anvil when the cup is dented by the firing pin. Again, numerous mechanical details apply to the shape of the firing pin, force, thickness of cup, loading pressure of the explosive, etc. Primers enclosed in a rimmed outer cup are called *battery cup primers.*

Since percussion primers are used for initiation of explosives as well as ignition of propellants in small guns and gas cartridges, and also for purely pyrotechnical initiation such as of pressed delay trains and ejection charges, the formulations vary with the purpose, though some formulas are said to be equally applicable to detonation or ignition. In other cases, low-violence and low-gas formation are claimed to make the primer especially useful for pyrotechnic ignition.

A number of impact-sensitive inorganic compounds have been used or recommended as prime ignition material in the past. They are nitrogen sulfide, nitrogen selenide, and certain chlorites. The properties of chlorites are described by Levi.[181] Anhydrous sodium chlorite, remarkably, explodes by itself on impact. Several heavy-metal chlorites exist, also cationic complex salts. Lead and silver chlorites combined with sulfur explode on touch with a glass rod, according to Millon,[182] and these and similar mixtures can become spontaneously explosive.[183] The apparently more tractable combinations of chlorites with sugar have been tried for primers. The explosive behavior of cationic complex salts will be briefly discussed in the Note to this chapter. In the literature, one finds also molecular compounds of lead nitrate and hypophosphite,[184] and of styphnates and hypophosphites of lead[185] or of iron.[186]

The most frequently used modern percussion primer mixtures, as shown in Formulas 7—11, are combinations of a not too brisant prime explosive salt such as lead styphnate (normal or basic) containing inorganic fuels and oxidizer salts for increased heat output.

Useful for pyrotechnical fire transfer are the *M*39 (Army) mixture, Formula 7; mixture PA-101, Formula 8, said to be particularly well-suited for the ignition of delay compositions; FA-70, Formula 9, used in numerous pyro primers such as the *M*29 and in commercial small arms ammunition primers, e.g. Winchester # 8-1/2; NOL-60, Formula 10; and FA-959, Formula 11. Red phosphorus in combination with barium nitrate and other materials has been used in several formulas in percussion primer mixtures and is also the subject of a patent.[187]

The requirement for stability at and functionality after exposure to temperatures of 400—600°F has led to new formulations that include a patented high explosive[188] named Tacot (du Pont) that neither melts, degrades, or decomposes up to 375°C (about 700°F).

Percussion primers may be integral components of an item such as a gas-producing cartridge, the outer shell of signal star assemblies, and of small flares. The latter are fired from a special firing device or pistol. On the other hand, the primer and the spring-loaded firing pin mechanism may form a discrete unit that in the field is connected to a safety fuse or detonating cord or is screwed in a canister or other container for pyrotechnical material. Such firing devices are described in the manual on booby traps[39] and other places. They are actuated by pull (*M1 and M2 Weatherproof Fuse Lighters*)*, pressure (*M1A1 Pressure-Type Firing Device*), release of tension (*M3 Pull Release Firing Device*, will also fire on increased tension), or release of pressure (*M5 Pressure Release Firing Device*). These are suited for operation by lanyard, trip-wires and a great variety of nasty arrangements of the harassment type linked to the opening of a door, picking up of a "bait," etc. Such applications would all be explosive. Pyrotechnic munitions use a well-known type of hand grenade fuze of the pressure release type, described later in connection with specific devices.

Other military prime ignition devices that also operate by mechanical force, but which are based on the two-component safety match principle, will be discussed in the following chapter on matches.

Electrical ignition of prime explosives or of prime ignition mixtures has numerous advantages and some disadvantages over mechanical initiation. The advantages:

Absence of moving parts;

Instantaneous action, and hence possibility of multiple, simultane-

* Also *M*60 *Fuse Lighter* and various similar items with built-in delay action.

ous, or accurately timable firings, important in commercial blasting and demolition work of numerous charges as well as in control or destruction of spacecraft;

Ignition in small, remote spaces with the electrical impulse instantaneously relayed, e.g. from a current-creating impact to the tail fuze of a bomb;

Capability of using extremely small energy (below 100 ergs and as low as 10—20 ergs) and, conversely, capability of dispensing with prime ignition materials altogether where relatively high currents are available;

Relative ease of hermetic sealing by having the electrical leads cast into molten glass ("glass-to-metal seal") in ceramic ("ceramic-to-metal seal"), or embedded into thermosetting plastic;

Achievement of highest reliability and, in line with it, ability to be subjected to partial nondestructive testing such as continuity (integrity of bridgewire and connections).

Some of the disadvantages:

The need for some kind of electrical power source;

A requirement for electrical circuitry and connectors creating possible sources of failure;

The delicate operations in affixing a tiny bridgewire to the base plug by welding or soldering;

Accidental firing of the more sensitive types by stray currents, radio frequency (RF), or body "static."

It is obvious that in small individual units such as hand-held or hand-thrown signals, percussion priming will be the preferred means of initiation. On the other hand, the delivery of photoflash cartridges from a dispenser on an airplane, the initiation of numerous (though not all) types of gas cartridges and of many delay-timing and switching devices is most often performed by an electrical impulse. Also, all blasting caps that are not exploded by the flame of safety fuses are electrically set off.

Carbon bridges, consisting of nothing more than a dab of water-based colloidal carbon black between two closely-set conductive areas in a molded plastic plug, are the resistors in the most sensitive electric primers and they use one of the most heat-sensitive, stable prime explosives, i.e. normal lead styphnate with some lacquer, as in the *Electric Fuze Primer Mk 121.*

The majority of electric primers, squibs, and actuators use a cylindrical base plug ("header"), which contains one or two wire leads (sometimes more for a redundant system) molded into an insulating plug material made of a thermosetting plastic or a low melting glass. If the latter, the glass bead is fused into a metal cylinder, and metal alloy and glass are carefully matched in thermal expansion properties to assure perfect bonding. After grinding the surface of the plug, the resistance wire is affixed by soldering or welding. This flat surface permits pressing-on of the pyrochemicals. In other arrangements, the bridgewire is affixed to protruding conductive leads in such a way that it is possible to surround the resistance wire by a bead of priming material. The latter arrangement is also used in so-called electric matches in which the priming bead and successive layers of ignition materials form a small bulb resembling a matchhead.

In order to avoid the tedious bridging of the plug, attempts have been made to replace the wire with a conductive path created by sprayedon metal and more recently by forming a metal layer through vacuum depositing. On the other hand, it is now possible to perform wirebridging in continuous automatic operation.[188a]

Electric primers are often characterized by the amount of current that may safely pass through the bridgewire without setting off the priming charge, as well as by the current that will reliably fire the charge. The no-fire current, if applied in nondestructive testing, may do nothing to alter the all-fire characteristics of the item, but in other cases it can degrade the priming charge and thus cause irreversible changes by a "cooking-off" (chemically degrading) of the chemical adjacent to the heated wire. The higher the all-fire charge is above the minimum stated amount, the quicker is the action. Again, since the electric primers are subordinate components of pyrotechnical devices, only a few typical items and formulas will be presented here.*

* This is a field that is in constant flux and some of the information that follows may already have been superseded, but it is included at least as general information rather than as a catalog of extant systems.

Electric Matches[189] are small resin-impregnated paper strips on which conductive brass strips are laminated with a wire loop affixed over one end. Two layers of priming material—Formula 12—followed by a chlorate/charcoal/lacquer mixture for flame and fire transfer and protective coatings of clear lacquer form a bulb resembling a bookmatch tip. Ignition is achieved by a current of 500 mA for a minimum of 50 msec. This is only one example out of a series of electric matches, which are manufactured to various current input requirements and also with different chemical compositions.

The M59 is a plastic molded button-shaped electric igniter used for the ignition of the propellant powder expulsion charge in the *M112* and *M123* photoflash cartridges. The pyrotechnic mixture—Formula 13—will tolerate a no-fire current of 1.3 A for 60 sec and will fire with 1.9 A in 50 msec.

The Mk 1 Mod 0 Squib is a relatively new rocket igniter and all-purpose squib (MIL-S-17923 NOrd) designed to permit firing at all altitudes and under wide extremes of temperature and humidity, as well as after prolonged severe exposure. It fires with 2.5 A dc in 0.2 msec with minimum brisance and excellent flame output and is nonfunctioning at 0.2 A dc or with a condenser voltage of 25 V and 4 μF. Its ignition charge is given in Formula 14. A black powder charge (A5) furnishes the fire transfer capability.

The older M1A1 Squib uses a combined priming and flame-forming composition (Formula 15) basically identical with the charges in the *Mk 1*. It is somewhat lower in sensitivity and will not withstand extensive temperature and humidity cycling tests.

The Mk 3 Actuator, as well as related and derivative items, is a typical completely sealed electric actuator with a flush-mounted resistance wire on which the prime explosive charge—in this case pure milled normal lead styphnate—is pressed. It is used among others in the *Mk 35 Mod 0 Explosive Switch*, the purpose of which is the opening of a normally closed circuit and the closing of another circuit. Its firing characteristics are: no-fire at 100 mA for 5 min and all-fire at 580 mA for 2 msec. The actual firing current is a condenser discharge of 3.75 μF at 40 V. The prime explosive together with a small pressure-sustaining ball powder charge performs the mechanical action—in the case of explosive switches—of moving a piston.

These few examples of typical electric primers could be extended to the sizable number of squibs, igniters, or actuators found in commercial catalogs, in data sheets, and in commercial or military compilations of the latest designs of devices used for the prime ignition of rocket igniters, fuse trains, etc.[190,191]

The tendency to move up to higher and higher firing (and no-fire) currents has brought out the fact that pyrotechnic fuel-oxidizer mixtures, not normally considered as prime ignition charges, can supersede prime explosives or chlorate mixtures where safety considerations demand such relative inertness and where adequate current is available. There is, of course, nothing surprising in the fact that compositions of zirconium with the more active oxidizers respond to relatively small current inputs.[192,193] The next step is indicated in two patents. One[194] claims formulas of boron, barium chromate, and potassium perchlorate; another[195] rather unorthodox mixture of magnesium, tellurium, and tellurium dioxide emphasizes a high ignition temperature and the application of 3—5 A or higher firing currents. The latter is in line with a "one ohm, one watt, no-fire" squib with an all-fire current of 4.5 A as demanded in recent years. Even higher no-fire/all-fire currents can be useful with a great variety of pyrochemical combinations in which materials of low ignition temperature or low decomposition temperature are omitted. If, however, more active materials are preferred, while relatively high and prolonged no-fire currents are prescribed, then a heat-dissipating design arrangement and choice of materials must be resorted to. Replacing a bridgewire by a vacuum-deposited layer of chromium or gold tends to dissipate heat from a no-fire current input. It has the additional advantage that the tedious technique of affixing the bridgewire is omitted. A header made from beryllium oxide and coated with a metal film has been claimed to permit exposure of lead azide to 10 A or 10 W of continuous exposure without firing while being charged from a 1 μF charged capacitor.[196] Beryllium oxide has the remarkable property of being a ceramic insulator with a (relatively) very high heat conductivity.

The principle of using an electrically conductive prime ignition mixture and thus eliminating the provision for a resistor wire or coating of the base plug can be achieved in two ways: by the addition of conductive powders such as acetylene black or of a noble metal powder to lead azide in varying percentages;[197] and by the use of mixtures containing at least 20% of lead dioxide that make the pyrotechnic composition moderately conductive. This latter scheme has been the subject of patents by Ciccone[198] and Peet and Gowen.[199] The former claims a combination of fine and of coarser grained zirconium with barium nitrate and lead dioxide. Where the mixture is to provide a form of propulsion rather than ignition (or perhaps both), PETN (pentaerythritol tetranitrate) is added as a gas-former. The second patent is similar but claims that the additional use of zirconium hydride (ZrH_2) acts as a desensitizer in a conspicuous manner causing an increase from a firing energy of only 2000 ergs to a nonfiring energy of 800,000 ergs when part of the zirconium is replaced by the hydride (Formulas 16, 17, and 18).

In this connection, the problem of prevention of accidental firing by means other than selection of highly insensitive chemicals can only be touched upon. It consists in the use of "filters," shielding, or use of integral attenuators in the form of special composition headers[200] made from phosphatized carbonyl iron powder. None of the proposed systems of protection is satisfactory in all respects.

Mainly of interest in the explosives field is another approach to the problem—namely, the avoidance of primary explosives or mixtures altogether. By igniting secondary explosives under high confinement, the enormous pressure buildup can lead to shock-wave formation. This is enhanced by creation of a gap or change of continuity of some kind between parts of the explosive column. Space between two charges may be filled with uncompressed or porous explosive. These designs have been the subject of a patent[201] and of other research effort.

Note to Chapter 11

In this chapter, the use of certain cationic coordination compounds, also summarily called "complex salts," was mentioned and some additional remarks on the pyrochemical behavior of such compounds may be in order.

Notwithstanding the extensive special literature on complex salts, such as the books by Sutherland[202] and Bailar[203] and the constant flow of current information, the explosive properties of their pyrochemically active compounds are rarely or only casually mentioned and come to light in the generally available publications mainly on the occasion of serious accidents.[204] The hazardousness is obviously quite general, though there are degrees of explosive behavior or of violence of mere deflagration, but it is not always realized that such substances may be sensitive primary explosives or powerful high explosives.

The compounds in question contain the cations cobalt and chromium, both in the trivalent state; bivalent copper; bi- or trivalent iron and others such as in the later-discussed alkaline earth compounds. At least, these are the most stable ones besides those of platinum and palladium, which are of lesser technical interest for obvious reasons.

The anions that furnish the oxygen for the pyrochemical action are nitrate, perchlorate, iodate, and periodate, and—exceptionally—chlorate, nitrite, and permanganate.

Complexing compounds (also called ligands or coordinating groups) are ammonia and ethylenediamine, urea, the thiocyanate group, hydroxylamine, and hydrazine. The last two yield the most active and explosive compounds, while urea is probably the least active. Coordinating groups may enter the molecule in mixtures together with pyrochemically inert ones, so that a great number of variations is possible. How many groups enter the molecule and how their nature affects the number of anions present is all well-known to the chemist but of little interest here. Pyrochemically, it is clear that larger organic ligands such as pyridine, though they may form stable complex cations, are of little use since they will make the compound too underbalanced in oxygen. There are no more than twelve active oxygen atoms present in a perchlorate of a trivalent metal and less if an acidic ion such as a halogen or the thiocyanate ion enters the complex.

The pyrochemically active complex salts have three interesting properties: they can "stabilize" an otherwise deliquescent metal perchlorate by making it nonhygroscopic, as will be shown for colored flame later; they can yield high explosives and even primary explosives "tailor made" to special requirements; and they can introduce a catalytically active metal salt or oxide in nascent stage in a pyrochemical system such as a solid propellant while being exothermically decomposing by themselves, thus contributing rather than detracting from the energy output.

Amiel[205] suggested in 1934 the use of bis- or trisethylenediamine copper chlorates or perchlorates for use in primers. Tomlinson *et al.*[206] probably made the first fairly systematic study of the explosive properties of certain chromium (III) and cobalt (III) complex salts. Médard and Barlot[207] concentrated on the di- and trihydrazine metal (II) nitrates, especially of nickel, as primary explosives. Use of complex salts as low detonation pressure explosives, reported by Abegg and Meikle[208] and performed by Fronabarger, Hoppesch, and Rittenhouse of UMC Industries was based on the authors' suggestions, as was earlier, classified work for Picatinny Arsenal whose subject was the increase of the burning rate of a solid propellant by admixture of such compounds.

The properties of the heavy-metal chlorates and perchlorates containing hydrazine, and their unusual sensitivity, were studied as early as

1910 by Salvadori,[209] later by Friederich and Vervoorst,[210] and more recently by Maissen and Schwarzenbach.[211]

Iodates and periodates, largely unknown quantities in pyrochemical reactions with few highly specialized exceptions, seem to have attracted much more interest outside of the United States as one can see by perusing the list of references in Supplement I to Mellor, Volume II. Explosive salts containing these anions have been described in Russia by Lobanov.[212]

Stability at elevated temperature and difficulty of preparation of many of these compounds are problems with explosive and other pyrochemically active complex salts.

chapter 12

Matches

> There was one day in the year—among the Romans it was the first of March—when it was the duty of every family to put out its sacred fire, and light another immediately. But to procure this new fire, certain rites had to be scrupulously observed. Especially they must avoid using flint and steel for this purpose. The only processes allowed were to concentrate the solar rays into a focus, or to rub together rapidly two pieces of wood of a given sort.
>
> *Fustel de Coulanges*, 1956[213]

History of Matches

The story of fire-making culminating in the development of the modern match might well fill a small book by itself, including such topics as the sacred fire on the hearth, the varieties of primitive methods of fire-making by different ethnic groups in the world,[29] the numerous devices of chemical and pyrochemical nature developed in the 18th and 19th centuries, the abolition of white phosphorus matches, etc. The salient points of these developments are treated in quite some detail in the general and technical encyclopedias and in a limited number of special articles,[214,215] the finest and most detailed ones being the already-cited series by Crass.[20]

Only an extremely condensed history of matches will be given in this chapter. It appears to be least confusing to separate the art of fire-making into three branches: *mechanical* (to which may be added chemical but not pyrochemical methods), developing into the flint lighter; *phosphorus*-based devices culminating in the modern SAW match; and finally, *chlorate*-based mixtures leading to the present day safety match.

Here is a quotation from *The Voyage of the Beagle*,[216] a work of the young Charles Darwin, still interesting and enjoyable:

> They then proceeded to make a fire, and cook our evening meal. A light was procured, by rubbing a blunt-pointed stick in a groove made in another, as if with intention of deepening it, until by the friction the dust became ignited. A peculiarly white and very light wood (the Hibiscus tilaceus) is alone used for this purpose: it is the same which serves for poles to carry any burden, and for the floating outriggers to their canoes. The

> fire was produced in a few seconds: but to a person who does not understand the art, it requires, as I found, the greatest exertion; but at last, to my great pride, I succeeded in igniting the dust. The Gaucho in the Pampas uses a different method; taking an elastic stick about eighteen inches long, he presses one end on his breast and the other pointed end into a hole in a piece of wood, and then rapidly turns the curved part, like a carpenter's centre-bit.

The scene is Tahiti and the year 1835, which shows that fire-making then (and throughout the greater part of the 19th century) was still primitive over large areas of the globe. In variations of the cited methods, the drill may also be turned by a string around it and the latter moved with a bow by a sawing motion or in pumping fashion. The burning lens was used in antiquity and played a role in ceremonial fire-making for relighting the sacred fire on the hearth,[213] as did a concave mirror in the form of a gold cup used by the priests of the Incas in Peru. Such fire was "purer" than the one created for ordinary purposes by sparks from hitting stone (flint) against stone, later steel against pyrites, and finally steel against flint, letting the spark fall on easily ignitible tinder, sulfur, or carbonized linen. The only practical survivor of the latter method is the modern "flint" lighter with its much hotter spark, which will ignite a gasoline drenched wick or a dry wick impregnated with a nitrate. But it took a long time until the tinder box, and correspondingly, the flint lock on a gun were replaced by match and percussion primer.

In parts of India, fire was made by adiabatic compression of air in a "fire pump"—wooden tubes with a wooden piston that contained some tinder in a hollow at its end—basically identical with the ignition in the modern diesel engine.

We may add here the *Feuerzeug* (lighter) of Johann W. Döbereiner (1823), which was a bulky hydrogen generator (zinc block dipped in sulfuric acid) with a piece of platinum sponge attached, which, by catalytic action, first became bright red-hot and then ignited the gas. Similar devices appear here and there on the market utilizing methanol and a metal screen coated with platinum catalyst, which becomes hot enough to light a cigarette pressed against it when air carrying methanol vapor is sucked through the grid. The use of semipyrophoric substances, which had gained some popularity during the 19th century as "pyrophorus" and mentioned in Chapter 7, is another unorthodox approach to fire-making.

The real story of matches starts in 1680 with the laborious manufacture of white phosphorus by Hankwits in London and the application by Robert Boyle of the then enormously expensive material to ignite a sulfur tipped wood splint. Between 1780 and 1830, numerous contrivances approaching the shape of an individual match were introduced, especially in France and Italy. They lead eventually, from about 1835 on, to real matches as we know them now. These were called "Congreves" in England after the inventor of the war rocket and "loco focos" in America—both names suggestive of the untamed and hazardous character of the early matches. By reduction of the phosphorus content from 20% to 5½% and adjusting the amounts of the other ingredients such as sulfur or rosin, chlorate, various fillers, and a binder, the quality of these matches was greatly improved, though no basic change in formulation had taken place by the time the phosphorus match encountered its decline toward the end of the 19th century. However, the obnoxious sulfur in the splint was later replaced by beeswax and finally by paraffin.

The poisonous qualities of white phosphorus matches showed themselves rarely in use except when the tips were chewed up by children or swallowed by people wishing to commit suicide. According to the U. S. Dispensatory,[217] an infant was killed by eating the heads of only two of such matches.

Workers exposed during the manufacture of the matches to the fumes of white phosphorus, which entered the body mainly through defective teeth, were subjected to an uncurable destruction of the jaw bones, the so-called phosphorus necrosis. With the awakening of social conscience in the protection of workers, all civilized countries either eventually prohibited use of white phosphorus in matches or taxed them out of existence as was done in Russia in 1892 and more stringently in 1905. The United States followed the latter example with an internal revenue tax of two cents per one hundred matches, which became law in April 1912, effective on 1 July 1913.

While it was possible to make a match that ignited by friction on any solid surface without use of white phosphorus, none could equal the ease of handling of the phosphorus match. A combination of potassium plumbate and red phosphorus with binder and filler called *Schwieniger Masse;* mixtures of a sulfophosphite with potassium chlorate; scarlet phosphorus—nonpoisonous but more reactive than regular red phosphorus—and other materials appeared between 1890 and 1905 to replace white phosphorus in friction matches or "strike-anywhere" (SAW) matches as they are called today. The problem was finally solved satisfactorily by an invention of two French chemists, Henri Sévène and Emile David Cahen, whose U. S. patent (1898)[218] was acquired by an American manufacturer, William A. Fairburn, and offered to other companies for the production of safe, nonpoisonous SAW matches. Their significant ingredient is the nontoxic compound, tetraphosphorus trisulfide (P_4S_3), called "sesquisulfide" in the trade. Originally, a single tip bulb was used, but this was later changed to the double-dipped (tipped) head described below.

We come now to the most important types of matches, the ones based essentially on the discovery of potassium chlorate by Berthollet* ca. 1786. What might be called the first two-component pyrochemical fire-making device was based on the property of chlorate to produce a flame in the presence of a fuel when moistened with strong sulfuric acid. Chancel's instantaneous light box or *briquet oxygéné* (1805) consisted of a match tip of potassium chlorate, sulfur, binder, and other ingredients on a wooden splint. Pressed against asbestos saturated with concentrated sulfuric acid in a glass vial, this device furnished a flame, often accompanied by sputtering and spraying of acid to the detriment of clothing, if not worse. However, the invention in one form or another stayed popular for many years. For a variant of this type, the invention of Samuel Jones (1828), we again let Charles Darwin[216] speak, with a description of his visit to a Brazilian plantation in July 1832 during the early months of his famous, long circumnavigation of the world:

> I carried with me some *promethean* matches, which I ignited by biting; it was thought so wonderful that a man should strike fire with his teeth, that it was usual to collect the whole family to see it: I was once offered a dollar for a single one.

This match consisted, according to other sources of information, of a small glass capsule containing somewhat diluted and blue-colored sulfuric acid and coated with a sugar/chlorate/binder mixture, the whole wrapped compactly into a three-inch-long piece of paper. For the less courageous, breaking the vial with pliers was recommended and a kit sometimes contained such pliers.

The true friction match—the immediate precursor of the safety match (though in fact a recalcitrant strike-anywhere match without

* Claude Louis Berthollet (1748-1822) must not be mistaken for another important French chemist, Marcellin Pierre Eugène Berthelot (1827-1907). Both were interested in gunpowder and explosives, which adds to the possibility of confusion.

phosphorus)—was the invention of John Walker (England, 1826 or 1827). It contained chlorate and antimony sulfide and could be ignited with some skill with a piece of folded "glass paper." More sensitive *lucifers* followed, which contained an addition of sulfur. But soon after the discovery of red phosphorus (1844), all this changed. Pasch in Sweden and Böttger (1845) in Germany prepared striking surfaces containing the new nontoxic material, but the world-wide acceptance of the system started with the availability of purer materials and improved formulas by J. E. Lundstrom in Jonköping, Sweden (1855). Strangely, the United States was slow in accepting the safety match. The big kitchen match, which could be taken out of the pocket and struck with ease with one hand on a shoe sole or on one's trousers, was more suited to the American temperament and was perhaps less hazardous in use at a period when other matches tended to be hard striking, explosive, and sparking.

The final major step in safety matches was the invention of the paper match, now called book match, ascribed to Joshua Pusey (1892). Somewhat more difficult to handle than the wood stick match, these inexpensive, advertising-carrying matches only gradually gained acceptance, but now dominate the American match industry as will be seen in the following sections of this chapter.

Economics

In 1958, the latest year for which a census of manufacturers has been published,[219] the value of all matches made in the United States was $135 million, a figure ten times the value of all fireworks items produced in this country. This makes the match industry the largest pyrotechnical producer and the only economically significant one. Even if nowadays matches fulfill a dual role as flame producers and as a vehicle for advertising, there is every reason to regard them as pyrotechnical items representing the most intricate and interesting type of pyrochemical action.

About three-quarters of the 400 billion (400,000,000,000) matches made in 1958 in this country were of the cardboard variety, the safety-type book match.* Because of their highly mechanized high-speed manufacturing methods and the tie-in with advertising, book matches are the cheapest and most popular form of flame producers in this country. Since they cost as little as one cent per one hundred lights, they

* According to an unofficial release, production of *book* matches in 1964 as reported by 14 companies was 311 billion lights against 287 billion in 1958.

are often given away as a service and in connection with their advertising message on the cover. When the price is higher, the only difference in what the customer gets is a better quality of paper and a more elaborate art work and printing effort on the cover.

Book matches that are sold in markets or are given away with cigarettes are in part paid for by large-scale advertisers who themselves have nothing directly to do with the distribution of these so-called resale matches. Similarly, a large company will absorb part of the cost of the advertising on those books that bear a distributor's name and address, such as a service station or an automobile dealer. Banks, hotels, trucking lines, and many other businesses (not to omit the commissary of a certain state penitentiary!) will contract for matches with individual designs of their own, or sometimes a very small establishment will be satisfied with a standardized "stock design" on which name and address are imprinted.

While paper matches became cheaper to manufacture and more and more accepted by the increasing number of smokers, wood for match splints has become very expensive and the wooden match is falling into disuse.

The large "strike-anywhere" (SAW) or "kitchen" match has retained some of its popularity longest, but with the prevalence of the pilot light in gas stoves, and the electric stove in rural electrification and in city use, mainly the little guild of pipe smokers appreciates its virtues.

Manufacturing Procedures

Book matches are punched from 0.038-in-thick, lined chipboard in strips of one hundred splints of 0.125-in. width each. In an eight hour shift, a single machine can produce about 20 million match splints and deliver them half an hour later as completed, strikable matches, ready for cutting and stapling into "books." In this half hour, the tips of the punched-out splints are first immersed in molten paraffin wax, without which no persistent flame and fire transfer is possible. Immediately, the tip composition is affixed by dipping the ends of the strips into a thick but smooth fluid suspension carried on a cylinder rotating in a relatively small tank at the same speed as the match strips move in the clamps of an endless chain. While an evenly rounded match tip is formed, cold air is blown against the matches, which then enter a dryer. There, however, the main object is not so much the speedy removal of the water in the match composition as the prior congealing of the matchhead, which takes place at about 75°F and a relative humidity of

45—55%. More about this curious way of drying will be given below in the discussion of the formulas.

The matchcover board is an 0.015-in.-thick coated or lined chipboard or sometimes a fancy grade of a variety of decorative and more expensive boards. It is provided with a striking strip, printed on by a roller-coating process from a thin slurry of a composition described below. The cover is printed with an advertising message. It establishes the curious relation of the merchandising of an item of great intrinsic utility and of a paid message.

Wooden matches can be made by a veneering method whereby aspen wood is "peeled" off a section of a log and cut into splints that have a square cross section of 0.10—0.15 in. depending on the length of the match. The alternate method consists of cutting rows of round splints from selected blocks of white pine wood by means of individual dies that resemble somewhat a large darning needle of which the eye is the cutter.* The wood blocks for this operation must be free from all knots, as straight-grained as possible, and hand-fed in a certain direction into the splint-cutting machine, all of which adds to the cost both of material and labor. These match splints are circular in cross section except for two adjoining concave grooves caused by an overlap in the cutting action, which not only prevents formation of useless splinters but also aids in the fire transfer to the splint at the acute edges. The splints in both types of operation are forced into holes in cast iron plates and are thus transported through the various dipping operations. For quick access in case of fires, which are not infrequent (though easily controlled) with the strike-anywhere type of match, the chains are not enclosed in dryer compartments, but rather the whole room is temperature and humidity controlled.

A third type of commercial match popular in Latin countries is the wax "vesta" with a center of cotton threads or a rolled and compressed thin and tough paper surrounded by and impregnated with wax, each match a miniature candle of long burning time (about one minute). Similar in performance but approaching a wooden splint in physical strength is a wax-impregnated paper match invented by the author[220] that, however, cannot quite compete economically with the plain book match.

Two processes precede the affixing of the matchheads on wood or paper splints. The first one is glow-proofing of the splint by impregna-

* *Collier's Encyclopedia*[27] has an excellent photograph showing this operation.

tion with ammonium phosphate or a mixture of it with boric acid. (In paper matches, this impregnation is conveniently done during the fabrication of the paper.) This suppresses continuation of glowing of the carbonized splint after discard and prevents the burned part with still hot tip from falling off and singeing clothing. The second impregnation is the soaking up of paraffin wax into the stem for a certain length to assure flame forming and fire transfer to the wood. Head formation is similar to the process described for book-matches, except that for SAW matches a second, smaller tip is affixed to the larger bulb. The rollers in the dipping tank over which the splints travel are grooved, the first roller deeply for the base tip, the second roller shallow-grooved for the SAW tip. The same equipment, simply leaving out the second dipping, can be used for safety matches. Immersion of the tips in diluted formaldehyde solution (sometimes also employed with paper matches) aids in congealing and subsequent proper drying of the match-head.

All wooden matches are packaged in rectangular boxes either of the the larger size for kitchen matches or in the "penny box" for safety matches, which in Europe are made from wood veneer, but in this country from thin cardboard. Depending on the type of match, the boxes are either provided with a striking strip like the book matches or with a plain "sand-line" consisting of an adhesive line on which abrasive material is superficially applied or (lately) a homogeneous mixture is used.

Non-Standard and Military Matches

Because match manufacture is a series of high-speed and highly mechanized operations, any variation that involves dimensional or procedural changes is a major undertaking warranted only if a continuous high production is forthcoming. Thus, a customer may be accommodated (for a price!) if he desires an unusual shade of color of the tip or even of the striking strip, but major variations in the formulas and, especially, in the size of heads, cannot be tolerated. Hence, specialties, which occasionally appear on the market, must be classed as fireworks or special production items, made laboriously and at relatively high cost (as much as five cents per single match) by hand-dipping with very limited mechanization. Such matches produce a colored flame, give off perfume or fumigating vapors, or furnish a persistent glow or flame for the purpose of burning in a strong draft. In order to do these things effectively, an enlarged elongated bulb is preferred. Matches burning with a colored flame, popular as harmless fireworks items, were and probably still are manufactured in Germany on a fairly large scale and are called "Bengal Matches."[175] During World War II, the facilities were used for diesel oil igniters made on the same equipment. A high content of gas-forming fuels and especially the added effect of a small amount of phosphorus sesquisulfide make matches wind resistant. A match with a sizable amount of powdered charcoal will burn without flame but with strong glow if properly formulated and designed. Such a match would be commercially unacceptable because of the fire hazard but is excellent for igniting safety fuse.

An interesting variation of the regular match is the "pull-match." It is a paper match, considerably thinner and narrower than a regular book match, since it needs very little stiffness when being used. The tip part of the match is enclosed in a strip of corrugated paper glued to a flat cardboard (such as a box of cigarettes) and the inside of the corrugated board is covered with striking material. On pulling the match fast enough out of the corrugation, the tip passes and engages the striker and becomes lit.

A match that has aroused interest because of the exaggerated claims for its performance and that has become somewhat of a hardy perennial in journalism is the "eternal match." Based on the patents by Rezsö König and Zoltán Földi and numerous followers,[221] this *repeatedly ignitible* item consists of a small pencil-like rod, the center part of which is a safety-match composition, while the outer layer consists of a slow and cool-burning mixture whose essential, effective ingredient is always metaldehyde (sometimes combined with other volatile organic substances). These are expensive chemicals to burn as flame-former in lieu of paper or wood and paraffin; in addition, they evaporate in storage. Repeated striking of the pre-used match fouls the striking strip so that the ignition becomes progressively more difficult.

The fact that a prime ignition device can be blown out and re-ignited is a remarkable phenomenon for the collector of odd pyrotechnical facts, but otherwise the item is in no way competitive with normal match production.

A variation of the pull-match is the commercial *Pull-Wire Fuse Lighter*[222] that also has the military designation *M1 Friction-Type Fuse Lighter*,*[38] in which, however, the role of striker and match is reversed.

* Even though there are enough integers in the decimal counting system to number every piece of hardware in the world unequivocally, the military persists in starting

The match mixture is located in a metal cup, and a length of wire, corrugated and covered with striking mixture near the end, is threaded through a hole in the cup, with a handle for pulling the wire through it. The match cup is enclosed in a length of stiff paper tubing for insertion of a delay fuse, held in fixed position in the tube by a metal ferrule with sharp protrusions. On removal of the wire with a fast pull, this cheap device will fairly reliably produce a spit of flame within the tube and transfer the fire to the fuse train.

In the *Portable Flame Thrower*, the original electrical ignition has been replaced by a mechanically actuated safety-type match. The fire-producing mixture (Formula 19) is in a tiny metal cup and is actuated by a blunt steel pin coated with phosphorus mixture, which is pushed by a lever action into the cup.

Match buttons, ignited by hand with a striker such as the one on the wooden top of a railroad fusee or on a separate strip under the lid in the *M1 Fire Starter*,[38] are formulated simply according to MIL-STD-585 (Formula 20), and the striker according to MIL-STD-537 (Formula 27), though fusee manufacturers also use other proprietary formulas. A hot and relatively slow and intense flame, affording transfer of the fire to the substratum into which the match button is embedded, is the aim of this arrangement and formula. Such formulas sometimes contain a nonhydrophilic binder such as shellac or nitrocellulose (Formula 22). However, in one case, a match of this kind made with an unsaturated polyester as binder developed explosive properties in storage, the reason not being ascertainable.

Exceptionally, there may be a requirement in which a match mixture, after being hit by a phosphorus-coated striking pin, disintegrates into hot sparks. For this type of fire transfer, a low binder, low ash formula (Formula 21) is indicated.

A borderline item between a commercial match and a specialty item is the "self-lighting" cigarette. The idea of combining a "smoke" with its integral ignition source goes back to 1835 when "loco foco or self-lighting segars"[223] were sold; since then, various patents have been issued for self-lighting cigarettes, several distinguished mainly by their amateurish claims to the use of theoretically possible but most impractical ignition mixtures. The latest and fairly workable item solves the problem by use of a very small dab of tip composition applied to a potassium nitrate impregnated circular paper strip.[224] A recent (1966)

each category of items with the number 1, hence the confusing array of M1 or Mk 1 items—squibs, fire starters, flares, etc..

U. S. patent appears to follow a quite similar approach.[224a]

Formulations for Commercial Matches

While the above-mentioned formulas are simple and effective for their specific purposes, the commercial match requires a complex, very carefully compounded and balanced formula for both the striking strip and the matchhead. Formulas for book matches and wooden matches are quite similar and so are the formulas used by different companies. There are no secret formulas anymore. Knowledge of a great many details and experience in adapting and improving a formula in the face of small changes in materials and external conditions with the aim of producing a superior and nearly perfectly uniform match is the real secret of a match company.

In order to understand the chemistry of safety matches and hence some of the finer points of their formulation, it must first be realized, as has been pointed out in Chapter 10, that the reaction between potassium chlorate and red phosphorus, which is the basis of its prime ignition action, is extremely hazardous and unpredictable. Their separation in match tip and striking strip is therefore not only a brilliant piece of inventive creation, but requires great ingenuity to strike a balance between effective action and safe action. The next point to consider is that the prime ignition in its entirety must be confined and followed by orderly progression of a flame and that any kind of flashing or dropping of glowing particles must be avoided. This latter task falls to the various siliceous ingredients in the matchhead, which form a sintered residue containing all the nonvolatile residues from the reaction. Similarly, the phosphorus in the striker must be prevented from burning off and causing a shower of sparks. This is done by choosing the proper grade of binder and its right ratio to the phosphorus. Since ease of ignition, spark-forming, and excessive wear of the striking strip all go together, while a more adhesive or stronger grade of binder decreases wear and—literally—tears and desensitizes the act of ignition, head and striker formulas must be not only "just right" in themselves but matched in behavior. The addition of finely powdered glass adds to the resistance toward "striking through" and burning off of the striking strip.

A similar but far more complex role is played by the binder in the matchhead. The glue must be strong enough to bind the powdered ingredients into a firm bulb, but its amount must be small enough to permit easy abrasion, which precedes ignition. The amount of glue is also limited by its pyrochemical function as a fuel: while the match is vastly underbalanced in fuel versus oxidizer, a further excess of fuel beyond certain limits would prevent the tip from flaming altogether, and the smouldering bulb would not reach the flame point of the wax in the stem, so that the match tip simply fizzes out.

In modern high-speed match production, the type of binder in the matchhead is curiously limited to one certain kind and to very few similar grades of it. High-grade animal-hide glue with the specific capability of quickly forming a reversible gel on cooling below 85°F from its more concentrated solutions will cause all components of the formulation to stay in their place once the dipping process is completed. Should the binder instead have a chance to form a skin in the top layer of the bulb, as happens with high molecular organic colloids in aqueous solution, the resulting "case-hardening" can convert the matchhead into a miniature bomb—the chlorate/sulfur/binder mixture can burn explosively under the slightest confinement. Before temperature and humidity control of the dryers of matches (or, with open machines, of the room itself) came into general use, the—unintentionally!—exploding match was no rarity. This was rarely the result of faulty formulation but mostly derived from matches made under hot and humid conditions when the congealing of the matchhead at the start of the drying process was delayed or never took place. The same might also occur in winter with use of artificial heat, though under these circumstances the result is more likely a poorly lighting, "hesitating," and insensitive match.

To continue with the specific functions of the chemical ingredients in matches, we consider next the two purposes in the small percentage of sulfur in the head formula: It acts as an easily ignitible and hence somewhat sensitizing fuel and flame-former, and its combustion product is the pungent but harmless and not nauseating sulfur dioxide (SO_2), which masks the much more unpleasant odor of burning glue. Thus it is a "perfume" of sorts, whose place can also be taken by powdered rosin.

Both matchhead and striker formulas contain an acid-neutralizing but insoluble salt such as calcium carbonate (whiting, ground limestone). This aborts any tendency to destructive oxidation of the sulfur in the head or of the red phosphorus in the striker. The latter process can be greatly accelerated by catalytic influence of certain heavy metals, especially copper.[225] Since copper-bronze powders are well liked as decorative components of the artwork on the cover of book matches, improper formulation of the striking strip may lead to complete destruction of the striker.

The least understood component of the matchhead formula is the potassium dichromate once generously present in many formulas, if some recipes in the literature can be trusted. It controls the burning rate and facilitates the ignition following the "prime" reaction between chlorate and phosphorus; even fractional percentages exert a noticeable influence. It can be replaced by lead compounds (lead thiosulfate) and, in general, by compounds that furnish finely divided oxides of metals that can occur in several valence states. This parallels the influence of similar compounds on the lowering of the decomposition point of molten chlorates, and we may call these additives "catalysts" even though their precise functioning is obscure.

Formula 23 is a typical, slightly generalized, commercial matchhead formula on the basis of which anyone with a bit of experimentation and adaptation can produce a well workable though not necessarily salable match. Not without good reason has the refreshingly blunt author of the old National Bureau of Standards circular on matches[226] said: ". . .if they [people who ask NBS about how to make matches] plan to produce matches for sale, they are hereby advised to seek a more promising way of earning a living." However, since many older formulas that have found their way into encyclopedias and books of recipes seem to be of dubious usefulness, some elaboration will be added in order to guide manufacturers of fireworks or ordnance items who may be faced with the task of using match formulas for some special device.

Match glue used in the tip must be a high-quality animal (hide) glue free of defoamer or only very slightly treated, so that during processing of the match paste a somewhat aerated structure is created and preserved in the final head. The potassium chlorate must be finely powdered, since coarser crystals would cause sputtering. The sulfur should be crystalline, finely ground "flour," not the commonly sold "flowers" of sulfur, which is too active and acidic. The siliceous materials—aside from the diatomaceous earth, which is nearly indispensable for "soaking up" the molten potassium chloride residue—may be powdered (bottle) glass only or a mixture of various siliceous materials, the cheapest of which is plain finely ground silica sand (SiO_2).

Nothing has been said as yet about the gay colors of our American (and some foreign) matches that are possible because the basic composition is nearly colorless. (The use of only fractional percentages of potassium dichromate as a catalytic burning-rate regulator and sensitizer creates some problem with white matchheads. To the distress of nontechnical supervisors, chemists are not smart enough to know how to "bleach" dichromates.)* Soluble aniline dyes in fractional percentages create the standard colors—red, blue, green, or any other special shade.

Formula 28 is the striker or "friction" formula for the commercial safety match. Nowadays, the binder is always insolubilized either by a special hardening process,[227] a formaldehyde treatment, or use of a casein solution in ammonia. This prevents the staining formerly experienced when matchbooks were carried in shirt pockets and subjected to perspiration or exposed to rain. The striking strip can also be made with hydrophobic binders such as nitrocellulose or various plastic emulsions, but the quality of such a very moisture-resistant striker is generally inferior as are all special formulas designed to make the striking strip adhere well to foil, plastic, metal, or other impervious surfaces.

It is not difficult to vary the color of the striker by omission of the carbon black or charcoal and tinting or lightening it with various pigments such as water-dispersible titanium dioxide, hydrated iron oxide, etc. Even bolder colors are possible but not practical.

The typical European match (Formula 24), because of its use of dark oxide pigments, leaves no chance for esthetic variety, but there are brightly colored matches outside the United States and matches from the Far East are now often in delicate pastel colors or white. Formula 29 is a European striker formula apparently for "penny box" application. The large amount of antimony sulfide in the formula would be shunned in this country, because antimony is somewhat of a strategic† and critical† material, fairly expensive and tending to settle out because of its specific gravity. It can replace the red phosphorus in part because of its low ignition temperature and its reactivity with chlorates.

Formulas 25 and 26 are typical compositions for SAW matches, but the two fulfill completely different functions: The tip, which is easily ignited by rubbing over any smooth or rough surface and also by scratching, stabbing, or (less reliably) breaking up (crushing), contains a rather large amount of tetraphosphorus trisulfide (P_4S_3), familiarly called "sesquisulfide." This is the easily ignitible agent in the tip. The bulb is "loaded" with fuels that provide a billowing flame, and (with the addition of some sesquisulfide) become the major contributor

* A true incident, Universal Match Corporation, ca. 1936!
† For these and other special terms, see Glossary.

to the moderate wind resistance of this type of match. If blown out immediately after ignition, a well made, large SAW match should "come back" with full flame once or twice. The chlorate content of the "base" is kept deliberately low in order to desensitize the bulb. Since the large SAW matches are packed several hundred to a box and half of them, facing in the same direction, touch each other, a heavy impact can cause a sizable amount of friction, which might be sufficient to ignite a high chlorate formula of this type.

Properties of Matches

Mechanism of Performance. The act of striking a safety match consists of bringing matchhead and striking strip in intimate contact by rubbing so that the reaction between chlorate and phosphorus can take place. It is not really an act of creating heat by friction as is often thought, though of course it is open to argument whether the mechanical action creates a minute "hot spot" from which the reaction spreads out or it is mainly a matter of intimate contact of chlorate and phosphorus. When one rubs a match very lightly over the striker in complete darkness, greenish sparks of burning phosphorus can be observed. Such mechanical force would appear to be no stronger than necessary to break through the surrounding glue layers. Normal, stronger rubbing will permit the tiny flash from the chlorate/phosphorus reaction to ignite the glue/sulfur/chlorate system, aided by the catalytic influences of the heavy-metal additives—dichromate, lead thiosulfate (in some American matches), and manganese dioxide and others in European matches.

Safety Aspects. Safety matches can be set off by impact or strong friction. This in no way detracts from the safety aspect since it takes great skill to ignite a safety match in the absence of a phosphorus striker. Such ignition can be performed by rubbing the match over a glass plate or a piece of fairly smooth and dense cardboard with just the proper pressure to create enough heat to ignite the match mixture by friction but without breaking up the fragile match tip. A blow with a hammer will explode a safety match, and sulfuric acid will ignite it like other fuel /chlorate mixtures as discussed in Chapter 10.

On the other hand, considering that the safety-match tip mixture has both explosive and prime ignition characteristics, matches in normal confinement (as in a matchbook) are extremely safe. Experiments have been performed in which matches in regular cardboard shipping cases were ignited by setting off electrically a single book in the standard carton of 50 books, 50 cartons to a case. In every instance, the flash from the ignited book spread only to the other books in the same small carton and often only to a few of these books! Neither is there enough oxygen present to propagate the initial fire, nor does the heat from all the matchheads in relation to the large mass of paper amount to much. The story is different, however, when cut-off matchheads are accumulated and set off in tight confinement, as many deplorable accidents involving young rocketeers have shown.

The question of heat output and flame temperature of a single matchhead is sometimes raised. According to the author's estimate, a single matchhead, without involvement of paraffin or paper, should furnish about 14 calories or 0.05 Btu. Arditti *et al.*[228] give the heat of combustion of matchheads of various provenience as 965 kcal/kg or less, which, using the weight of the tip of an American paper match as 22 mg, comes to a somewhat higher figure than the author's. In any case, this heat is only one-tenth or less of the heat of the burning paraffin, not including the paper. The same authors[228] have measured the "initial" temperatures of several foreign and domestic matches as between 1350° and 1930°C.

Claims concerning matches "going off" or "blowing up" spontaneously can generally be reduced to the fact that violent striking, ignition of a cracked bulb, and careless or absentminded handling of the lit match caused ignition of the whole book. Also, if more than one single book is carried in one pocket, interlocking may cause ignition under unfavorable conditions.

Matches ignite spontaneously at 180—200°C when gradually heated in an oven. A government specification[229] stipulates a minimum ignition temperature of 170°C (338°F). Strike-anywhere matches ignite under the same conditions at 120—150° (248—302°F), in line with a self-ignition temperature of about 100°C for the phosphorus sesquisulfide.[67]

Toxicity. Small children are inveterate explorers of objects in their reach and apparently the slightly bitter and salty taste of matchheads is no deterrent to their liking of chewing and sucking on matches. The toxicity or innocuousness of matches thus becomes a frequent subject of parental concern and even the medical practitioner confronted with a frantic parent may not always have the answer. While potassium chlorate may cause disturbances if ingested in larger quantities, the quantity contained in one match is very small (about 9 mg) and the amount a small child can suck out of one or even a few books should be below a toxic limit. Osol and Farrar[217] give 1 g as the maximum therapeutic dose of potassium chlorate for an adult. The other ingredients are either harmless, or present in such minute quantities that they can be regarded as harmless. Therefore, while safety matches cannot be recommended as part of a steady diet for infants, anxious parents need not be unduly disturbed if they find that their child has chewed up a book of matches.

Since the layman cannot easily understand that two different kinds of phosphorus exist with entirely different physiological effects, the word "phosphorus" may evoke the spectre of phosphorus poisoning where there is truly not the slightest need for concern. No book treating toxicology should nowadays include "Matches Containing Yellow Phosphorus"[230] except under historical aspects. Similarly, it is way off the mark to characterize yellow phosphorus by "Odor. Like matches."[35] However, it is worth mentioning that as late as 1939, the existence of white-phosphorus-containing matches was reported.

It is generally agreed that red phosphorus is nonpoisonous, though Osol and Farrar add the qualification "when pure." People have handled daily for decades the red phosphorus in preparing striking mixtures without any ill effects where no precautions as to contact with it are taken. This should be reason enough to doubt contrary statements such as by Sax,[231] which should be reexamined.

Humidity Resistance. During World War II, the armed forces of the United States requested development of a match that would be ignitible after submersion in water for 6 hours.[232] This task was fulfilled by partially coating kitchen matches with a nitrocellulose lacquer of adequate thickness and proper flow characteristics so that the coating was reasonably uniform and did not accumulate near the tip so as to make striking difficult or explosive. The author, though by no means very knowledgeable in this field, had no difficulty in producing his formula WM-49 (Formula 30), which is an underplasticized NC solution to which some cellulose acetate-butyrate was added for arresting the backflow because of the gelling properties of the latter. Millions of matches characterized by a deep green coloring added to the lacquer were manufactured during the second half of World War II (by Universal Match Corporation) and they withstood the water immersion at ambient temperatures for as long as 10 hours. Another company in this country independently developed the same kind of match, and a related development seems to have taken place for the Canadian government[215] with a match that can be "ignited in a rain or wind storm."

Actually, match tips cannot be made highly humidity resistant in themselves because of the eventual breakdown of the water-soluble chlorate, the softening of the glue under prolonged extreme humidity, and the relatively small amount of binder permitted in a matchhead formula. By means of a few tricks, safety matches can be made to withstand a limited exposure to extreme conditions to comply with EE-M-101H (Oct. 31, 1964).[229] The test consists in exposing the matchbooks in a strictly specified manner over water at 106°F for 8 hours. It is not practical to coat safety matches with a lacquer because of difficulty in striking, which increases with the effectiveness of the coating.

There is an interesting sequel to the lacquer-coating, waterproofing business, which affords some glimpses into the whole problem of protection of moisture-sensitive items against severe ambient conditions: All heavily-coated World War II kitchen matches eventually became unstrikable (after 5 to 10 years under ordinary indoors conditions in St. Louis, Mo. climate) because of decomposition of the SAW tip. However, matches coated with a much thinner lacquer soon after World War II (1946-47) are still in good condition today (1965) after about 18 years under identical storage conditions as with the heavily coated ones. No deterioration of the lacquer coating itself and of the base composition took place in either case.

Since the SAW tip is vulnerable to prolonged high humidity exposure because of its P_4S_3 content, the conclusion seems inevitable that where a certain water vapor transmission exists, *no matter how small*, and where the exposure fluctuates, the "better" protection may actually prove to be disadvantageous since just as the influx of moisture is slowed down, so is the reverse action—the drying out when low-humidity conditions set in. Unfortunately, no plastic or wax-coating has a water vapor transmission of zero. But even hermetic sealing, say in a glass ampoule or a metal box welded or soldered together with perfect sealing, may not protect a pyrotechnical item from destruction by moisture. Chemicals, plastics, and especially paper or felt parts, all contain some moisture in equilibrium with ambient air and when this equilibrium is disturbed in the enclosed space by uneven heating of the package, some vulnerable chemical component such as an igniter may be damaged, the whole package virtually "stewing in its own juice."

To sum up this side-glance on the problem of climatic surveillance, only an analysis of the problem of vulnerability, duration, periodicity, and extremes of exposure can decide on the best protection.

As far as matches are concerned, a heat-sealed plastic envelope will protect a book or box of matches from getting soaked in rain and during a prolonged though not excessive period of very high humidity. Single SAW matches thus enclosed in a cartridge belt type plastic envelope can be torn off individually and struck *without opening the envelope*, a simple and efficient though somewhat costly scheme.

Note to Chapter 12

Water-vapor transmission figures are given by a number of different methods and units, one being the method ASTM D697, which expresses g/24 hr/m²/mm thickness/cm Hg at 25°C. Figures quoted in the *Modern Plastics Encyclopedia*[233] for this method are in the first column of Table 4, while the second column refers to g/100 in²/24 hr/mil according to ASTME 96 taken from *Material in Design Engineering*.[234]

Table 4
Water-Vapor Transmission of Plastic Films

Polyethylene	0.02 to 0.08	0.3 to 1.5
Cellulose acetate	1.4 to 8.8	
Polyvinylchloride	0.35 to 2.0	0.5 to 0.7
Polypropylene	0.06	0.25 to 1.0
FEP fluorocarbon	0.002	
Nylon		0.9 to 1.0
Cellophane, coated		0.2 to 1.0
Fluorocarbon CTFE		0.025*

* Also, Aclar (trademark) fluorocarbon resin 0.015 from an advertisement.

Other tables including data on permeability of nitrogen, oxygen, and carbon dioxide, and conversion figures for the various permeability units, appear in an article by Lebovits.[234a]

A calculation—and it is only a mathematical exercise—shows that if a single matchbook is enclosed in a 2 mil (0.002 in.) thick envelope of the best grade of polyethylene with 100 cm² total surface area and exposed to 98% RH, assuming an average interior RH of 55% (1 cm Hg of water vapor pressure differential), the influx of moisture per week would be 28 mg or 0.7% of the weight of the enclosed paper and chemicals. This proves the above-raised point that prolonged, uninterrupted, severe exposure cannot be counteracted by plastic enclosure.

In practice, any lengthy extreme exposure would be in a warehouse where the individual envelopes are packaged in larger units, thus greatly diminishing influx of moisture. In conclusion, it might be said—and this applies to a variety of items, not only to matches—that sealing-in in plastic film after removal of interior moisture, canning of the sealed units as few as practical per package, and protecting the cans from chemical attack by humidity should be the best protection for small, moisture-sensitive pyrotechnical objects. Simpler and adequate for many purposes is the use of so-called scrim bags, which consist of laminates of metal foil and heat-sealable plastic and textile backing. In this case, water-vapor transmission should be restricted to minute pinholes in the foil and to the heat-seal edges.

Part III

Light

chapter 13

Underlying Phenomena

Spectacular light effects have been the chief aim of pyrotechnics throughout the centuries. The "fire-art" was mostly the art of making fireworks. While these have become less important as amusement, the progress in techniques and the selection of more effective chemicals have greatly improved pyrotechnic light production for military uses as important means of signaling and illumination.

In order to understand the relations between pyrochemical reactions and light, one can cite the laws of radiation as a starting point. But not all the ramifications of light emission from solid particles or gases in the complex entity of the flame from a flare or the fireball of a flash are known. The phenomena of luminescence and of spectral emission at high temperatures in complex flames are still rather obscure and even the relatively well explored field of radiation from solids has only sdoradically yielded data at temperatures in the important 2000—3000° C range. Many more figures are needed, e.g. emissivity at desirable wavelengths or discrete ranges of the spectrum, visible and invisible—all of greatest concern for the theoretical pyrotechnician.

It is gratifying to find evidence in a voluminous report *Symposium on Thermal Radiation of Solids*[235] of the large amount of work, both experimental and theoretical, on radiant emission of solids related to space technology. But closer inspection of these informative reports yields few applicable data for our specific area of investigation.

Thus, while optimal conditions of luminous efficiency or purity of color cannot be achieved on the basis of calculations, the theories of radiation might be a guiding factor—albeit as yet uncertain—in arriving at superior light sources.

By "light" we understand within the framework of modern pyrotechnics not only the narrow spectral range of radiation that constitutes visible light (0.40—0.78 μ wavelength) but also the spectrally adjoining portions of the ultraviolet (UV) and—even more important—the infrared (IR). Sensing devices that respond to invisible longer wavelength emission[236] make it possible to utilize such radiation from a flare for a variety of tactical uses.[237]

Our consideration of the laws of radiation starts with the concept of a "black body," defined as a solid that emits and absorbs (but does not reflect) radiation of all wavelengths in certain proportions, the absolute and relative amounts governed by the Stefan-Boltzmann Law, Planck's Law, and Wien's Displacement Law. Even though the ideal black body is a theoretical concept, radiant emission from a hole in an otherwise enclosed heated cavity, such as a tube furnace, and from certain dark substances, such as carbon, oxidized iron, and the metal platinum, approximates the black body emission.

We can summarize the black body laws as follows: the total amount of radiant energy emitted is proportional to the fourth power of the absolute temperature. Coincidentally, the relative amounts of UV and of visible radiation as a percentage of the total emission increase with temperature. Thus, an emitter of visible light benefits twofold from high temperature—the total increase in radiation with rise in temperature, which is the dominant influence, and the relative increase in shorter wavelength emission.

Table 5 indicates the visual impressions from heated solids, and since measurements at high temperatures are often hard to perform with pyrotechnic items, the melting points of some solids are juxtaposed. They can be used occasionally to give at least the minimum temperature reached though they are, of course, no substitute for actual measure-

Table 5
Visual Temperature Phenomena of Solid Bodies

°C	*Subjective Color Impression*	*M.p. in °C of Various Materials*	
480	faint red glow	Zinc	419.5
		Tellurium	449.5
500-650	dark red	Antimony	630.5
650-800	increasingly bright red	Aluminum	658
800-1000	bright "salmon" red	U.S. Silver Coin	890
		Pure Silver	961
1000-1200	orange to light yellow	Gold	1063
		Copper	1083
1200-1300	white	"Tophet" C Alloy	1350
1300-2500	brilliant white	Nickel	1453
		Platinum	1769
		Aluminum Oxide	2400
		Iridium	2430
2500-3000	dazzling white	Tantalum	2850
		Wolfram	3400

Table 6

Wavelength Ranges and Radiant Emission

Type of Radiation	*Wavelength Range, μ*	*Black Body Radiant Emission*											
		In % of total at °C						*In W/cm² at °C*					
		500	1000	1500	2000	2500	3000	500	1000	1500	2000	2500	3000
Intermediate and far infrared	20 to ∞	3											
Near infrared	1.5 to 20	97	95	80	64	49	38	2	14¼	44¾	96	167	246
Nearest infrared	0.72 to 1.5	0.2	5	19	34	45	49	—	¾	10¾	51	153	318
Visible light	0.72 to 0.40	trace*	0.01	0.4	2	6	13	—	trace††	¼	3	20	83
Ultraviolet	less than 0.40	—	trace*	trace†	0.01	0.1	0.5	—	—	—	0.02	0.3	3
Total radiant emission in W/cm² (approximate)								2	15	56	150	340	650

* Less than 1/10,000 of 1%.
† Less than 1/1000 of 1%.
†† About 0.002% of 1 W.

ments. Since the radiation laws are within certain limits independent of the nature of the radiating solid (or liquid), an experienced observer can also estimate a temperature between 500 and about 1000°C quite closely by the color and intensity of the glow. However, it is hardly possible to make valid statements from mere viewing in the white light regions.

An important fact to know is the smallness of the amount of visible radiation in percent of the total radiant energy emitted. This is shown in Table 6, which derives from rounded-off figures taken from a slide rule type *Radiation Calculator*.[238] Thus, the very bright reddish glow of a piece of coal at 800°C represents actually only about 0.001% of the total emission of radiant energy, the rest being heat radiation. (The amount of UV is even smaller than that of the visible.) An exceedingly bright object in the temperature ranges of the flames of flare candles emits only a few percent in the visible range if the emission conforms to black body radiation laws. Table 7 (from Table 13.1 in Shidlovsky's book[3]) reveals somewhat different data on relations between temperature and visible radiation. Again, it should be noted that these are strictly theoretical figures for a true black body.

Table 7
Brightness and Luminous Efficiency of a True "Black Body" at High Temperatures

°*K*	*Brightness, stilbs**	*Luminous Efficiency, lm/W*
1600	2	0.2
2000	44	1.5
2200	136	3.2
2300	223	4.4
2400	350	6.6
2600	779	9.4
2800	1552	13.9
3000	2872	19.2
3500	9432	34.7
4000	2.34×10^4	50.3
5000	8.41×10^4	74
6000	1.98×10^5	84

* Brightness (luminance) of 1 stilb is emitted from a surface of 1 cm² with an intensity of 1 cp.

Even before we enter the area of actual observations from flares, it must be pointed out that as far as actual materials are concerned the black body laws are riddled with deviations. The most common of these is the *grey body* concept according to which a solid substance emits energy in the same relative proportions as a black body, but in smaller amounts. The factor is called the *emissivity*, and an emissivity of 0.10 or 0.33 would signify that the radiant energy is one-tenth or one-third of that expected from a black body. This emissivity in turn varies with the temperature and also with the wavelength, and Figure 1 from the above-quoted NASA report[235] shows a typical "bundle" of curves of emissivities vs. wavelengths. Table 8 (from Table 13.2, Shidlovsky[3]) gives the emissivities of refractory oxides, important in pyrotechnics, though much below the pyrotechnically more interesting temperature level. It shows that in the shorter wavelength ranges, emissivity can approach black body conditions.

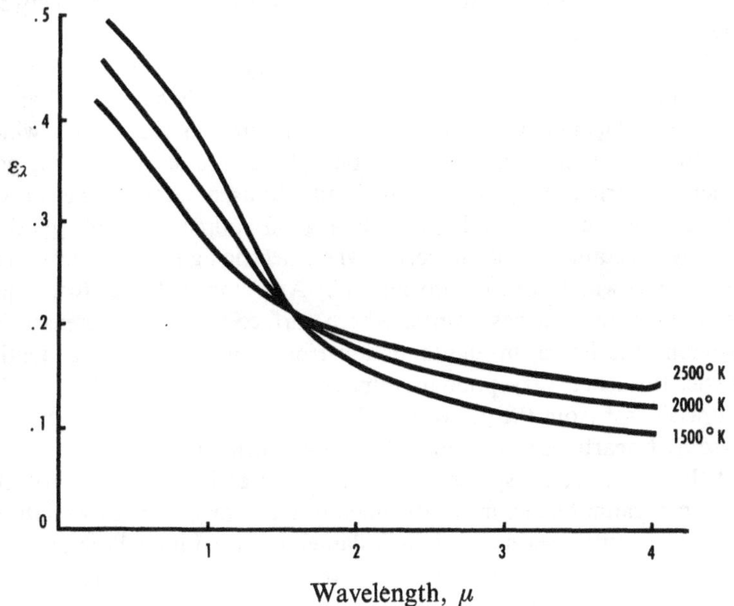

Fig. 1 Spectral Emissivity of a Typical Refractory Metal

Table 8
Variations of Emissivity with Wavelength and Temperature

Material	*Formula*	*Temperature, °K*	*Emissivity expressed as % of black body for wavelength in mμ*						
			750	700	650	600	550	500	450
Aluminum oxide	Al_2O_3	1600	24	25	31	40	53	81	90
		1900	31	33	38	50	65	89	99
Magnesium oxide	MgO	1500	—	23	—	35	45	65	—
		1900	37	41	—	53	61	65	83

Nichols and Howes[239] have studied the behavior of refractory oxides and have found that certain oxides at specific temperatures emit vast amounts of light in a limited spectral range (they claim as much as 85,000 times the expected amount!) and that such radiation depends on previous heat treatment.

Some oxides of the rare earth metals show for solid bodies quite an unusual behavior at elevated temperatures, which should be mentioned here even though its applicability to pyrotechnic uses appears to be remote: Erbium oxide glows with a green light, thulium oxide emits a carmine red glow when heated, but turns yellow and then almost white at higher temperature, and yttrium oxide emits a deep red color between 900 and 1000°C.[67]

The best known practically (though not pyrotechnically) significant deviation from black (grey) body conditions, is the behavior of thorium dioxide (ThO_2) in combination with 1% cerium dioxide (CeO_2), which is used in the Welsbach mantle. It permits the use of illuminating gas, which in normal mixture with air in the Bunsen flame reaches only a maximal temperature of 1600°C, as a good source of light—good at least by the standards of the period when electric lighting was just coming into its own. Prior to its invention by Auer von Welsbach (ca. 1880), luminously burning gas flames, whose surface area was increased by flattening the flame through emission from a narrow slit in a steatite nozzle, furnished very poor illumination (by modern standards) of a yellowish cast from the glowing carbon particles in a flame containing those hydrocarbons that tend to luminous burning.

It has long been suspected that white light emission from pyrotechnic flames cannot be entirely attributed to black or grey body radiation, but the scientific treatment of such flames seems to have been avoided in the broader studies on flames, as shown by the scant evidence in the book by Gaydon and Wolfhard[240] It merely skirts the problem in a chapter on the combustion of solid fuels. However, recent studies in this country and observations cited by Shidlovsky point toward the role that *luminescence* plays in pyrotechnic flames. Luminescence is the emission of visible radiation caused by excitation of atoms or molecules and their return to lower energy state if this process is not caused by high temperature. Shidlovsky quotes Vavilov[241] as defining luminescence as an excess over thermal radiation when this excess "has a definite duration, considerably exceeding the period of the luminous oscillations."

If such excitation of the atoms or molecules is the result of chemical reactions at ambient temperatures, we speak of *chemiluminescence.* In recent years, an amazingly powerful chemiluminescent organic compound has been developed at the Naval Ordnance Test Station and is now manufactured by the du Pont Company. On exposure to air, it glows brightly enough to permit tactical application. The project coded "Tiara" is at present still under security restrictions,[242] but it has been disclosed that the most effective chemiluminescent material known is a relatively simple liquid aliphatic amine, which can be solidified with the help of a waxy material.

Another luminescent phenomenon is *electroluminescence.* Novel types of EL phosphors stimulated by minute electric currents glow brightly in various colors.[243] These two examples are interpolated here because they are potential competitors to pyrotechnic illumination or signaling.

While light measurement or any deeper involvement with the optical aspects of pyrotechnics is not contemplated in this book, a short review of units of measurement is indicated. That the profusion and confusion of units must be bewildering has been plaintively noted in the book by Middleton.[244] The books on optics by Sears[57] and Hardy and Perrin,[58] the discussion of the subject by Shidlovsky, and informative papers and tables from *Temperature*[245] have been consulted. Units of special interest are:

The *light intensity* (I), called candlepower (cp), now based on the new candle (candela) and equal to 1/60 of the intensity of light obtained from a 1-cm^2 surface of a black body at the temperature of solidification of platinum (2046.6°K). For practical purposes there is no difference between the old and the new unit.

The unit of *brightness* (or luminance), the stilb (sb), is the light emitted with an intensity of 1 cp from a surface of 1 cm^2. The same unit is also applicable to the *illuminance*, i.e. the illumination referring to the incidence of light on a material surface. Here the reader encounters the lux (lm/m^2), and the "phot" (lm/cm^2), and the illogically termed foot-candle, which is an illuminance of 1 lm/ft^2.

Familiarity with these and other units is desirable for purposes of photometry, i.e. the measurement of light intensity from a pyrotechnic light source as well as of illumination—for example, the brightness necessary for recognition of objects or for the ability to photograph an illuminated area.

While radiant emission in the black body concept (which postulates an opaque surface) is proportional to the surface area, an application to the light emission from a flare candle, taking the flame surface as an emitter, is probably not permissible, as discussed in the next chapter.

The pyrotechnist, in designing white light sources for utilitarian purposes, is interested in achieving the maximum of light output by means of optimal conversion of caloric energy into visible radiation. As a first step, it is relatively easy to determine the amount of composition consumed per second and its calculated energy per unit weight (kcal/g) or volume (kcal/cm³) expressed in the equation below as H. In such calculations, metal powder in excess of stoichiometric relations must be carefully considered, since it furnishes a large additional calorific output by burning (coincidentally) in air. After photometrically determining the light intensity I (candlepower) and from it the candle seconds per gram of composition (L_0), we derive from the conversion formula F (lumens)$=4\pi I$ (candles) the equation that directly relates light output to chemical energy input:

$$\frac{L_0 \times 3.003}{H \times 1000} = \mathrm{lm/W}$$

Shidlovsky, who furnished this formula, gives as an example a figure of 18,000 cp sec/g from a composition with 1.65 kcal/g and arrives at an output of 32.7 lm/W. For the Mk 24 illuminating candle of 44,000 cp sec/g and 3.2 kcal/gm heat output, described later, the corresponding figure is 41 lm/W.

Merely from the fact that even a true black body would yield less than half this amount (Table 7) at an estimated flame temperature of about 2500 °C, it becomes obvious that in intense pyrotechnic radiation thermal grey-body emission is heavily augmented by selective radiation and luminescent phenomena. This can be experimentally demonstrated by a comparison of light emission from binary mixtures where various alkali salts act as oxidizers. Table 9 (from Table 13.7, Shidlovsky[3]) shows these relations. He uses a fixed ratio of 40% metal fuel and 60% of the nitrates of sodium, potassium, or barium. It is of course not

Table 9
Influence of Oxidizer on "Specific Integral Light Output"

Composition 40% *Mg*, 60% *Oxidizer*	*B.T. sec/in.*	*Candle seconds per gram*
$NaNO_3$	2.3	15,200
$Ba(NO_3)_2$	3.2	13,000
KNO_3	2.9	10.600

possible to equalize all significant factors such as amounts of fuel or of alkali, heat output, and burning rate for a fair comparison in one set of experiments.

If one analyzes the figures, the large contribution of the alkali metal in the oxidizer to the light output is evident—sodium nitrate is much superior to potassium, while heat output per gram (not shown) is about the same. Barium nitrate compares well with sodium nitrate but the heat output from this combination is about one-third higher than with sodium because the fixed weight ratio means a larger excess of magnesium.

On the other hand, the beneficial luminescent influence of the sodium competes with heat radiation from magnesium as shown in Table 10 (from Table 13.8, Shidlovsky[3]) on the specific influence of sodium.

Table 10
Illuminating Characteristics with Variation in Component Ratio

Composition % Mg/%$NaNO_3$	*B.T. sec/in.*	*candle sec per gram*	*kal/g*	*lm/W*
30/70	5.4	9,800	1.3	22.6
40/60	2.3	15,200	2.0	25.0
50/50	1.8	20,000	2.6	23.0

The hypothetical conclusions are strongly supported by recent work in this country. Figure 2 shows a composite of four traces from studies at Denver Research Institute.[246] Each trace plots continuously the radiant energy increment at wavelengths in the visible as well as in the near IR region for stoichiometric magnesium/alkali nitrate mixtures. The oxidizer salts are combinations of 10, 55, 85, and 100% sodium nitrate with 90, 45, 15, and 0% potassium nitrate. Though variations in burning time and total luminous energy output would demand some corrections, the differences in distribution of visible and IR output leave no doubt about the meaning of these curves. The mere fact that all the curves except the one for smallest sodium content in the formula deviate so drastically from a black body distribution curve, which would rise smoothly from small emission in the visible to a peak in the IR region, is proof of the specific luminescent action of the sodium.

A practical application of these facts, undoubtedly found empirically much earlier, is the work of Hart and Eppig[247] in 1947 at Picatinny Arsenal that led to the adoption of mixtures of magnesium, sodium nitrate, and a binder as high-intensity white light sources. These are still in use and are the best we know at this time. More will be said

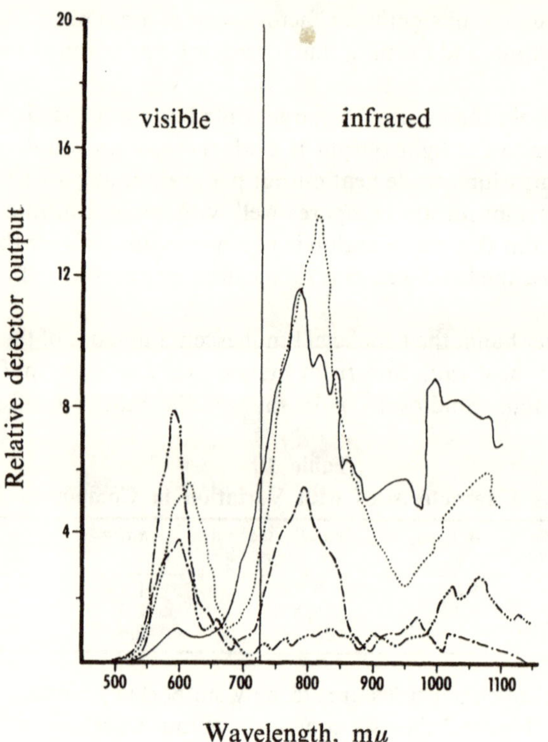

Rel. % $NaNO_3$	*CP*	*BT, sec*	*Total watts/ster*
10 ————	241	11	132
55	519	7	122
85 —·—·—·	598	9	96
100 ···—···—	667	$9^1/_2$	82

Fig. 2 Influence of Alkali on Radiant Energy Distribution

about them in the next chapter.

Where the flare is designed to act as an IR emitter as is the case in augmentation flares for drones in the testing of IR-seeking missiles, no advantage is gained by using flares whose spectral distribution deviates from grey body radiation—at least, not as long as the deviation is toward the shorter wavelength ranges. Formulas designed specifically for optimal emission in the desired wave ranges have not as yet been released for general publication in this country* and are not mentioned by Shidlovsky.[3]

Of even more recent interest are flares or rocket grain ingredients that are apt to produce a large number of ions from the exposure of alkali metal to high temperature. Presence of alkali metal additives in rocket exhaust at 3000°K is said[248] to cause in excess of 10^{14} ions/cm^3, and from mixtures of aluminum with the nitrates of cesium or potassium, electron densities of the order of 10^{12}/cm^3 have been reported.[249] Shidlovsky quotes from Reference 250 that a cesium plasma may be obtained from a mixture of aluminum and cesium perchlorate. The subject of the direct dispersion of alkali metals and other substances at high altitude will be discussed in the concluding passages to Chapter 19.

Color in a flame, as used in pyrotechnics, results from the spectra of excited gaseous metal atoms, molecules, or ions.[251] Salts of a certain limited number of metals are vaporized and the gaseous molecules or their first partial dissociation products lead to band spectra. On further splitting to neutral atoms of the metal, atomic lines are produced, and eventually metal ions create ionic spectral lines. The latter are undesirable for color production in the flame. So-called C-type chemiluminescence, wherein a small number of molecules emit an abnormally large amount of radiation, plays an important role in colored emission of red or green flares. An excellent discussion of the mechanisms of pyrotechnic color production is found in reports by Douda.[252,253,254]

The metals that are suitable as color producers in flames are mainly found in the first two columns of the periodic system—lithium (red), sodium (yellow), potassium (violet), rubidium (red), cesium (blue), calcium (orange-red), strontium (red), barium (green), radium (red), and copper (blue or green). To these might be added boron and thallium (both green) and the weak color producers zinc (bluish white), indium (pale blue), and tellurium (greenish).

Of these elements, only sodium, strontium, barium, and copper have found practical application in the form of their salts in order to create deep color effects. The others either lack intensity of color, are too expensive, are toxic, or do not furnish compounds of practical usefulness in conventional devices because of hygroscopicity.

In the following chapters, pyrotechnic light production will for purely practical reasons rather than as a logical arrangement be divided into the creation of white light other than flashes, for military uses with

*However, for additional information, see Chapter 50.

emphasis on high intensity candles; flashes, both for military and civilian uses for photography; colored lights for ordnance and civilian utilitarian purposes; and finally, the broad field of fireworks.

chapter 14

Production of White Light

General Remarks

Pyrotechnic light production has two purposes: illumination and recognition. In other words, it either is important to light up an object or an area to make it visible to the eye or to a camera, or one desires to see or scan by instrument the light source itself as a signal and, in the case of fireworks, as an esthetic design of light.

Illumination for photographic purposes is best achieved by very brief high-intensity flashes of light, which will be treated in Chapter 15. Signaling by white light, while important in certain spotting charges, is less common than signaling with colored flames as described in Chapter 16, while white and colored light production for amusement will be found in Chapter 17 on fireworks.

The major part of this chapter will be devoted to high intensity, long-burning white parachute flare candles, which are used to illuminate large areas of the ground for the purpose of showing bomb targets and bomb damage at night, illuminating combat areas, acting as anti-infiltration devices, or aiding in emergency landings of aircraft. Points of discussion will be the optical requirements, types of hardware, chemistry of compositions, and techniques and allied problems of manufacture.

Optical Considerations

The light of a full moon on a clear night has been given as yielding an illuminance of the ground of 0.025 foot candles (lm/ft^2), which may be called the minimum of useful light for a military operation while 0.05—0.1 foot candles are required for some night operations.[255]

If we restrict our requirements to the (geometrically) normal aspect of light to target, without involving the cosine law with regard to hilly ground or slant range of light to ground, the required candlepower, assuming identical atmospheric conditions, will depend on the height of the flare above the target. Since a parachute flare descends, but renders in normal practice a fixed light output during its burning, the most unfavorable—i.e. highest—point of the flare above the ground determines the needed minimum output. For the above-outlined minimum illumination at an illuminance of 0.025 foot candles, a flare of 1 million candlepower will satisfy this requirement from an altitude of 4000 ft and for a 5000-ft operational radius. The same candle would have to drop to 2000 ft for 0.05—0.1 foot candles at an optimum radius of 4000 and 2500 ft, respectively. More details on formulas and charts for illumination can be found in a Picatinny report.[256]

The gist of this thumbnail sketch is that a candlepower of 1-2 million over the longest possible or practical burning time of 2-3 minutes is needed for many operations. Air-dropped parachute flares of this size exist as described below, as do smaller less powerful units (but designed for operation at lower altitude) in the form of ground-launched howitzer or mortar-fired illuminating shells.

Standardized High-Intensity Flares

In order to provide powerful illumination from the air by means other than search lights, end-burning flare candles suspended by parachutes are ejected from tubular containers, that have been released from external bomb racks or, for high speed airplanes, from special dispensers. A parachute/candle assembly can also be ejected in flight from the rear end of a howitzer or a mortar shell. Still another way of projecting a parachute flare into the air is from the ground using a trip flare, which is constructed like an antipersonnel mine of the "bouncing betty" type.[257] Hand-held devices are normally too small for useful ground illumination but serve as signals.

The described methods of launching a flare indicate also in some measure the expected candlepower, altitude, and time of burning, since the air-launched canister is the largest; artillery ammunition much smaller; and ground launchers—at least as customarily designed—the smallest and achieving lowest altitude. To these might be added the so-called airport flare—a powerful ground-burning flare for improvised airport illumination. Most parachute flares descend after deployment of the chute at about 10 ft/sec; sometimes, much faster.

Surveys of existing munitions (for illumination) are found in technical manuals,[33,41,43,258] but without pyrotechnic details such as formulas. The quoted candlepower figures may be much too low compared with those now obtained from the same munitions due to improvements since these manuals were compiled. Dr. Hart[255] quotes the doubling and tripling of light outputs between 1948 and 1958 on numerous flares.

The *Mk 6 Mod 6* flare has a cardboard-encased candle with 17 lb of illuminant, which is about 15 in. long and 4 in. in diameter. It burned for at least 3 minutes and gave 1 million candlepower when produced in the 1950's. Formula 31 shows the now-historical illuminant composition that is inferior to the modern *Mk 24 Mod 3* candle of similar dimensions (Formula 32). The latter, which started as the *XA2A* experimental flare in whose development the author was involved, has a more complex fuzing system whose principle is described in Chapter 24. Some Army flares, such as the *ANM26*, *M138*, and *M139*, exceed in size or candlepower or both, those named above, but are also considerably heavier.[255]

Illuminating shells differ from airplane flares mainly in the material for the flare casing, which is a substantial steel canister with attached swivel, though sometimes, as in the *M301A2 Illuminating Shell for 81 mm mortars*, a "boxboard casing" is used. In large canisters, a center tube permits passage of a cable, which delays deployment of the parachute. This center post, naturally, complicates the pressing of the compositions.

Illuminating shells exist for 60, 81, 105, and 155 mm Army shells and for large Navy shells, such as the *Navy 5"/38 Illuminating Projectile.*

They vary in light output from 350,000 to 1,000,000 cp. Burning times are generally 60 seconds or less. Since an illuminating shell such as the *M118* for guns or howitzers has the rather high rate of drop of 35—40 ft/sec, it must be launched—i.e. expelled from the shell—at a minimum of 2100 (better 2500) ft, so that its 60 sec. B. T. can be useful and the candle burns out while still substantially above the ground.[41]

Modern formulas for illuminating shells are similar to that for the *Mk 24* flare except for the higher amount of binder (Formulas 33, 34, and 35).

Recent development of illuminating flares has tended toward longer burning (5 minutes) and more powerful flares of larger diameter. Unfortunately, at the present "state of the art," they are very heavy and require the use of larger parachutes. In order to increase the efficiency indirectly, attempts are made to slow down the descent of the item.

Thoughts have also been given to a better weight to useful illumination ratio by using a candle that stepwise becomes smaller in diameter as it burns and approaches the ground.

Lower-Intensity and Nonstandard Illuminating Flares

Illumination from flares on the ground, with the exception of the above-cited emergency airport flare, need not be powerful; in fact, extreme brilliance may defeat the purpose. This is the case where a sentry wishes to detect a person. Depending on the relative location, the subject is either silhouetted or seen behind the light source, in which case the observer may be blinded by the glare[259] if observer, flare, and the observed are more or less in one straight line.

Hand-held underwater flares for use by divers differ from other types by their battery operated, self-contained initiator protected by a rubber diaphragm. The flares, according to a commercial pamphlet,[259a] burn for one or two minutes and furnish 80,000 cp. They can be used at 100-ft depths and lower, and are useful for underwater inspection and exploration—not to forget their popular appearance in some television programs.

A U. S. patent[260] claims a composition purportedly useful as an underwater flare, though it appears to be overly rich in fuel under conditions where ambient air is absent and the water is not likely to partake in any useful reactions.

In World War II, a flare was designed to be lit under water for silhouetting of submerged submarines.[15] Its interesting formula is given as Formula 36, but I do not know if it performed well. Obviously it was not tactically useful.

For other purposes of temporary illumination, such as in caves, one may use parade torches and similar items described under Fireworks (Chapter 17). Smoke formation must be held to a minimum.

Chemistry of Illuminants

Through the knowledge we have from theoretical considerations, we may postulate that a flare of superior white light output should have the following characteristics:

1. The major fuel-oxidizer reaction should produce the highest possible heat output per unit volume of compressed composition;
2. The heat of combustion in air of excess fuel should be high;
3. The melting and boiling points of the products of reaction should be extremely high and their decomposition temperatures lie above their boiling points;
4. The solid refractory main product—oxide—should disperse readily in air but must have a very high boiling point. No

sintering—partial melting and agglomeration of particles—should occur near the reaction zone;

5. Volatile reaction products or by-products should be present in sufficient quantity to carry away the reaction products from the reaction zone, but ingredients that tend to convert refractory products into vapor should be absent;
6. Emission characteristics should be optimal for the visible part of the spectrum and luminescent phenomena must contribute to maximum emission of visible radiation.

Of all fuel-type metals investigated so far, magnesium powder best fulfills the role of the principal emitting source. It vaporizes in the reaction zone, producing gas-colloidal oxide. Aluminum is less effective, though its calorific output is slightly advantageous. Fine particles such as flake aluminum burn with great brilliance as do titanium, zirconium, cerium, uranium, and their hydrides. However, such particles tend to burn in the solid phase without evidence of vaporization, as has been shown by Harrison[261] for titanium and zirconium. As far as aluminum is concerned, only if the oxidation takes place at a temperature near its boiling point (2327°C), which is about 1200° higher than that of magnesium (1105°C), can vapor phase oxidation take place. Korneyev and Vernidub[262] noticed that with preheating of aluminum and oxygen to 400°C, oxidation is continuous and 81—99% complete, while the unpreheated system reacts in a pulsating manner as aluminum vapor periodically breaks through the oxide film covering the droplets of metal, leading only to 37—56% completion of the reaction. It may be mentioned that while magnesium in powdered form seems to burn completely and in the vapor phase, as evidenced from its ash and flame formation, a burning magnesium ribbon will leave as much as 18 % of unburned metal in that part of the ash that is sintered.[263]

Since aerosolized reaction products surround the flame as dense white smoke, a complex relation exists between smoke formation and light. At our present state of the art, a superior white-light source cannot do without the presence of white smoke from magnesium oxide and sodium oxide (formed from nitrate). By reflection of the light, such smoke can have some beneficial effect, but of course it will be detrimental for illumination. This is a physicochemical as well as a topical problem. The position of the flame, the ambient air currents, and the convective draft from the heat of the reaction combine to keep the flame visible or to obscure it. However, it has been indicated from theoretical consideration that the magnesium oxide within the flame itself is *not* opaque to visible radiation.

The choice oxidizer for high-energy white flares is sodium nitrate. It fulfills several important functions: The reaction with magnesium leads to formation of magnesium oxide; the heat of reaction per gram of components is high because of the high oxygen content of the nitrates and the low equivalent weight of the sodium; and the sodium exhibits luminescent properties that add significantly to the useful light production.

In older formulations, barium nitrate appears as the oxidant but with sodium oxalate added. Light output and other pertinent data for experimental flares containing both sodium and strontium nitrate in varying proportions are found in an NAD Crane report.[264]

White light is sometimes produced with barium nitrate to which a much smaller percentage of strontium nitrate is added in order to neutralize any appearance of color. This has been the object of patents by Schladt.[265] Lower-energy white lights, especially for fireworks, may also use potassium nitrate or chlorate.

During World War II, the Germans utilized the oxidizing properties of sulfates to stretch their short supply of nitrates.[266] Plaster of Paris replaced part of the sodium nitrate in float and in ground signals with magnesium as the fuel (Formulas 37 and 38). Water was added, and while such water becomes chemically bound by the plaster forming the quite stable dihydrate of calcium sulfate, one might assume that such mixtures were only suitable for quick consumption under wartime conditions.

Binders are added to the fuel—oxidizer mixtures to aid as internal lubricants while in liquid state during the compacting phase. Small additions of castor oil, linseed oil, or paraffin wax can be used. Shellac and other natural gums, rosin, resinates, stearic acid, or dextrins are found in many formulas, especially for small items and fireworks. In large flare candles, materials of strong binding force, which solidify and bind by catalytic action without external heating, are now the preferred binders. The ones most often specified belong to the class of alkyd resins and are unsaturated polyesters that contain copolymerizing styrene monomer. The best known in pyrotechnic practice are certain types manufactured by the American Cyanamid Company under the tradename *Laminac*. Other cold-curing resins such as epoxy resins can be used.

In the Combined (CIOS) and the British (BIOS) Intelligence Objectives Subcommittee Reports after World War II, the preference is for "synthetic phenolic resin" in tracers[267] and other compositions[268] and of polyvinyl chloride ("Igelit") both in colored and white signals.[269] A phenol-formaldehyde resin, apparently of a noncuring lacquer type, is described by Shidlovsky under the name *Iditol.* He also mentions Bakelite A, which, however, is only recommended where high-temperature curing to about 150°C is safe.

While the binder adds to mechanical strength and to the important bonding of the materials to the casing and (where in existence) to a centerpost, its heat output and concomitant gas formation also play a part in flame formation and perhaps in light output.

Formulas 32 through 35 show the above-described preferred flare ingredients in suitable proportions and particle sizes for proper burning rate and light output. It has been found convenient to refer to this type of mixture by the code name "Rita."* The burning time (inverse burning rate) decreases with an increase in metal fuel and with a decrease in particle size. However, within narrow percentages ranges, other influences may prevail. A change in particle size of the magnesium to fineness below 200 mesh can increase the burning rate by a factor of 3-5, i.e. the burning time will be from 2½—4 sec/in. against 12 sec/in. for an experimental *Mk 24* type series.[270]

A curious fact taken from single-experiment data from the same report is that the candlepower of experimental *Mk 24* flares stayed about the same, notwithstanding the increased burning rate, so that the candlepower second figure and the candlepower seconds per gram composition became less than one-fourth of the standard value. However, for a flare with a diameter of 1.65 in. instead of 4.25 in. the fine magnesium caused both increased burning rate and increased candlepower and the product of both was only slightly below that of the standard item. The report by Lottes of NAD Crane[264] shows a similar constancy of candlepower second figures for a different type of experimental flare of small diameter. Many factors enter the picture and it is obvious that in large diameter candles at high burning rates, strongly adverse conditions prevail that prevent good luminous efficiency.

A Picatinny report[272] shows that the burning rate increases up to a magnesium content of about 65%.

In order to evaluate, calculate, or estimate the energy output as a a basis for luminous efficiency in flares, we must examine the reaction

* Naming it Rita seems *not* to have been an act of gallantry as was Adolph von Baeyer's naming of barbituric acid after a girl named Barbara whom he courted during his studies on derivatives of urea.[271]

mechanism.

The conventional way of presenting the $Mg/NaNO_3$ reaction is as follows:

$$5Mg\ (s) + 2NaNO_3\ (s) \longrightarrow 5MgO\ (s) + Na_2O\ (s) + N_2 \qquad (a)$$

The ratio of magnesium to sodium nitrate is 41.6:58.4 and the heat of reaction is calculated as 2.0 kcal/g of the mixture.

Because of the high temperature at which the reaction takes place, two corrections might apply to this calculation: The heat of sublimation of the sodium oxide (not known) should be considered rather than the standard solid state assumed, and the reaction might be reformulated to go to metallic sodium:

$$6Mg\ (s) + 2NaNO\ (s) \longrightarrow 6MgO\ (s) + 2Na\ (s) + N_2 \qquad (b)$$

While the stoichiometry changes to 46.6:53.4 of magnesium to sodium nitrate, the heat output taking sodium in the solid state is identical with that for formula (a), i.e. 2.0 kcal, and when allowance is made for evaporation of the sodium, 1.9 kcal/g (actually a reduction of 8% of the heat output).

The next calculation, that of the excess magnesium burning in air, is simple, the figure being 5.9 kcal/g magnesium for the amount above stoichiometry.

It seems futile to engage in calculations of complicated and uncertain reactions between the resin and nitrate since in the presence of an excess of metal fuel one might as well consider the resin as burning in air, using a figure of 7.0 kcal/g of *Laminac* given some years ago to the author by the manufacturer. The figure used at Picatinny Arsenal is $-\Delta Hf^0 =$ 67.7 kcal for the formula C_5H_5O [s] from which, in good agreement with the author's figure, one calculates the heat of combustion as 7.1 kcal/g if water is in standard liquid [l] state or 6.7 kcal/g for H_2O [g].

The final result of these calculations is that the heat outputs from the *Mk 24* formulation, if we disregard entirely the fate of the binder, can be calculated as amounting to 2.9 kcal/g for reaction to sodium vapor, 0.1 kcal higher for sodium in standard state. If the combustion of the resin is included as part of the useful heat, 0.3 kcal/g of composition is to be added. As to the resin, one might have to take into account that at the temperature of the flame, the combustion products from the resin would hardly be those of calorimetric determination, i.e. CO_2, H_2O, N_2, but rather CO, H_2O, and N_2 (the dissociation of water below

2500°K is less than 4%).

We have now all the basic data for a scrutiny of the properties of the high-powered prototype flare *Mk 24*, an experimental "7½ in." candle, and the *81 mm Illuminating Shell M301A2.* These data were chosen because they came from closely controlled experiments with *Mk 24* and 7½ in. candles from an NAD Crane report.[273] The light output and burning time of the 81-mm canister are nominal figures from recent production requirements. Table 11 shows a number of measured and derivative figures which should be of help to the flare designer. Of these, the candlepower per square inch of burning surface area is probably the key figure.

Table 11
Light Output from High-Intensity Flare Candles

Candle	*Mk 24 Mod 3*	*Exp. "7.5 in." Candle*	*M301A2, 81 mm*
Diameter, in.	4.25	7.00	2.835
Burning area, in.^{2}	14.2	38.5	6.3
Candle length, in.	16.75	7.0	4.0
Candle Weight, g	6804	7711	622
Apparent d, g/cm^3	1.74	1.74	1.57
Burning time, sec	170	61	60
Burning time, sec/in.	10.0	8.7	15
Mg %, nominal	58.0	58.0	55
$NaNO_3$ %, nominal	37.5	37.5	36
Binder %, nominal	4.5	4.5	9
Candlepower (million)	1.77	4.05	0.50
cp sec (million)	300	247	30
cp sec/g (thousand)	44	32	48
cp/in.^{2} burning surface (thousand)	125	105	79
Heat output, kcal/g*	3.2 (2.9)	3.2 (2.9)	3.4 (2.8)
Efficiency, lm/W	41 (46)	30 (33)	42 (51)

* With and without consideration for the heat of combustion of the binder.

The efficiency of the light output in relation to the released heat energy was calculated according to the Shidlovsky formula shown in the preceding chapter. While Shidlovsky is inclined to consider the surface of the flame as the emitting surface, there are indications that the flame is, or at least behaves as if it were, transparent and that the emission under otherwise identical conditions would be proportional to the burning zone rather than the flame surface. This does not mean that the geometric surface area at which the reaction in each layer of the cylindrical candle originates *is* the actual emitter. If it were, the steel-encased flare canister would make an extremely poor envelope as the flame front receded into the steel shell. This is definitely not the case for high-intensity flares, though there would undoubtedly be a limit to the depth into which the flame front of a candle in a steel canister could recede without decline in light output when viewed in profile.

Technical Considerations

Consideration has been given to replacement of the envelope by other related materials such as plastics for paper, or light metals for steel. The former may have some advantage because of greater strength and moisture resistance. Certain plastics appear also to contribute to light output, though indications are that such influence is restricted to candles of smaller diameters. As to the use of metals such as magnesium or aluminum as flare canisters, their low melting points militate against the enticing notion of utilizing the fuel value of the envelope as contributing to light output. Such canisters simply melt away in irregular patches and expose the unburnt portion of the candle.

The technical problems involved with both cardboard and steel cases include the question of loading. Under the conventional system of loading a powder mixture directly into the case and consolidating it by hydraulic pressure, it is necessary to subdivide the charge into several parts. As a rule of thumb, the height of the consolidated charge should be no larger than the interior diameter of the tube or shell, but shallower increments in larger numbers are often used. This adds to uniformity of compaction of material.

Where the envelope is made of cardboard, there is not only the danger of bulging and breaking, but also the influence of frictional forces on the cylinder wall, which cause it to collapse at the lower end in an accordion pleating that weakens the structure and shortens the tube length. The use of large, cumbersome, expensive, hinged split molds and the consequent need for time-consuming handling adds greatly to the cost of loading these flares.

In steel-encased flare candles, the loading is somewhat simplified because of the rigidity of the metal case, but other problems arise. The direct contact of metal wall and candle substance may lead to erratic burning as a result of the heat conductivity of the metal. In addition, cracks, insufficient bonding of the candle, or contraction may create accelerated burning with build-up of pressure, which can result in actual detonation of the candle. The most unfavorable conditions exist if a center post increases the possible areas where flaws may occur and when the burning rate of the compositions is particularly high.[270]

Swabbing the inner surface of the metal canister with a catalyzed

Laminac solution, giving it time to convert to a thin protective coating, will prevent erratic performance of standard illuminating canisters. A more substantial coating is provided by applying and curing a proprietary plastic material called *Coast Proseal.*

As a rule, loading pressure for all military items is as high as practical. By maximal compaction, the largest amount is packed into the smallest space with the justifiable expectation of greater efficiency; the effects of sudden acceleration, deceleration, vibration, or temperature variations might also be minimized. However, one should also think of possibly lowered ignitibility, the stresses exerted during pressing on the casing, and the proven possibility of the break-up of larger crystals (350 μ) at 10,000 lb/in.^{2} [273a] with consequent effects on burning. Therefore, the compaction pressure for the large candle does not exceed 10,000 lb/in^2. It takes little arithmetic to calculate that even below this limit the *Mk 24* candle needs at least a 60-ton press.

Specific Hazards of Production of Large Flares

The nature of the chemicals involved in high-powered flare compositions and the large amounts processed at one time and used in one flare candle create specific hazards, which have been assessed only after grievous experiences. Unfortunately, such experience is rarely shared, analyzed, and critically evaluated, but rather obscured by the pyrotechnic fraternity, a shining exception being the specifically applicable report of a fatal accident at a Naval installation.[270]

The large particle size of the magnesium in the above-discussed flare formulas, the relative inertness of nitrates (compared with chlorates and perchlorates), and the large amount of liquid resin of friction-reducing (and thus phlegmatizing) character should make the "Rita" formula one of the safest. Instead, open trough-type mixers (mix muller) with moderate charges (less than 50 lb) have blown up with detonating force. Candles in the mold in the process of compaction have broken molds and damaged pressing equipment, and burning candles have exploded during tests with serious consequences.

In some of these cases, a reason can be given for the most likely causes of accidental ignition. If the sequence of addition of materials to the mixer is such that fuel and oxidizer come in contact before at least one of the two has been thoroughly wetted, the hazard of ignition by friction or by another obscure cause is greatly increased. For this reason, the preferred sequence of mixing is the combination of metal powder and binder, the latter sometimes diluted with a volatile solvent such as acetone or trichloroethylene. Though there may be some qualms about addition of a highly flammable or a potentially pyrochemically-active chlorinated solvent, the advantages of fast and uniform wetting of the magnesium together with the high volatility of the solvents may outweigh the more remote hazards and other disadvantages (residual solvent strongly inhibits the hardening reaction of some resins).

In one case of an explosion in the mold, it was found that it occurred during the last pressing after the first fire had been added. Any accidental ignition due to excessive friction, as might occur when steel canister and pressing ram are not perfectly centered, would of course be easily explained when the much more friction-sensitive first fire (Formula 168) is involved. Such hazard is greatly lessened by a reversal of the pressing procedure, starting with the first fire, but this is often impractical and is impossible where a permanent bottom closure is present.

While these findings are valuable, they should not detract from the point that fuel-oxidizer mixtures of any description (with perhaps the sole exception of standard coarse-grained thermite) can explode violently in loose or compacted state and not necessarily in what may be called confinement. The rapidly released gases or gasified solids, under a self-accelerated burning rate and pressure build-up, make the term confinement meaningless for more than minimal, thinly spread quantities. And since complete control over conditions in or around a mixing installation is beyond human foresight, a potential explosion must be made harmless, at least as to life and limb of personnel, by proper shielding and barricading, and other safety precautions. An essentially complete automation would be impractical (though it can be achieved for simpler processes such as photoflash bombs), and a satisfactory compromise approach can well be reached that permits the direct handling of the completed mixture (subdividing, weighing, loading into mold) without undue hazard.

The explosion of candles during burning, especially of experimental variations or new items, is in some respect a more serious problem because it is even less common and not expected to occur at all. The author has observed this phenomenon with a fairly small candle in, of all things, a large (1000 ft^3) vacuum chamber under highly diminished pressure. The resulting shock was sufficient to cause the large chamber to become dislocated from its mountings.

White Signals

Visual signals indicate the location of an item in flight or at the target, or they constitute a distress signal or a prearranged message— "One if by land. . . . !" Pyrotechnic light signals are normally designed for night use, in which case their candlepower need seldom be high. For daytime use, however, in order to achieve a contrast against the sky as in missile tracking, a candlepower of 100,000 or more is needed. This suffices for small missiles such as the Sidewinder or Bull Pup, which employ such flares as the *Tracking Flares Mk 21 Mod 0* and *Mk 27 Mod 0*. For tracking at high altitude, flash charges are preferred, as shown in the next chapter.

Enhancement of the radiation from the propellant exhaust of drones by target flares may be called "visual" signaling even though it is only for the benefit of a sensor in an attacking missile guided by IR radiation.

A peculiarity of both tracking and target flares is their preferred mode of initiation, i.e. by fire transfer from the hot propellant gases. This is called "parasitic ignition."

Spotting charges may be designed as both day and night signals by a combined light and smoke formation.

A perhaps unexpected use of signal flares consists of misleading a hostile recipient of the message. Faber[1] relates that in World War I the sending in the air of light signals for the purpose of starting artillery barrages was at times thoroughly confused by the shooting off of miscellaneous fireworks items by one party in order to confuse the enemy. In World War II, the dropping of signal flares from airplanes to indicate targets ("Pathfinder System") was counteracted by flares lit on the ground or launched from the ground in rural areas in order to decoy the attacking bombers ("Antipathfinder System").[48,268]

Quantitatively, the most abundant use of pyrotechnic signals is probably in *tracers* for projectiles. Since most tracer formulas pertain to colored flames, this whole subject will be treated in Chapter 16.

In the class of training weapons and simulators, one finds mostly small flash charges. These, however, can become quite hazardous, since the distance from the trainee may be perilously short if he fails to avoid setting it off. *The M118 Illuminating Booby Trap Simulator* averts this hazard since it consists of a slow (30 sec) burning illuminant (Formula 39) ignited through match and starter mixtures by means of a pull wire—in this case, a wire strung across the infiltration path.

A typical formulation for a white signal flare is the one used in the *M18A1 White Star Cluster Ground Illumination Signal* of five fr.e-falling stars propelled into the air from a rifle grenade launcher.[33] The formula (# 40) is of the class covered in the earlier cited Schladt patents.[265]

Some white light signals based on phosphine gas (which ignites spontaneously when reaching the surface of water) were described in Chapter 4, viz. the *Drill Mine Signal Mk 25* and the *Marine Location Marker* (*night*) *Mk 2.*

For marking the location of objects in water, the *Signal, Smoke and Illumination, Aircraft: AN-Mk 5 Mod 4* (formerly *Signal, Drift, Night, AN-Mk 5 Mod 4*) and the *Signal, Smoke and Illumination, Aircraft: AN-Mk 6 Mod 2* (formerly *Light, Float Aircraft, AN-Mk 6 Mod 2*) are used.[258] They are wooden bodies floating on the surface and containing a rather unusual candle composition. It produces both light and smoke, thus being useful in daytime and night time. The 12—15 in.-long flame is visible in clear weather at night as far as 7 miles and the smoke in daytime up to 3 miles. The burning time of the single candle in the drift signal is 12 minutes and since the float light contains four sequentially burning candles, its burning time is about fourfold. A smaller *Drift Signal, Mk 4* has a burning time of only 3 minutes.*

Formula 41 shows the composition of the compressed lead-sheathed pellets for the three items above. It is neither comparable to the regular pyrotechnic flame former—i.e. fuel burning luminously in reaction with solid oxidizer and air—nor with the later-described smoke-formers in which the agent is evaporated and definitely *not* ignited. A thermitic reaction between magnesium and manganese dioxide furnishes enough heat to evaporate the phosphorus, which then burns at the orifice of the container. Thus, no part of the candle proper is exposed to water and, because of the small orifice and the pressure on the inside, infiltration is unlikely. Should the flame be doused by waves or spray however, the spontaneous re-ignition of the hot phosphorus vapor in air remedies the deficiency at once.

Testing of Flares

Notes on photometry and chemical analysis will be restricted to the barest minimum. Together with other techniques and aspects of pyrotechnics these subjects would go far beyond the scope of this book.

The measurement of the candlepower of a flare is not a difficult pro-

* An improved design of the *Marine Location Marker* is the *Mk* 25, *Mods* 0, 1, 2, *and* 3 described in the revised version of OP 2213.[43] A significant difference from the older items is the use of water-activated batteries and electric squibs for the initiation.

cedure but it is still somewhat involved. It is described in the specification MIL-C-18762(NOrd) (1955). It requires the erection of a test tunnel painted a flat black color on the inside and provided with baffles to prevent excessively high readings because of reflected light entering the barrier layer-type light-sensitive cells. The "fireplace," where the candle is burned, is painted white—a condition easily maintained because of the ample deposits of magnesium oxide from the flame. An exhaust fan eliminates obscuration of the flame by smoke. Provisions must be made for burning the flare in at least two vertical positions, flame up or down. Since the light output (and burning time) of illuminating shells varies greatly with the circular motion caused by the spin of the projectile (at 3000 rpm the burning time is reduced to one-half), a turntable may be required to simulate operational conditions.[273a]

In order to bring the photocell to a response that approximates the sensitivity of the human eye for the different parts of the visible spectrum, color-correcting filters are interposed that attenuate the ranges in which the normally used selenium cell is more responsive than the eye, i.e. below 0.53 and above 0.58 μ wavelength. Standard lamps are used for calibration.

The procedures are not claimed to measure accurately the actual candlepower but rather permit a fairly precise measurement of approximate values for purposes of routine checks and for acceptance tests. More elaborate are the provisions if one is specifically interested in the spectral quality of the light or in areas of the nonvisible portions of the spectrum. By use of a multiplicity of optical filters with several photocells, or by use of other sensors such as lead sulfide and lead telluride cells or bolometers and cut-off filters, some information can be obtained. Only by use of a scanning device that permits a continuous variation of the wavelength for which output is recorded can a true picture over either the visible or other discrete portions of the total spectrum be obtained. Monochromator spectrometer data are discussed in an NAD Crane report,[274] which also shows the graphic trace of a curve relating radiant energy *vs.* wavelength for a pyrotechnic flare compared with various nonpyrotechnic light sources. Continuous measurements of visible and infrared radiation over wider wavelength ranges are something quite new and, as shown in Figure 2 in the preceding chapter, are able to clear up some of the problems concerning the still-obscure parts that different emission phenomena play in relation to each other in pyrotechnic flames.

Quality control of flare production might involve a chemical analysis but the great time lag between the performance of such an analysis and the flow of produced goods restricts such analysis to spot tests or to special inquiries. However, thinking along lines that deviate from classical quantitative analysis will eventually lead to more rapid quality control.

Calorimetric tests are probably of very small value in any pyrotechnic device that burns in air. Such testing is discussed in Chapter 34 in connection with processes where the heat output as such is of interest.

chapter 15

Flash Charges

Many fuel—oxidizer mixtures can react instantaneously with explosive force under conditions leading to self-acceleration of an initially slow and orderly reaction. Pressure build-up is the primary accelerating factor. It may take place with consolidated charges, not necessarily in complete confinement, if cracks or defective bonding precipitate a reaction rate at which the gases cannot properly escape, whereupon the increased pressure leads to still faster reaction—a cycle completed in milliseconds. The same may occur—in fact is more likely to occur—with loose powder mixtures. If such an explosive reaction and dispersion is the desired effect, confinement of the charge alone is not relied upon, but an explosive stimulus is applied from a central burster; if the charge is large, a certain amount of black powder will help in uniform formation of a fire ball. Since all the energy of the reaction is liberated in less than 1/10 sec, the candlepower from a properly formulated flash charge is very high. Illumination of the ground from a large distance and over a large area can be achieved for the purpose of photographic reconnaissance. Smaller flashes are suitable as signals and for close-range photography.

Illuminating Flashes

Illuminating flashes for photography were first used by amateurs and reporters. For on-the-spot action, magnesium powder was dispersed and ignited in hand-held contraptions. For indoor picture taking, the metal powder and oxidizer were kept in small paper bags, which were suspended by a string and ignited by an impregnated strip of paper fastened to the lower end. In the writer's own experience, this quite unsafe device seemed to have done nothing worse than engulf the room in dense smoke and lead to pictures of dubious quality and odd poses. Attempts were made to reduce smokiness by the use of unusual chemicals such as thorium nitrate as the oxidizer[66] or the oxides of molybdenum or tungsten in combination with magnesium or aluminum powder.[275] Equally obsolete is the method of creating the effect of lightning on stage by blowing the brightly flashing spores of lycopodium (club mosses) into an alcohol flame.[7]

In passing, we must also mention here the modern photoflash bulb, mainly since it represents a pyrochemical phenomenon in a highly unusual design. The flash action is safely produced within the confines of a plastic-coated glass bulb; the oxidizer is pure oxygen but at diminished pressure; and ignition is by electric current. Thin aluminum ribbon or wire has long been the light-giving metal fuel, but it is being replaced by zirconium, which permits the use of a much smaller bulb. Even more advantageous seems to be the pairing of the fuel metal with fluorine as the oxidizer or with the more easily handled oxygen difluoride (OF_2)—both still in the experimental stage.[276]

Because the ignition of the aluminum in flash bulbs requires a powerful igniter mixture, special first-fire mixtures containing zirconium and potassium perchlorate have been developed and are reported in the literature.[277,278]

An unusual flash powder consists of a mixture of powdered aluminum and pyrophoric iron.[279] The mixture is made from ferrous oxalate and aluminum heated to 475°C. The iron ignites spontaneously on exposure to air and is said to ignite the aluminum powder.

Illumination for photographic reconnaissance at night in wartime is performed by powder mixtures or, exceptionally, by metal powders alone dispersed in air by an explosive charge. Depending on the altitude at which the unit is released, a relatively small "cartridge" or a much larger "bomb" is used. Table 12 shows five U. S. and two German items and pertinent data taken from several sources.[48,255,280]

The candlepower from a flash charge observed with proper instrumentation measuring candlepower *vs.* time will rise steeply and decay rapidly. For the *M112A1 Photoflash Cartridge*, the following data have been obtained: time to peak 2.2 msec; after 15 msec, candlepower is half of peak; after 28 msec, only one-tenth. Generally, the useful light output is measured by integrating the candlepower *vs.* time curve over the first 40 msec from start. From the viewpoint of coordination of the flash with the aerial camera, a minimum time to peak and a reduced burning time for the period of adequate light emission are desirable goals for further improvement.

Because of the destructively explosive force of the flash powder, the cartridge must be ejected from its dispenser and not exploded before a certain minimum time has passed. The *M112* cartridge is furnished with delays of 1, 2 or 4 sec; the larger *M123* with 2, 4, or 6 sec. As with

Table 12

Photoflash Cartridges and Bombs

Type of Munition	*Altitude, ft*	*Charge, lb*	*Total weight*	*Peak cp, million*	*cp sec, million*	*cp sec/g metal*	*Formula*
Cartridges							
M112A1	3,500	0.50	1.0	120	1.4	16,700	Type III, Class A Formula 42
M123A1	8,500	1.75	4.0	400	5.0	15,700	Type III, Class A Formula 42
Bombs							
M120A1	30,000—45,000	85	165	3,200	90	5,800	Type III, Class A Formula 42
T93		230		4,500	111	1,100	Type III, Class A Formula 42
Dust Bombs							
German W.W. II		33		450	63	4,200	Flake Aluminum
〃 〃 〃		66		800			Atomized Aluminum
(U.S.) M122	20,000	75	110	1,300	45	1,300	Mg/Al Alloy 65/35

flares, the data in the older manuals[33,43] as far as output, loading density, etc. are concerned may be lower than achieved in more recent times. In addition, the reported figures contain some obvious misprints or errors and the delay-time variation, e.g. of the 1-sec delay, has been improved from an earlier ± 0.3 sec to ± 0.1 sec. This, however, is still regarded as excessive and complicates the shutter synchronization problem. Photoflash bombs rely on mechanical fuzes.

Aerial night reconnaissance photography was started by the British in 1918 and went through numerous phases because of the difficult tie-in of illumination and camera action. The historical aspects of the development, with emphasis on the problem of relating the interval of highest light emission with photographic exposure, are related in a BuWeps report.[281]

During the course of the development, the preferred materials were magnesium, aluminum, or magnesium/aluminum alloy with the oxidizers barium nitrate and potassium perchlorate. The metal alloy was used in World War II because of the severe shortage of powdered aluminum and magnesium, the latter being laboriously produced by a modified lathe-cutting operation that was woefully inadequate to provide the smaller particle sizes, whereas the alloy is an extremely brittle, relatively easily comminuted solid. It was used in mixture with either 40% potassium perchlorate or 54.5% barium nitrate. Later, a mixture known as the Type III, Class A, specification MIL-P-466A was preferred and is still the standard composition for both cartridges and bombs (Formula 42). It contains atomized aluminum as the fuel.

The efficiency in terms of cp sec/g metal for flash charges is much inferior to that for flares. Since a much larger part of radiant emission is in that part of the infrared that can be made photographically useful with special IR sensitive film, efforts have been made along these lines.[281]

Another important point is decrease of efficiency both with increase in the weight of the charge and when only metal powder without oxidizer is dispersed. The latter type, the so-called "dust bombs" were developed because large amounts of the explosive flash mixture make the airplane carrying photoflash bombs extremely vulnerable to destruction. The flash charge and with it everything close by may explode when the bomb is perforated by bullets or shell fragments. Bombs that contained only metal powder used either aluminum powders or (in this country) a magnesium/aluminum alloy (65:35) as in the *M122 Photoflash Bomb*.

The use of metal alone causes a slower time to peak light output—42 msec in the case of the *M122*, which is longer than the time to completion of the main action with regular flash powder. The inefficiency of the dust bombs led to development work on segregated compartments for fuel and oxidizer both surrounding the burster coaxially. However, these efforts did not lead to a better utilization of the powder.

The problems encountered in the creation of a fire ball from a flash charge, its development, duration, size, shape, useful light output, etc. depend on many factors and are undoubtedly even more complex than the problems of ordinary pyrotechnic flames. Chemical ingredients, geometry of the total assembly, the quality of the burster, the degree of confinement of the charge (i.e. the strength of container and its seals) prior to the burst all have a bearing on performance. From a preview of the subject in the forthcoming *AMCP Manual*,[30] this specific subject and its literature has been thoroughly treated therein. It has been given short shrift here because very few people engage in photoflash charge manufacture and still fewer in a theoretical study of its functions. Undoubtedly, further studies could lead to spectacular improvements in luminous efficiency.

Because photoflash powders are hazardous in any quantity, one might regard their processing and the manufacture of cartridges and bombs as the most hazardous in pyrotechnics. It is just this definite, unequivocal danger that leads to such extensive precautions and special management that the hazard becomes greatly reduced. A safety factor lies in the possibility of mixing the ingredients either in the final container by vibration or with the help of an attached canister or mixing device and of disconnecting the two units by remote control only after the mixed powder has come to rest in the cartridge or bomb. This makes the production actually safer than that of flares where loose powder has to be handled after mixing by pouring, pressing, etc. with only partial remote control action.

Signal Flashes

The majority of flash charges used as a signal or an imitation or symbol for some flash from a weapon are relatively small. They comprise those that are incorporated in training weapons such as the *M110 Gunflash Simulator* used for the training of artillery observers and troops in maneuvers and as a decoy, since its effect closely resembles the flash of certain guns (Formula 43); the *M115 Projectile Ground Burst Simulator*, which combines a pyrotechnic whistling sound with a loud report and flash (Formula 44); the *M117 Booby Trap Flash Simulator* (Formula 45); and several air-burst simulators fired from a grenade launcher or a pyrotechnic pistol. All these are described in a technical manual.[33]

A similar but larger item is used with a special firing device to simulate a tank mounted gun, i.e. the *Gun Firing 90 mm Blank Cartridge with Detonating Cap* (Formula 46).

To these items must be added the devices used in the motion picture industry for "special effects." Most of these are clever mechanical arrangements but pyrotechnics plays a part, especially in battle scenes. A recent (1966) book is devoted to this subject.[281a] A noteworthy fact is that a manufacturer of such items advertises, among others, red, green, and yellow flash powders.[281b]

The most ambitious and also least innocuous training devices are the ones which are meant to simulate an atomic bomb. The *EX*1 *Mod 0 Nuclear Air-Burst Simulator* is an air-launched bomb weighing in excess of 500 lb and containing various flash, smoke, and sound charges. With a total of 300 lb of a modified magnesium flare mixture, red phosphorus, and high exPlosive, it produces an impressive fire ball and "mushroom cloud."

Signal-type flashes are employed in the calibration of the electronic tracking system of missiles where small cartridges, especially suited for effective light output at very high altitudes, are ejected at intervals. Picatinny Arsenal's "Daisy" cartridge weighs only 3 oz, furnishes 23 million candlepower at 80,000 ft, and is visible at a distance of 1000 miles.[282] There are similar cartridges of slightly different size and charge weight. A hazard with these items is the fact that they may damage the missile if their explosion is not properly delayed following ejection. They also cause radio-frequency interference.

A peculiarity of the class of high-altitude flash charges is that (undoubtedly in connection with spectral emission) they are as effective in the highly rarefied atmosphere as at low altitude.[283,283a] As shown in Formulas 47—50, the spectral effects are produced by calcium either as metal or as compound, and additional advantages are gained by the presence of a sodium compound. Information on these developments is found in several Picatinny Arsenal reports.[283-286]

Flash charges are also used to indicate the point of impact of practice bombs and of small caliber spotting rounds shot from a small arms weapon with a trajectory identical to that of a larger gun. Formerly, black powder was used for such purposes, providing both flash and smoke. Nowadays, a zirconium-based flash composition of 1.5 million candlepower and exceptionally long duration (700 msec) has been successfully employed in a 20-mm spotting round of about 6.5 cm^3 volume.[30]

Another mixture, more modern than black powder but obviously also a "smoky flash" (though all flash charges end up as a more or less evanescent smoke puff), is described in a recent Picatinny Arsenal report.[287] It consists of a mixture of equal parts of red phosphorus and magnesium, ejected and ignited by meal powder.

chapter 16

Colored Lights

This chapter will be devoted to colored lights used for military signaling and for civilian emergency flares, while the following chapter will contain the colored flames, flares, and spark effects of fireworks for pleasure. The division is necessary because not only the choice of colors but also formulations and manufacturing techniques are different. Both branches can of course benefit from each other and some overlapping is obvious.

Color Selection

Briefly, colored *signals* must be unequivocally recognized and named as to the color they represent by all observers who are not color-blind. The color should not lose its quality when viewed from a long distance or under other unfavorable conditions such as through haze or in daytime. Fireworks, on the contrary, need neither be very intense in luminosity nor restricted to definable colors. They are viewed from a conveniently close spot on clear nights and delight the eye by a wide variety of delicate tints of, say, pink, carmine, or lavender—none of which would be a suitable "red," coded as such.

Holmes[288] in an extensive World War II study on *non*pyrotechnic signal lights summed up his findings as follows:

> One outstanding result of the experiments is that blue is not seen *at low intensity*. Blue and purple cannot be regarded as satisfactory signal colours. Blue would be safe if the illumination were very high . . . *Red is the ideal signalling colour* . . . if *orange* lights are not also seen by the observer. *Green* . . . is a safe colour . . . *confusion with blue* . . . Yellow and white colours are safe *if used separately* [italics added].

It was also brought out in the printed discussion of this article that "flashing" (not in the sense of a flash charge but showing the light just long enough that attention can be focused on it) gave better recognition in cases of doubt than prolonged steady lighting.

These views seem to be well applicable to the pyrotechnic color recognition and identification problems. Red, green, and yellow are the foremost recognized signal colors, though one occasionally finds the designation "amber" and there are blue signal lights. The latter seem to be somewhat maligned and deprecated,[289] apparently not so much on optical grounds but because it is difficult and perhaps impossible to create a blue pyrotechnic flame of great depth of color.

In a more limited way, a problem exists in the proper *naming* of colors,[290] and perhaps also in the bias of some observers with otherwise normal color recognition (such as the writer) toward the color yellow, whereby the slightest admixture of yellow to blue, especially in light shades, makes the person call this color "green" rather than "blue."

As will be shown below, it is possible to extend the variability of color messages by combining several signals as simultaneously-seen discrete lights of the same or of different color.

Colored signals as ordnance or utilitarian commercial items comprise a variety of devices, which function either on the ground or more often by being propelled into the air from the ground, from a submarine, or from an aircraft displaying one or more "stars." If of short duration these are unsupported, but if the burning time is longer they are suspended from a small parachute. Such signals are useful for communication, identification, or as distress signals or spotting charges.

Compositions for Colored Lights

In compiling a list of compositions for colored flame production, formulas have been taken from specifications for certain items and a number of such formulas could be found in a rather unusual source, i.e. a manual on the toxic hazards of pyrotechnic items.[50] Others have come from the writer's notebooks, a few from patent disclosures or from special reports. Since there is now little overlapping between the formulas for ordnance items and for fireworks, the latter, which are appended to the following chapter, are only of passing interest to the designer of military flares and stars.

Formulations for *red* and *green* flame are remarkably similar. The basic coloring agents for red and green are the elements strontium and barium, mainly in form of the relatively volatile chlorides.* It will not do, however, to use the chlorides as such since the formula would simply be too "congested" if these salts were combined with a fuel—oxidizer heat source. In other words, there might not be enough of all three major ingredients, or, as one might say facetiously, the pyrotechnician

* They are mentioned as ingredients of some novelty—items resembling wax candles and burning with a colored flame. Bennett[313] cites formulas for such candles. Their fuel may be paraformaldehyde or similar nonsooty burning agents.

is stumped by the fact that there is only one hundred percent in a formula. At least one chemical must therefore do double duty, and the best solution to this problem would seem to be to utilize as oxidizers the chlorates or perchlorates of strontium and barium, which become halides after consumption of their oxygen. This is only practical with barium chlorate, which had been used until some time after World War I in the United States in military formulas, but which according to Fedoroff (Vol. 2)[49] is not used anymore. Strontium and barium perchlorate are much too hygroscopic for practical use.

Two ways are open to utilize perchlorates as superior colored light producers: One consists of finding cationic complex salts of the alkaline earth perchlorates in which the coordinating ammino, ethylenediamine, etc. group converts the salt in an anhydrous, nonhygroscopic, stable compound.

Smith and Koch[291] have reported on the use of the compounds $Me(NH_3)_2(ClO_4)_2$ where Me stands for Ca, Sr, and Ba. A sample received about ten years ago from the G. Frederick Smith Chemical Company with the fórmula $Sr\ en_4(ClO_4)_2$ (en=ethylenediamine) is still in excellent, barely agglomerated condition, and when ignited with a match yields a flame of beautiful deep red coloration. Douda[251] has reported on the properties of tris(glycine) strontium perchlorate with the same goal in mind, but standardized colored flares have not as yet come forth from these interesting efforts.

The other approach is to embed the hygroscopic perchlorates in a plastic material, which thus becomes fuel, binder, and protective coating and also opens the way for new production techniques. This scheme has been tried as shown in a U. S. patent[292] for lithium perchlorate embedded in nylon for the purpose of formulating high-energy propellants, and it is applicable to colored flame production.[291a] A similar idea is incorporated in another U. S. patent[293] that deals with colored flame matches or more exactly match splints made from cellulose acetate in which the perchlorates of copper or strontium perchlorate, dissolved in triethanolamine, are dispersed. Again, the color effect is most pleasing, but the technique seems to have remained confined to laboratory samples.

We must now return to the less esoteric aspects of colored flame production. A practical approach is the use of the nitrates of barium and strontium as oxidizers followed by their partial conversion into halides by combining them with the oxidizers potassium chlorate or perchlorate. Eliminating the chlorate for military purposes because of its hazardous behavior in combination with fuel-type ingredients, potassium perchlorate is best suited because of its stability, nonhygroscopicity, the negligible tinctorial effect of the potassium (as long as the quantity is not too large), and the capability of the residual chloride to react in metathetical interaction with the residue from the strontium and barium nitrate to yield the desirable alkaline earth chlorides. However, this secondary conversion can only be a partial one and is even more limited by the fact that the amount of perchlorate must in many cases be small because of its accelerating influence on burning time. It is therefore necessary to introduce a specific halogen donor of high chlorine content. The favorite materials are organic compounds: hexachlorobenzene [C_6Cl_6 (JAN-H-257)] with 75% chlorine; polyvinylchloride [PVC (MIL-P-20307)] with 64% chlorine; and the rarely-used dimer of hexachlorocyclopentadiene, Dechlorane (trademark of Hooker Chemical Co.) ($C_{10}C_{12}$), with 78% chlorine. Apparently not used as color intensifiers are Parlon (trademark of Hercules Powder Co.), a chlorinated polyisopropylene with 67% chlorine, and the later-discussed smoke ingredient, hexachloroethane (HC) [C_2Cl_6 (JAN-H-235)] with 90% chlorine. Note should be taken that hexachlorobenzene (C_6Cl_6) is different from benzene hexachloride ($C_6H_6Cl_6$). The latter occurs in five isomeric forms, one of which is a well-known insecticide (Gammexane, Lindane).

Copper powder or copper oxide is occasionally added to green formulas as an intensifying flame colorator. The conventional binders are used but sparingly in order to minimize weakening of the flame by luminous(sooty) burning of organic binders. This is especially important in flares of low luminosity where, incidentally, one must also consider that a cardboard tube may have undesirable burning characteristics. The luminosity of the flame is boosted by the addition of varying amounts of magnesium powder, which increases the visibility at greater distance or in daytime, but contributes nothing to the coloration, rather it tends to overpower it.

Yellow flame presents no color problem considering the very strong tinctorial power of the yellow sodium light. Because of its nonhygroscopicity, the oxalate is nearly always used to introduce the sodium. There would seem to be no reason to abstain from the use of highly purified sodium nitrate if the flare could be fully protected from the influx of high humidity during storage. It should be noted that according to Douda[252] halogens depress the color radiation for the alkali salts such as sodium. Furthermore, a series of formulations involving mainly magnesium and sodium nitrate, with or without added sodium oxalate, point toward an improvement in color purity in the presence of the oxalate.[253]

The most difficult color to produce in satisfactory depth is *blue*. This problem is less serious in the fireworks art, but the materials used there are looked at with suspicion in modern ordnance formulation. The essential color producing element is copper, which emits a blue light in the presence of chlorine (in its absence, green). The large number of compounds of copper one finds enumerated in the fireworks literature allows the conclusion that none is very effective.

Formulas 51—82 give a fair cross section of recent developments in colored-flame formulation. The novice in the field must be baffled by the fact that white and colored flares are so closely related in composition considering the differences in generation of visible radiation of wide or of narrow wavelength bands. The reason lies in the uncomfortable fact that the formulation of colored light sources is a series of compromises between visibility and color recognition. A consequence of this similarity is that the techniques of mixing, pressing, and fusing are identical. However, with colored flares of low candlepower one must avoid the presence in or near the flame of objects that may be interfering light emitters such as residues of some first fires, cardboard that contains luminous substances (sodium), or residues from fuse trains.

As to instrumentation for measurement of light output, the means are also identical with that for white light. However, if one is interested in the purity of the color, the measurement of chromaticity coordinates, and other data specifically pertaining to color, the photometric measurements must be modified by simultaneously employing several sensors with or without a variety of filters which permits the passage of light in limited wavelength ranges.[3,253]

Standardized Items

Many standardized items of this class are found in technical literature.[30,43,257,258] Because of the varieties encountered, only a summary and partial enumeration can be presented. The simplest consist of a large shotgun-type shell, about 4 in. high and 1½ in. in diameter, with a percussion primer in the rimmed base. During World War II, when the writer's company fabricated these in enormous quantities, only the rimmed base was metal, while the body of the shell was made of cardboard. Since then, one-piece impact-extruded aluminum shells have been adopted and carry the added designation -A1. (-A1B2 signifies a steel body). The content of the shell is fired from the *Pyrotechnic Pistol AN-M8*, commonly called the "Very Pistol," though the latter is a different item of smaller diameter, designated *Very Signal Pistol Mk 5 Mod* 0.[43] The content of the shells is either a single star of the series *AN-M43—45* in red, yellow, or green; two stars of the same or different colors, *AN-M37—42*; or a colored "tracer" and two stars of the series *AN-M53—58*. The latter scheme would permit 18 combinations, but not all of them seem to have been used. Such signals were employed during World War II to identify incoming aircraft as friendly when they displayed the "color of the day." The formulas used are found under Nos. 51—59. Of these Nos. 51, 54, and 57 belong to end-burning, fusee-type flares, but they are grouped with the stars and tracers since they form a family of formulas with identical ingredients representing the "state of the art" during World War II.

The unsupported pellet assembly in a thin-walled, perforated aluminum cup is ejected by black powder and ignited with quickmatch and pressed-on coarse black powder. The fire transfer causes enough delay so that the stars or the tracer are ignited at some distance from the point of ejection. The pellets reach an altitude of about 250 ft and burn about 7 sec in the case of the single or double stars. Where there is a tracer, it burns first for about 3 sec and the two stars, which ignite subsequently and simultaneously near the apex of flight, burn for another 3-4 sec. all being visible for about 5 miles at night or 2-3 miles in daytime. The candlepower of the stars is in the 12,000—30,000 range with the red stars generally being on the high side of light output.

A longer burning item of similar construction and launching is the *M11 Aircraft Red Star Parachute Signal.* The "star" in this case is case is actually an end-burning candle since otherwise the burning time of 30 sec would be unachievable.

Some signals are designed to be launched without the need for a special bulky pistol, i.e. from a rifle or carbine. A single parachute supported star is contained in the *M17* to *M21 Ground Signals* or a free-falling cluster of five stars in the *M18* to *M22* series. These signals differ from other series by using white, green, or "amber" and by omitting red.

The term "ground signal" means that the signal is *launched from* rather than burned *on* the ground, a somewhat confusing terminology. Even the word "hand-held" does not imply performance on the ground, as the following items show. There seems to be no unequivocal term for a truly hand-held (or stuck-in-the-ground) torch light.

The "hand-held" *M125, 126* and *127 Illumination Ground Signals* require no external launcher at all. Their assembly of five green stars (M125), single red star (M126), or single white star (M127) is propelled into the air up to 700 ft by rocket action. Launcher tube and initiation device are the outer casing and cap, and the "rocket-motor" contains a compressed black-powder pellet.

The *Mk 13 Mod 0 Day and Night Distress Signal* is for observation at ground level. It is double-ended for use as a hand-held red flare or an orange smoke emitter. Another larger hand-held signal with a sturdy wooden handle and emitting red, blue, or white light for 1-2 min is the *Navy Light Mk 1 Mod 0, 1, and 2.*

The Signal, Smoke and Illumination, Marine: AN-Mk 13 Mod 0 (formerly: Signal, Distress, Day and Night: AN-Mk 13 Mod 0) leads to the well-known commercial *railroad fusee*, which is also a military item under the designation of *M72, Railroad Warning Fusee; red 20 min.* This is the most popular size (about 16 in. overall with spike and cap). Shorter and longer ones of 5—30 minutes B. T. exist. The 30-min flare as well as yellow and green flares of this type are commercial items.

The red-burning *Railroad* or *Highway Fusee*, described in detail especially as to testing procedures as a commercial item in a Bureau of Explosives leaflet,[294] is a low-intensity (70—85 cp*) flare of slow burning time (about 2 min/in.) During burning, a prodigious molten and dropping-off residue is formed, part of which tends to form a sintered scoria or chimney. The fusee is ignited by a safety-match-type "button." On top of the wooden plug that is part of the protective cap is a layer of match striker material protected by a water-resistant wrapper. Formula 63 shows the composition of a good commercial item; closely related ones have been reported by Weingart.[7]

As a commercial item, mass-produced by at least half a dozen manufacturers, the fusee is a very inexpensive (about $25—50 per gross) flare, but within limitations, serviceable and well-functioning if protected from exposure to very high humidity for prolonged periods. There seem to be no statistics on the manufacture and consumption of this item. Beside its use by truckers and brakemen as a warning light, it has military and semi-military applications such as in clandestine operations for the outlining of drop zones or of landing strips. The

* TM9-1370-200 gives, apparently erroneously, 0.85 M cp and so does the author's book *Modern Pyrotechnics.*

railroad fusee has in the past been the subject of a number of patents,[295-301] some of which make informative reading.

Very small red signal flare cartridges with a compact, fountain-pen-sized "launcher" have been merchandized for civilian use as emergency signals, but have also aroused considerable military interest as shown by the *EX79 Mod Signal Kit*, which can be carried in flight clothing.[302,303] An item of this kind has also been the subject of U. S. patents.[304] Tests have shown that such a miniature cartridge rises to 275 ft and burns for about 3 sec, being well visible in daylight. Also, a 12-gauge shotgun flare shell has been designed whose flare reaches 500 ft and is extinguished by the time it drops to 150 ft.

The most abundant manufacture of pyrotechnical ordnance components is undoubtedly that of the so-called *tracers*. They are very small flares pressed into a cavity at the base of a small-arms or artillery projectile or into a separate assembly fitted into the base. Since the tracer (if properly functioning) burns in "cigarette fashion," it may serve as a delay that destroys the projectile after a definite interval.

Tracer mixtures must be compressed very highly (up to 125,000 lb/in.^{2}), must not explode when ignited under the high pressure of the propellant gases, and should not blind the gunner. Thus they present special problems. Red is the preferred color, but others, including white, have been used outside the United States, as shown in Formulas 64—69. Burning time depends on the function of the shell, 3—4 sec sufficing for an armor-piercing (AP) shell, while 6—10 sec may be required for antiaircraft ammunition.[257] Information on the chemistry of tracers and tracer ignition has come out of Frankford Arsenal,[305,306] Picatinny Arsenal,[307] and has also been found in the patent literature.[308,309]

In Formulas 70—80, additional modern or fairly modern (i.e. W.W. II or later) recipes are presented with short notes on their specific properties and applications. In general, the historical position of a formula can be gaged by the prescription or absence of certain ingredients. Natural gums, asphaltum (gilsonite), castor oil, and hexachlorobenzene usually mean that the formula dates back to the forties. PVC, Laminac, and ammonium perchlorate appeared in the fifties or later.

Dye Markers

The dispersion of dye in water as a marker, indicating for instance the location of the initial point of contact with a submarine, is pyrotechnical only insofar as a delay fuse and a bursting charge dissipate the dye. The *Depth Charge Marker, Day, Mk 1, Mods 1, 2, and 3* (Mil-M-18541A) is thus classed with pyrotechnic devices, in line with the late Dr. David Hart's (Picatinny Arsenal) remark that in ordnance everything that cannot be easily classed is thrown into pyrotechnics. The active principle in the sea marker is uranine, the sodium salt of the well known yellowish-green dyestuff fluorescein (40% in *Mod 1*, 86% in *Mod 2*) mixed with a water-soluble inert compound to facilitate dispersion. In another version of the item (*Mod 3*), a stearated chrome yellow pigment that forms an undissolved thin layer of brightly colored material floating on the water is used.

Other items that act by the spreading of soluble dye are the smaller *Slick Marker Cartridge AN-Mk 1 Mod 0* and *Slick Marker AN-M59*. It is said[43] that the latter, by the use of 1 kg of dye, forms a slick 20 ft in diameter persisting for 2 hours and visible from 3000 ft at a range of 10 miles.

chapter 17

Fireworks

This chapter, except for some minor alterations, was written by the Reverend Ronald Lancaster, M.A., and all the formulas pertaining to fireworks have come from this contributor.

According to a survey in 1960,[310] the total retail value of firework items in the U. S. A. amounted then to 16 million dollars, of which about 13 million dollars was domestic. 60% of the total production and imports is used on the Fourth of July, the rest mainly in the South at Christmas and New Year. In Great Britain, fireworks are mostly used on the Fifth of November (Guy Fawkes' Day), the total expenditure being in the region of 4 million pounds sterling annually. Germans have their celebrations at New Year's Eve (Silvester) and spend over 30 million marks per year.[22] These figures do not seem to include utilitarian items such as warning flares, railway torpedoes, or special items used in the motion picture industry.

The condition of the firework industry seems to vary considerably in different parts of the world and very much reflects the attitude of the state or country. This is very obvious in the United States where fireworks are strictly regulated in some states, often only locally. Restrictive legislation is mostly provoked by unfortunate accidents to smaller children and this in turn may be partly a consequence of permissive attitudes of parents and of negligent supervision. But even display fireworks that are used with special permission and donated by civic groups or private enterprise have become rare in the United States with the above-noted exceptions. We could not say in Britain that the public has lost its interest in fireworks, though the increasing tendency to live in flats certainly creates a problem. For many years now, the British and some other European governments have exercised a sensible control over the manufacture and sale of fireworks that amounts to a combination of the restriction of the use of certain dangerous materials and a gentleman's agreement over the sale of some potentially dangerous fireworks.

The automation of firework production is something that has come only slowly. Paper "caps" are produced on automated machinery at a rate of 25 million a day, and machines are used to roll and label tubes, fill certain compositions, etc. The writer has not yet seen any machine that completely automates the production of any given firework and suspects that even if time and money were put into such a machine it could only be used for mixtures that were limited in sensitivity or simple in form. Filling machines and automatic presses are used in Europe, but friction occasionally causes explosions or fires. The machines are so arranged that they can be quickly cleaned and restarted.

Bombshells, particularly cylindrical shells, require great skill and knowledge in their manufacture and can only be made by hand. At the present time the high cost of manual work accounts for the high price of fireworks; thus, many countries such as Japan, Hong Kong, and Macao are able to compete very favorably with the Western hemisphere.

Recent years have seen many changes, not the least in the attitude of manufacturers. In the past many of the "old school" regarded their knowledge as a great secret, but really it was technique rather than composition that was the great secret. Firework makers used to think that the addition of a little of this or that made their fireworks so much better, but their opinion was frequently biased and had no scientific basis. As time goes on, formulations tend to become simpler rather than more complex, and this has obvious advantages.

A number of books have been written on the subject and quite a few formulas have been put into print. The best and most recent of these are Weingart,[7] Davis,[8] Izzo,[9] and Faber,[1] but their formulations do not give very much away. Any firework manufacturer knows that it is not only the quantity of a chemical that matters, but also the particle size of the ingredient, for this governs the burning speed of a composition more than anything else. In addition, many of the printed formulas are quite useless or too expensive from a commercial point of view even if they function properly; no manufacturer, for example, would take very seriously a star composition made with sodium nitrate, sulfur, and aluminum, etc., but there is one in print. Some formulations use picric acid in color production, but these formulas are only of academic interest. Aluminum, perhaps more than any other substance, presents problems to the firework maker as so much depends on the method of manufacture of this material. Atomized aluminum is used increasingly now and is more reliable, but it has somewhat limited use with our present knowledge mainly because it is more difficult to ignite a round particle than it is to ignite a flake.

In recent years ammonium perchlorate has been used increasingly, though many still regard it with great suspicion since experience is so limited. This material in supposedly safe mixtures has apparently given occasional surprises by exploding. Part of the problem lies in the instability of ammonium or cuprammonium salts and also the possibility of double decomposition that arises in the presence of moisture. Quite a good green flame, for example, can be produced with barium nitrate and ammonium perchlorate, but if double decomposition takes place, hygroscopic ammonium nitrate is formed. This does not normally happen, but the writer recently saw a commercial item that had absorbed moisture during storage. The balance had been upset possibly by the addition of polyester containing a promoter.

The omission of formulas would be disappointing to the amateur or to a reader who desires an introduction to this field; therefore, some formulas have been included. The writer faces the delicate situation of being connected with a large firework company as advisor, but feels that he has not been unfair to the industry generally and has not excessively expanded on what is already to be found in print.

As a general principle, light-producing fireworks for pleasure, in contradistinction to ordnance items, have the following characteristics: Compositions are not usually highly compacted, but merely tamped, shaken down loosely, or at the most charged with a mallet and tool or pressed. Some compositions are damped, consolidated, and cut into cubes or formed into cylinders. This contributes greatly to the ease of ignition and of fire transfer but limits the resistance to physical abuse and influx of moisture. While all pyrotechnic mixtures have some explosive potential, civilian fireworks formulas often employ (for better effects) materials and combinations of materials that are more hazardous than would be permitted in some military items. It is difficult to generalize on this, of course, and examples could be produced to support either side. Certainly the use of potassium chlorate has dangers since it forms impact and friction sensitive mixtures with all solid fuel-type materials.* Some commercial fireworks have a limited storage life, particularly if they contain materials that have hygroscopic tendencies, but it is amazing what will keep. The writer has frequently fired

* Brock describes some spectacular firework factory explosions that have occurred since the introduction of chlorates into fireworks. Consequently, the combined use of sulfur and potassium chlorate, an especially sensitive mixture, was forbidden in England in 1875.

fireworks that were twenty years old or more and has found them quite satisfactory.

Finally, fireworks do not aspire to enormous light outputs but feature spectacular, though actually only moderately bright, white stars or flashes and deeply colored effects because distances are small between source and observer.

Illuminating Fires

The textbooks suggest that extensive use of *tableau fires* was quite common in the past, but this is certainly not true of the present day. Simple light effects were produced from loose mixtures of the color-producing nitrate, additional oxidizer (chlorate or perchlorate, which also deepens the color), and some fuel-type material. Loose powder tends to burn very rapidly and therefore is not very economical, but heaps of red or green fire can be very effective when placed near trees or bushes. It is also quite useful to tamp the mixture into a thin-walled tube about 3 or 4 in. in diameter with the colored fire on top of an illuminating composition made with magnesium. Smoke is a problem in this as in all fireworks that create light, but a moderate amount of smoke enhances a stationary or slowly moving flame by reflecting light. Formulas 81—86 are suitable mixtures for this purpose. The so-called *Bengal Illuminations* are used for illuminating public buildings, a technique extensively used by the Germans for lighting up their castles, etc. The candles are thin-walled tubes about 2 in. in diameter and 12 in. long and are fixed horizontally behind a board so that the onlookers do not see the flame itself. This horizontal position also prevents the dross from burning the tube and thus altering the burning time which is quite slow— being anything from 30—60 sec/in. Formulas 87—90 are suitable for this purpose.

Lances

These fireworks are similar to the Bengal Illuminations mentioned above but are very much smaller and faster burning. *Lances* are usually about 5/16 or 3/8 in. in diameter and about 4 in. long. The tube is made of about three or four turns of a thin paper so that it will burn away with the composition. Surprisingly, these fireworks can present problems, for they sometimes form a long chimney-like ash that can obscure the color and make the lances lose their brightness. Compositions frequently have to be adjusted to fit in with materials, but Formulas 91—95 are typical. The lances are placed at 3 or 4 in. intervals on canework designs to form pictures or words. In order to achieve the necessary simultaneous ignition of the lances, a type of instantaneous fuse train known as "piped match" is used, consisting of cotton threads impregnated with black powder and covered with paper tubing. The only way to test a good lance is to see it burning horizontally in the dark at a distance of about 50 ft.

Torches

The slow-burning, hand-held *parade torches* consist of large lances of about 1 in. diameter and provided with a suitable handle. Parade torches must be formulated so as not to splutter, burn irregularly, or emit noxious fumes—conditions not always easy to meet, particularly since these are cheap items. Lance compositions would be suitable for use in parade torches, and starch or some similar substance could be added to make the composition burn more slowly. Formulas 96 and 97 furnish an additional effect with aluminum. It would appear that parade torches of a specifically pyrotechnic type were more commonly used in the United States than in Europe where it is much more usual to use a roll of "hessian" (a type of woven jute cloth) that has been soaked in paraffin wax and then fixed to a wooden handle. The well-known railway fusee has been described in the preceding chapter and as a utilitarian device takes an intermediate position between a firework item and the ordnance-type illuminants. Also in this category are the *marine handlights*, which are usually made of a strontium salt and magnesium, or the old *ships blue lights*, which were made of potassium nitrate, sulfur, and red orpiment (arsenic sulfide, As_2S_3). Ordinary colored *ships pilots' lights* are also lance type compositions that have been slowed a little.

Stars

Stars are cylinders or cubes of a hard firework composition. The older method was to damp the composition with a gum solution and then beat it down into a solid mass, which was then cut into squares with a long knife. The cylindrical stars were made by pressing a copper or brass tube into the composition and then ejecting the star with a suitable plunger fitted inside the tube. These methods are still used, except that the cylindrical stars are made in molds producing anything from 6 to 120 in a single operation. Adhesives have changed considerably over the years. Gum arabic solution that has been kept and allowed to ferment must have been the cause of many accidents when used with chlorates, and the writer has shuddered at the sight of gum arabic solution being used with a mixture of potassium chlorate and aluminum even in recent years. It is fairly common practice these days to add powdered dextrin to the dry composition and just add water to damp the mix sufficiently for it to bind together. Use is also frequently made of a solution of shellac in alcohol or some of the synthetic resins such as polyester or phenol formaldehyde. In recent years, there has been a great advance in the use of pressed stars. In some cases, the composition is pressed into a paper tube using a suitable mold or the material is pressed without confinement on a modified tablet press. Pressed stars are not easy to make, for some compositions are too sensitive, some will not press and hold together, while others become so hard that they will not light.

The ignition of stars can thus be quite a problem. It has often been said that cut stars light better because of their sharp corners and while this is true to some extent, much depends on the composition itself. It is fairly common practice to coat stars with gunpowder while they are still damp, and while this certainly makes ignition easier, it can make them more sensitive if the star contains chlorate.

The burning speed of a star is very important and will vary according to the purpose for which it is used. An exhibition *roman candle*, for example, will need a much fiercer composition than a *shell star*, because a roman candle star is projected through the air at quite a speed and the flame is sometimes extinguished.

In view of the fact that potassium perchlorate is more stable, it would be preferable to use it for making stars, but this is not always satisfactory as it tends to produce a much smaller flame than the chlorate and often needs the addition of charcoal to increase the burning speed. In general, therefore, it is better to use *chlorate* for roman candles and *perchlorate* for some shell stars. Red stars are normally made with either strontium carbonate or strontium oxalate. Yellow stars are made with sodium oxalate and cryolite. Green stars are invariably made with barium chlorate since barium nitrate gives such a weak color. Blue has always been a problem because copper salts are not pleasant to work with. Copper acetoarsenite (Paris Green) gives a good color and is much used, but it has the disadvantage of high toxicity, causing workers to have skin troubles and nose bleeds. Copper oxychloride is quite useful, but does not give the best blue and does not keep particularly well. It has always been considered that copper carbonate was almost useless, but in fact it does give one of the best blues when used with ammonium perchlorate (note the word of caution earlier in this chapter). Even with potassium chlorate one can get a really excellent blue that is quite cheap by using copper carbonate and the right chlorine donor.

Some of the existing literature suggests that zinc and steel filings are used in stars, but this is not generally true now. Zinc is occasionally used in smoke production but is not used much otherwise. Silver stars are usually made with aluminum and potassium perchlorate or sometimes potassium chlorate. Good silver effects can also be produced from barium nitrate and dark pyro aluminum. White stars are usually made with potassium nitrate, sulfur, gunpowder and antimony, but red orpiment can also be used and has the advantage of good ignition under the most difficult circumstances.

The well-known glitter effect is produced by aluminum and antimony combined with gunpowder. The glitter (a twinkling effect) seems to be produced by the sudden bursting of a molten blob of dross made by these materials. Sodium oxalate can be added to the mixture to produce a yellow glitter, but it also seems to enhance the glitter effect, perhaps because of the formation of carbon dioxide. Strontium oxalate can be used if a white glitter is required. Another important matter in the manufacture of glitter compositions seems to be the kind of aluminum used—the denser hammered powders seem to be preferable to the kind produced in a ball mill.

Golden streamer stars and willow effects are made with charcoal or lampblack or a mixture of both. A good quality of lampblack should always be used, but it has the disadvantage of being very dirty to handle and difficult to wet because of its lightness and grease content.

Contrary to general opinion, magnesium stars are not often used in commerce, mainly because they are expensive and do not keep very well. A very good pressed star can be made with strontium or barium nitrate in combination with polyvinylchloride (PVC) and magnesium. These compositions were developed by the Germans prior to 1939 and were used by them in the last war, becoming generally known after the release of the BIOS reports. Formulas 99—109 give examples of star compositions.

Roman Candles

It requires skill and experience to make a good roman candle since there are very many snags. At the bottom of the tube at the clay stopper there is a small amount of grain gunpowder. On top of this powder is the cylindrical star, which just comfortably slides down the tube, the star being surrounded with just enough gunpowder to go down the sides.

A slow burning "candle composition" is then placed on top of the star and gently rammed. This process is continued until the tube is full. The fit of the star in the tube must be just right—if the stars are too slack, they pop out of the top of the tube and land on the floor burning; if they are too tight, they come out with such force that they blow themselves out; and if the inner lap of the tube is not firmly stuck down, they come out like machine gun bullets. The amount of grain powder varies with the position of the star, of course, and it is important to use a good grade of grain. The "candle composition" is quite important and its burning speed is related to the grain charges. The writer has observed the great force used with gang rammers in some firework plants and is not surprised to hear that explosions have taken place with their use. Many seem to think that it is necessary to damp the composition and so partially granulate it, but this is not necessary, just as heavy ramming is unnecessary. Formula 110 shows a typical roman candle fuse.

Waterfalls

This very popular firework is best made with a mixture of potassium perchlorate and aluminum. Potassium chlorate was used in the past (Formula 111) but perchlorate gives a better effect and requires the addition of a slightly higher proportion of aluminum, including a proportion of coarse powder (flitter) to give a long drop. The composition is charged into thin paper tubes that burn away with the composition. Most people seem to fasten the tubes on wooden battens so that the cases burn in a horizontal position, but there is a tendency in England to fasten the tubes to a length of cord so that they dangle vertically and swing slightly. The above-mentioned composition has obvious disadvantages and so it is quite common practice to make this firework with barium nitrate and dark pyro aluminum. The effect is not quite as good but the mixture is safer. With barium nitrate there is a tendency to get a big glare at the top and a drop that is inferior. Sometimes *waterfalls* are made by using a bank of iron gerbs.

Rockets

The most spectacular fireworks are those observed high in the air against the night sky, which permits large crowds to witness the display. In order to elevate the fireworks to an adequate height, one can use a bombshell projected from a mortar or a rocket, and although a rocket cannot carry as much as a shell, it nevertheless tends to be more popular.

The firework rocket consists of a "motor," which is a stout, convolute-wound paper tube. A nozzle is formed by constricting the tube or by ramming clay into the tube, which rests on a plug with a central spindle. The latter forms the nozzle opening and also shapes the propellant powder, which is rammed or hydraulically pressed into the tube in small increments. By this procedure, about three-fourths of the length of the propelling charge has a tapered central hole of about one-third the internal diameter of the tube at the nozzle end. The last addition of the propellant is without hole and is followed by a layer of clay. A hole through the clay transfers the fire from the propellant powder to some gunpowder on the other end in the rocket head and actuates the various effects stored in it.

Since the propellant, which is one of several black powder type formulations (Formula 112), is tightly rammed or more frequently hydraulically pressed, it burns in a progressive, orderly fashion—faster at first, because of the large surface area caused by the long tapered hole and then linearly towards the end plug. It is desirable that the ignition in the rocket head take place exactly at the apex of flight, but this can be difficult in commercial manufacture.

Small rockets for sale in shops are sometimes made entirely of gunpowder, which is compressed into a tube. There is no spindle but it is necessary to have the clay washer at the nozzle end with a smaller hole in the center. These rockets are not as beautiful because the tail does not show so much, but there are obvious commercial advantages.

Rocket sizes are usually described in ounces and pounds. This is a relic from the past when the 4-oz rocket, for example, had an internal diameter that would just admit a lead ball weighing 4 oz. Weingart[7] and others suggest that the American system uses the same descriptions but has different sizes. The literature gives compositions for different sizes of rockets, but this is quite useless unless the size of the spindle and other details are disclosed.

While the utilitarian congeners of firework items are treated in other chapters, some practical uses of the firework rocket will be mentioned here. Small rockets can be used to carry packages of thin, small propaganda leaflets in the front compartment. Of more practical importance is (or at least has been in the past) the line-carrying rocket that throws a thin "leader" from ship to shore. The idea of such emergency equipment goes back to the eighteenth century. The paper rocket was later replaced by a sturdier metal-encased item, which for better aim and faster firing was set in motion and initiated at the same time by a blank cartridge and within a pistol barrel—the so-called *Schermuly Apparatus*. Brock[2] describes the historical development of this invention in considerable detail. In the United States, the *Lyle Lifesaving Gun* is used for line-throwing with the help of a special projectile fired by a blank cartridge.[311] In England, the Wessex *Self-Contained Rocket Line Throwers* are similar and have a range of up to 300 yards.

More recently, the cumbersome marine rocket has been replaced by a much smaller device with a plastic type propellant that is not much more than 1 in. diameter and 2 in. long. The whole unit is about 12 in. long and has no stick.

Shells

A rocket contains its own means of propulsion and is quite different, therefore, from the shell, which has to be propelled by a suitable gunpowder charge. Extensive use is still made in Europe and the East of hollow paper balls containing stars and bursting charge, and fitted with a 3-5 sec delay fuse. These shells are fired from mortars with diameters of anything from 2—16 in. and even 25 in. has been used in the past. Marutamaya Ogatsu of Tokyo is said to make the largest shell, which attains the remarkable height of over 1000 ft and is fired from a 36 in. mortar. In general, ordinary ball shells do not give the best aerial effects, and much better results can be gained from cylindrical shells, though these require great skill to manufacture and are very time-consuming.

Sparklers

These popular and relatively harmless amusements for children are made by dipping straight pieces of wire into a viscous fluid mixture and withdrawing them slowly. The process somewhat resembles the dipping of matches and the similarity is reflected in the relatively large amount of binder (dextrin, etc.) in Formulas 113, 114, and 115. It is common practice to make sparklers on wires with barium nitrate, aluminum powder, and steel filings, but the steel has to be protected by coating with a suitable material such as paraffin or linseed oil. The drying process has to be extremely slow and this tends to make mechanization difficult. Shower sticks are similar to sparklers and consist of wooden sticks coated with a mixture of potassium perchlorate and aluminum, etc.

Wheels

Revolving pieces are turned by fiercely burning *fountains*. The tube is similar to that of a rocket with a suitable choke made by constricting the tube or using a clay washer, but the composition is charged solid and has no central spindle. Small triangles are turned with tubes about

1/2 in. in diameter while the very large display wheels have tubes up to 1 1/2 in. internal diameter.

Saxons are very similar except that the ends of the tube are closed with a clay plug and the fire is made to come out of a small hole punched in the side of the case so that the fire is at right angles to the axis of the tube. The tube has a nail through the center to make it revolve.

Pinwheels consist of a very long narrow bore tube filled with a modified gunpowder mixture. The tube is wound around a central disk made of wood or plastic.

Fountains, *flower pots*, and *gerbs* are similar to turning cases, though usually having a less fierce composition. These items are based on gunpowder that has been modified by the addition of extra potassium nitrate, sulfur, charcoal, etc. Charcoal in its various mesh sizes is used for gold effects and lampblack is occasionally used to produce its own rather characteristic gold sparks. Aluminum will create a rather pleasant silvery-gold effect and cast iron turnings are very useful in spite of the fact that they will not keep unless suitably coated. Titanium is quite the most useful metal, having been introduced into black powder mixes in recent years. While it is generally known that finely divided titanium is rather sensitive, the coarse filings seem to be completely safe; in fact, the unreactive nature of coarse titanium would appear to make it safer than magnesium of fine aluminum. Formulas 116—118 and 122—124 are representative of this group.

Noisemakers and Novelty Items

These include *flashcrackers*, *powder crackers*, *jumping crackers*, *toy torpedoes*, *humming* and *whistling devices*. The utilitarian railway torpedo and the military version of the "salute" firecracker (M-80), as well as whistling pyrotechnic mixtures, will be treated in Chapter 21. The potassium picrate whistle is still used by some manufacturers and with its black tail it can be very effective in roman candles. Although many consider this whistle to be more dangerous than gallic acid whistles, there seem to be few records of accidents with this substance. It is usual to press the material with a hand press and a small amount of stearic acid or asphaltum can be added to the picrate.

The amateur pyrotechnician is urged to prepare flash mixtures in gram quantities, avoiding any rubbing or pounding action. Weingart[7] stated that the danger from large crackers was mainly in use, since the mixtures are only explosive in confinement. This is an unduly optimistic statement. Not only is the mere flash from a loose powder mixture quite damaging to the processor, but even the unconfined mixtures in larger quantities will explode with great violence on mere ignition. What constitutes a "larger" quantity depends on the innate properties of the composition. Half a pound of some mixture in an open tin may burn quietly or may "blow" with a loud noise; in this case, "open" burning is just not open enough to allow the escape of the gaseous products of reaction fast enough to avoid the pressure build-up that leads to explosive burning or even to detonation (shock wave creation, also called "high-order" explosion).

Ordinary pyrotechnic flash composition is made with potassium perchlorate and dark pyro aluminum and depends greatly on the quality of the aluminum. Weingart gives mixtures that also add sulfur or antimony sulfide, but this is dangerous and quite unnecessary. A small quantity of flash mixture will make quite a large bang. It is usual not to fill the tube completely. Another good flash composition is a mixture of barium nitrate, dark pyro aluminum, and sulfur. This composition is not quite as fierce, and it is usual to almost fill the tube and use tubes of larger bore. The small "bangers" sold in the shops in Europe are usually filled with rifle powder. Formulas 119—121 are flash compositions.

Small noisemakers called torpedoes consist of either a sensitive primary explosive or of Armstrong's Mixture in combination with small pieces of gravel enclosed in a paper forming a tiny bag. The primary explosive, which is very sensitive to explosion by friction or impact, is silver fulminate. When thrown against a hard surface, the item explodes with some noise but the explosive effect is local so that no damaging particles result from the explosion. For the same reason, so-called trick cigars using this type of explosive are rather harmless, though certainly not a recommended "practical joke."

Book matches are sold that emit a sharp crack after striking. This is caused by a tiny dab of silver fulminate (or perhaps some other prime explosive) in shellac applied below the matchhead at the back of the splint. By affixing this explosive onto the matches in the front row of the book only, one can remove a normal match from the back row, but the victim to whom the book is offered will normally use the front-row match. It is hardly necessary to add that no match manufacturer will bring such matches on the market, but it is easy to add the noise-making agent to the finished match.

Humming fireworks are made by charging a short thick-walled tube with a fierce burning gunpowder type mixture and blocking the two open ends of the tube with clay. A small hole is then drilled into the composition, but the hole is drilled at such an angle that the gases that come out on ignition will make the tube revolve at speed on its own axis. In the air this makes a humming noise.

There are several pyrotechnic indoor effects available, of which perhaps the best known one is the *Pharaoh's Serpent.* Textbooks give details of the nitration of pitch with linseed oil, but the manufacture of a good snake is not an easy matter. It is generally known that naphthol pitch gives the best results, but the material needs to be chosen carefully. After the pitch has been nitrated, it is ground and mixed with a suitable material to make it burn and then pressed into pellets. The material that is added must not be a powerful oxidizing agent or it will burn the ash that one is trying to produce. Picric acid has been used in the past, but this is undesirable as it tends to creep to the surface and stain the fingers. Pitch produces quite the best snakes, which are as much as fifty times the original volume and nontoxic. Mercuric thiocyanate, $Hg(CNS)_2$, has been used in the past, but it is as toxic as it is expensive. The *Snake-in-the-Grass* makes use of a small snake pellet inside a cone of ammonium dichromate to which a small amount of oxidizing agent and fuel has been added. Green chromic oxide is produced at first, followed by the black snake.

Snow cones are mainly fine magnesium powder with some other fuel and a small amount of oxidizing agent. The powder burns quickly, producing a shower of magnesium oxide that slowly settles in the room —much to the delight of the housewife and mother!

It is not easy to bring pyrochemically-created flame into the home. Flame-coloring salts are sometimes put on logs in fireplaces and yield a pleasant effect, but candle flames are hard to color because of the salt residue. Other flame bases also present problems because of the unpleasant fumes, but neither such considerations nor the dismal economic outlook have prevented inventors from covering this field. Chlorides and nitrates have been combined with paraformaldehyde, metaldehyde, and other organic compounds (likely to release irritating fumes into the air). Hexamine is somewhat better, but its yellow coloration tends to mask lithium or other salts. Two German patents[312] for such mixtures are found in Volumes II and IV of Bennett's *Formulary*,[313] and there are others, such as a Swiss patent.[314]

Conclusion

Military pyrotechnics has the advantage of having numerous research workers who, under the stimulus of war, are able to devise better production methods. These workers are also fortunate to have money at their disposal for research in general, while the fireworks industry is not so well endowed and enters a new phase mainly when new materials are available. Thus we may speak of the era of black powder with its predominance of movement and propulsion; the introduction of chlorates and pure chemicals in general, leading to color effects; the advent of magnesium and aluminum with improved light, spark, and sound effects; and more recently, the use of titanium, synthetic resins, and atomized powders for more stable and better defined, reproducible items. However, the increase in cost due to some of these latter materials can be a serious drawback in commercial production.

Modern advertising methods and the ease of communication ensures that the populace will have its senses stimulated by a wide variety of color displays and other excitement but, alas, this leaves the fireworks display just as another "show" rather than the big, wonderful, eagerly expected event. The individual fireworks item, especially the firecracker, is frequently misused, so that it becomes a nuisance to others instead of being used sensibly and for enjoyment. Thus there is a danger that the fireworks art may go the way of other specialized arts and occupations and be swept aside by changing attitudes and shrinking commercial value.

There will always be a few dedicated enthusiasts, and perhaps ingenuity will create new and exciting items with a strong safety factor in order to give the art a new boost. We may find new designs of such dazzling and overpowering magnitude that even the jaded taste of the present-day public will be satisfied.

Part IV

Aerosols
(Smoke and Dispersed Agents)

chapter 18

Screening and Signaling Smokes

Principles of Effectiveness of Smokes

"The word aerosol denotes an aerial colloid, a suspensoid with air as the medium."[244] Using the term aerosol for liquids dispersed from a pressurized can—hairsprays to whipped cream—constitutes, as Fowler would call it, a "popularized technicality" since the droplets are much too large to qualify as aerosols. Also, strictly speaking, the dispersed particles including the surrounding air should be called collectively the aerosol, not the dispersed phase alone. Information on the subject, besides the special references and personal experience, has been obtained from TM 3-215,[35] the article on chemical warfare in Volume 4 of Kirk-Othmer's *Encyclopedia*,[24] Shidlovsky,[3] and the British or Combined Intelligence Reports after World War II, specifically the one on smoke signals.[315] A vast amount of information is found in the voluminous four-part history of the Chemical Warfare Service by Finklestein.[316]

Tiny liquid droplets are often called "mist" or "fog" and solid particles "smoke," but for purposes of obscuration and signaling, we shall follow ordnance practice and call all of them "smokes." These aerosolized substances exist in a wide range of particle sizes of about 1 mμ to 10 μ radius. In smokes for obscuration or as signals, we are concerned with 0.5—1.5 μ optimal size.

Such small particles have a slow settling rate, acquire electric charges while moving through the air, and are subjected to Brownian motion and of course to air currents. Thus, a cloud of such particles in air acquires what may be called a fleeting persistence or "duration" (as distinguished from the duration of smoke emittance from a smoke producing device). It hovers over the area of its creation but is removed and dissipated by wind or the thermic updraft, generally before it can settle or before the particles can coalesce.

In military operations, the optical behavior of smokes and more recently the dispersion and dissipation of definite substances as aerosols have become of great importance. If light is thrown on a smoke cloud whose particles are approximately of the size of its wavelengths, and if the particles mainly let pass and reflect, but do not absorb, the light, the result will be the almost complete scattering of the incident light. In sunlight, optimal conditions prevail when the light is neither in the line of sight with the observer nor directly opposite but at an angle of 45°.[3] Be this as it may, the effect is either one of *screening* (i.e. from view) of objects and persons in or behind the cloud of smoke or the aim is to *notice* the smoke cloud, puff, or trail as such.

The tactical uses of screening smokes and the means of producing them are only of limited interest to the pyrotechnician, since his devices are unable to produce enough smoke except in a relatively small area. However, the second purpose of smokes—the emission of smoke clouds or trails as a means of signaling—is well served by pyrotechnic grenades, pots, or bursts, especially since various bright colors can be achieved beside white smoke.

Such actions as the envelopment of a ship and the hiding and protection of advancing or retreating troops require slarge-cale and prolonged smoke emission. In an emergency and under favorable wind conditions, a ship may create a smoke cloud by deliberate maladjustment of its oil furnaces. However, contrary to what the layman may take for granted, the highly light-absorbing black soot particles and other types of dark smoke are least effective, whereas white (i.e. colorless) particles are best.

A minor use of screening smokes is the protection of orchards from sudden temperature changes by preventing the warmth of the area from being "radiated away," while at the same time the smudge pots that furnish the smoke clouds add some warmth to the area. Another screening action by smokes being considered is the (probably impractical) protection of populated areas from nuclear radiation.[317,318]

Obscuring smokes are effectively and least expensively made by heating and evaporating of "fog-oil," a cheap petroleum product similar to thin motor oil (SGF #2).

From the viewpoint of theoretical effectiveness, i.e. actual obscuration in terms of lb/mile^2,[316] white phosphorus (WP) has been measured as being 47 times as effective as oil-vapor. The obscuring agent is the combustion product of the phosphorus—pentoxide (P_2O_5, now often written P^4O_{10})—in reaction with the moisture in the air. However, white phosphorus burns rapidly and the heat of its combustion causes updrafts that tend to negate the obscuring effectiveness ("pillaring"). By converting the phosphorus under hot water into very small pellets (0.5 mm in diameter) and coating these with a viscous synthetic (GRS) rubber solution, a slower developing and hence less easily dissipated smoke is produced from this "plasticized white phosphorus" (PWP). Both WP and PWP are disseminated by bombs, grenades, or by artillery and mortar shell barrages.

Where screening smoke must be delivered by explosive action, and where at the same time smoke screens without fire hazard are desired, a solution of 55 parts of sulfur trioxide (SO_3) in 45 parts of chlorosulfonic acid ($CLSO_3H$) is cheap and effective though its screening power is about one-half or less that of phosphorus. This "FS" mixture and the more expensive and otherwise less desirable titanium tetrachloride ($TiCL_4$) called "FM" become aerosolized droplets only in the presence of water and presumably display their power of obscuration best and fastest in humid air.

HC-Smokes

We come now to the only type of screening or obscuring smoke that can be truly called pyrotechnical—the so-called HC-type.

In World War I, Captain Berger of the French Army developed a mixture of zinc powder and carbon tetrachloride (CCl_4), which reacted strongly exothermically (0.6 kcal/g mixture calculated for the product zinc chloride in gaseous state) to form zinc chloride as volatile hygroscopic droplets and carbon. A modification of this mixture appeared in World War II in the United States as a Chemical Warfare Specification (CWS) with the designation *Filling, Type E* and used in the *AN-M8 HC Smoke Grenade* (Formula 125). Since it excluded the metal zinc, it was much more stable, but the use of a thin volatile liquid such as carbon tetrachloride in a pyrotechnic item was undesirable. Two other mixtures, *Type A* (Formulas 126 and 127) and *Type B* (Formula 128), introduced the solidchlorine donor hexachloroethane [(C_2Cl_6), called HC], a term also generally used to typify the smoke mixtures containing it. Though this is a substance of high vapor pressure and is somewhat difficult to break down into a fine powder, it is a great improvement over a liquid material. However, *Type A* with zinc and ammonium chloride was undoubtedly very sensitive to moisture. *Type B* was the forerunner of the present *Type C* (Formula 129). The latter contains equal or near equal amounts of zinc oxide and HC with aluminum added in various percentages from 5.5 to 9.0. This controls the rate of reactions so that under otherwise identical conditions the burning time varies from 147—55 sec. When the percentage of aluminum is low, a change in the amount is of greater influence than at the high end.[35]

The burning time also responds strongly to small amounts of additives such as zinc carbonate. Studies performed in the writer's laboratory [318a] also showed that the crystalline structure of the zinc oxide plays another significant part in the burning rate, a fact previously observed by McLain and Meyer.[318b]

HC-smoke mixtures react at very high temperature, bringing the steel canisters to red heat, and create a luminous flame that may have to be concealed in some way for tactical reasons. The smoke itself is greyish-white, attesting to the presence of carbon in the smoke. The amount of carbon is reduced when the aluminum content is low.

The reactions are complex and one can assume, as expressed in TM 3-215,[35] a primary reaction between aluminum and HC followed by a metathetical reaction of aluminum chloride with zinc oxide, leading to aluminum oxide and zinc chloride. Another way of hypothesizing on the sequence of events is the assumption of a primary thermitic reaction of aluminum and zinc oxide followed by interaction of the liberated zinc with the HC. In either case, the end result remains the same:

$$2Al + 3ZnO + C_2Cl_6 \longrightarrow Al_2O_3 + 3\ ZnCl_2 + 2\ C \qquad (a)$$

This corresponds closely to the formulation for the highest (9.0%) metal content for which a heat of reaction of 0.75 kcal/g mixture can be calculated (with $ZnCl_2$ at the standard solid [s] state, otherwise the figure would be 0.54). If we reduce the aluminum content to one-third, we arrive at the other extreme, i.e. a carbon-free reaction:

$$\tfrac{2}{3}Al + 3ZnO + C_2Cl_6 \longrightarrow \tfrac{1}{3}Al_2O_3 + 3ZnCl_2 + 2CO \qquad (b)$$

However, a mixture, with such low aluminum content yields only a calculated heat output of 0.32 kcal/g (0.10 for $ZnCl_2$ as a gas), so it is not surprising that there is a tendency to erratic burning.

New variants of HC-smokes have also appeared in U.S. patents,[319] which claim *inter alia* a mixture of magnesium, zinc oxide, HC, and ammonium perchlorate.

With the exception of the hand grenade, type *AN-M8*, which contains 19 oz of HC mixture (2 min B.T.) and is thrown, HC-smoke munitions belong to the largest of the strictly pyrotechnic munitions. The *M1 Smoke Pot* (*HC*) contains 10 lb, the *M5 Smoke Pot* (*HC*) 30lb of composition; in addition, they can be stacked for multiplication of the 5—20 min of smoke emission. First fires for the HC-smokes will be discussed in Chapter 23.

HC-smokes are useful for small-scale or supporting screening action. The fumes are somewhat toxic but only slightly irritating and cause no serious adverse effects for short periods of exposure.[316] All screening smokes can also be used for signaling, especially the smaller items. An example is the *M15 Smoke Hand Grenade* (*WP*) containing 15 oz of white phosphorus—a fairly heavy-walled, bursting type munition.

Testing of smoke mixtures for efficacy is difficult. Comparative simultaneous burning of several items and observations from a proper vantage point, assuming uniform air movement at the necessarily fairly widely spaced units is just about the best one can do. A really objective test is the generation of the smoke under controlled and measured influx of air, drawing it through a test chamber where the amount of light scattering between a lamp and a photocell can be registered. Such a test is described in an NBS report.[320]

Colored Smokes

While a recent study[321] has shown that detection and color identification is best for white smokes, it is desirable in daytime signaling to have a variety of brilliant colors in the form of aerosolized organic dyestuffs or of inorganic compounds. These may be dispersed by explosive action furnishing a puff of smoke or by evaporation and condensation of the coloring material. Obviously, a puff, streamer, or a billowing cloud of bright red or orange coloration is not likely to be mistaken for some incidental smoke from a bursting high-explosive shell or some burning matter ignited accidentally. White smoke over water emitted from the previously-described drift signals and float lights, which emit both smoke and flame from burning phosphorus, is apparently adequate since there the chance of confusion is less likely.

The simplest way of dispersing dyestuffs of good heat resistance is by tamping the dye, mixed with 20% of table salt as an extender and coolant, into a shell, drilling a center hole, and filling it with a high explosive such as 67/33 Baratol.

According to information from the Research Laboratory of Picatinny Arsenal,[322] the disadvantage of large-scale destruction of the dye by the heat from the explosion can be reduced and the efficiency of color display improved by reversing the arrangement of chemical components. In the colored shell *XM 152, 2.75" Warhead, HE*, the dye is in the center surrounded by salt, with the explosive charge in the outside layer. The smoke-cloud results from the expansion following the *im*plosive action.

Small smoke puffs have been made by mixing equal amounts of dye and small granules of EC* Blank Powder, an extremely fast burning, smokeless, only partially colloided powder containing barium nitrate. In tight confinement, such a mixture will on ignition explosively disperse the dye without the benefit of a detonator. Other cool-burning explosives for the same purpose have been patented by Davis[324] and by Tuve,[325] and the scheme of depositing the explosively dispersed dye on 90—99.5% of an inert powdery material has been claimed by Magram.[326]

It is possible to create colored aerosols by exclusive use of inorganic pigments or reactants but they are either impractical or generally less desirable than the ones using organic dyestuffs. Red iron-oxide pigment has been used in explosive smokes. Shidlovsky[3] also mentions red lead, ultramarine blue, and even cinnabar—very expensive and of high specific gravity. A purple smoke can be produced from an iodate with a reductant[327] but seems to be both expensive and high in weight/output ratio.

Graff has claimed mixtures that produce "puffs of smoke" of yellow color based on the volatilization of bismuth oxides; brown from copper and lead oxides; and orange based on lead oxides.[328] Such smokes may be desirable when a colored smoke must be produced from readily available laboratory chemicals rather than from the special organic dyestuffs or where the latter are to be avoided for other reasons (Formulas 132—135).

Attempts have been made to dissolve dyes in titanium tetrachloride and disperse them with the hydrolyzing vehicle, but this scheme suffers from the low solubility of the dyes. Better success was achieved with chromylchloride (CrO_2Cl_2) alone or with titanium tetrachloride.[329]

Inorganic black and yellow-to-brownish smokes can be developed by chemical interaction using hexachloroethane as the major ingredient. Mixtures of HC, iron powder, and magnesium, or of HC, magnesium, naphthalene, and anthracene were developed in Germany toward the end of World War II, and a recent German patent[330] is evidence of the continued interest along these lines. The formulas claimed include HC, iron powder, aluminum, magnesium, and minor additives. The black HC-smoke is named as one of the color varieties of the *Mk 3 Mod 1 Colored Smoke Grenade* and *Mk 7 Mod 0 Aircraft Emergency*

* EC stands for Explosives Company of England, the original manufacturer of this item.[8,34,323]

Identification Signal (Formula 130); the yellow (Formula 131) with iron powder seems to be more practical than the above-mentioned. Experimental mixtures of iron powder, HC, and potassium chlorate, developed by the Chemical Corps in World War II,[316] react vigorously with water and may ignite spontaneously if moistened. The color effect is due to dispersion of ferric chloride ($FeCl_3$).

By far the most important colored smokes are those in which the dyestuff is gradually evaporated from a mixture of dye and a pyrotechnic heat source. This involves the selection of suitable dyes and of cool-burning, gas-forming mixtures, which melt and evaporate the dye with a minimum of destruction. However, the "recovery" may be no more than one-third of the amount of dye in the formulation.

Dyes that in practice in the United States have been found to be useful because of their resistance to decomposition by heat and their capability to evaporate and recondense as a brilliant cloud or trail are enumerated in Table 13. They have been taken from the standardized smoke formulas, from a Picatinny Arsenal report,[331] which also gives a number of experimental formulas and their variation of burning time and duration with the formulation, and from an AMC Pamphlet.[32] A number of other dyestuffs are mentioned in the foreign literature, such as indigo, methylene blue, rhodamine B, etc.[3]

The heat-producing mixture is composed of potassium chlorate and powdered ("confectioner's") sugar or sulfur as the oxidizer/fuel mixture and varying amounts of sodium or potassium bicarbonate as diluent and coolant. Thiourea has been used as the fuel.[331a] In one instance, however, when this compound was used it seems to have made the mixture unduly heat-sensitive so that it "cooked-off" toward the end of a "hot-soak" at 160°F. In the U. S. patent literature, various special additives or substitutions to the basic colored-smoke formulations have been claimed.[332,333] Processing aids such as kerosine (or a very thin lubricating oil) reduce the dust nuisance, improve compaction, and lessen the hazard from accidental ignition by friction.

Calcium phosphate promotes free flowing of the chlorate and sugar. In some formulas, both the chlorate and the sugar are specified as containing 3% of magnesium carbonate for the prevention of agglomeration, which greatly aids in achieving good mixing. A little infusorial earth is effective in preventing lump formation of sulfur. Small amounts of such "inerts"—siliceous materials, iron oxide, graphite—are sometimes added because they promote an even progression of the reaction.

Besides confectioner's sugar (Specification JJJ-S-791F, Type 1, b),

Table 13
Dyes for Colored Smokes

Name	*MIL Spec.*	*Army 96-111- (obs.)*	*M.p. °C*	*Color**	*Remarks*
Auramine		-68A		Y	
Auramine hydrochloride	-A-3664	-68A	267	Y	
4-dimethylamino azobenzene			117	Y	Spec. XPA-PD-367 Sudan Yellow GGA, Rev. 1
2-(-4 dimethylamino phenylazo) naphthalene	-D-3613	-102	174	Y	
Indanthrene Golden Yellow GK	-D-50029A		385	Y	
Benzanthrone	-B-50074			Y	
Paranitraniline				Y	(Ref. 3)
Oil Scarlet BLD		-66		R	
9-diethylamino rosindone		-100		R	
1-(2-methoxy phenylazo)- -2-napthol	-D-3179		178	R	Sudan Red G 2-anisidine azo beta naphthol
1-aminoanthraquinone	-D-3698	-97	253	R	α-amino anthraquinone
1-methylamino anthraquinone	-D-3284A†	-78	170	R	
85% MIL-D-3284 } 15% Dextrin JAN-D-232	-D-3718	-66A		R	
Rhodamine B				R	(Ref. 3)
60% MIL-D-3698 } 40% [96-111-68]				OR	
Benzene azo beta naphthol				OR	"Butter Orange"[3]
1, 4 di(methylamino) anthraquinone	-D-21354		223	B	
1 methylamino-4-p-toluidino†† anthraquinone				B	Sudan Blue G[3]
Indigo				B	(Ref. 3)
Methylene Blue				B	(Ref. 3)
1, 4 diamino-2, 3-dihydro anthraquinone	-D-3668	-81		V	
20% MIL-D-3284 } 80% MIL-D-3668	-D-3691	-67		V	
1, 4-di-p-toluidino anthraquinone	-D-3277	-80	218	•G	
71% MIL-D-3277 } 29% MIL-A-3664	-D-3709			G	

* Y=Yellow; R=Red; OR=Orange; B=Blue; V=Violet; G=Green.
† Misprinted 3824 in AMCP 706-187.
†† Instead of "toluidino," a better designation is "tolylamino."

one occasionally finds specified the more expensive lactose (MIL-L-13751) and, exceptionally, other carbohydrates such as dextrins. It is also possible to replace the chlorate by other oxidizers such as ammonium perchlorate but this is unusual, and the free hydrochloric acid from its decomosition pmay destroy some dyes.

After all that has been said about the friction and impact sensitivity of mixtures of chlorates with fuels, especially sulfur, one wonders that such combinations are specified and can be processed on a fairly large scale. I have never witnessed an accidental ignition of the carbohydrate type mixtures though my experience, as far as production quantities are concerned, is limited. However, the sulfur containing mixtures are undoubtedly much more hazardous. They have caused fires and explosions both during mixing and during pressing, sometimes with fatal results.[607] Thousands of tons of such mixtures were processed during World War II and again more recently. This means that the large excess of fuel-type material (i.e. the organic dyestuff) and the addition of large amounts of pyrochemically inert and phlegmatizing sodium bicarbonate greatly reduce accidental ignition because of friction or for other reasons, *but do not entirely exclude it.* Especially during the initial phases of mixing, the distribution of the active materials may be favorable to such ignition. Recent (1966) production of *M18 Colored Smoke Hand Grenades* by the author's company has been performed by mixing the ingredients in a mix muller in the presence of trichloroethylene and drying the mixture prior to pressing. Aside from reducing and, hopefully, eliminating entirely the hazard of accidental ignition during the blending operation, this "wet" processing renders a much denser mixture. Small fires have occurred in the final stages of processing in the canister, attesting to the need for special precautions throughout the processing of sulfur/chlorate-type smoke mixtures.

Another point of safety that must be raised is the combination of potassium chlorate and ammonium chloride in a few formulas. This pair is regarded as hazardous since metathetical interaction would form the unstable ammonium chlorate, which might become the cause of spontaneous ignition. Shidlovsky[3] vouches for the harmlessness of these mixtures, specifically of a mixture for an apparently grey smoke containing 20-30% chlorate, 50% ammonium chloride, naphthalene or anthracene, and a little charcoal, His argument is similar to the one expounded above for chlorate/sulfur mixtures, i.e. the overpowering effect of the large amount of a substance of cooling influence—in this case, the ammonium chloride. It has been the author's experience that ammonium salts, aside from the question of stability, generally achieve little good in pyrotechnic formulations and make them especially difficult to initiate.

Formulas 136 and 137 give a summarized picture of the large number of former Chemical Corps (now MIL-STD) series of red, yellow, green, and violet amokes. Orange color, which is esteemed in distress signals such as the *Coast Guard Floating Orange Distress Signal, Model FOS-1* (former Specification 160. 022) of 4—5 min B.T. and its larger counterpart of 15 min B.T. (sub part 160.057), is easily made using a mixture of yellow and red smoke dyes in a ration of about 2:1. It might be added that the specifications for the cited Coast Guard Signals seem to be unusually strict, even going so far as to specify the "hue, value, and chroma" according to the Munsell Scale.[334]

Burning time of colored smoke mixture can be varied widely but is mostly kept in the range of 10—40 sec/in. Increase in bicarbonate and dye and reduction of chlorate, in relation to sugar, reduce the burning rate. In the theoretical chlorate sugar ratio is stoichiometrically 1.44:1.00 for the reaction to carbon monoxide

$$4KClO_3 + C_{12}H_{22}O_{11} \longrightarrow 4KCl + 12CO + 11H_2O \qquad \text{(c)}$$

and yields 0.7 kcal/g (water taken as liquid), while the complete combustion yields carbon dioxide and proceeds at a ratio of 2.85:1.00:

$$8KClO_3 + C_{12}H_{22}O_{11} \longrightarrow 8KCl + 12CO_2 + 11H_2O \qquad \text{(d)}$$

Its heat output is 1.0 kcal/g.

Shidlovsky[3] has calculated the same reactions using lactose, which contains one molecule of water, and gives 0.63 and 1.06 kcal/g, respectively. Since in smoke formulas with sugar, the chlorate:sugar ratio is generally between 1:1 and 1.5:1 parts by weight, it is clear that we deal with a large excess of actual and potential fuel in the formula and hence a chance for preservation of much of the dye. In formulas with sulfur, the stoichiometric ratio of 2.5:1 chlorate to sulfur is commonly observed, but here the much larger addition of bicarbonate comes into play. The reaction of potassium chlorate and sulfur furnishes about the same amount of heat as the sugar mixture, i.e. 0.7 kcal/g of stoichiometric mixture.

Only scant data have been published on the heat of sublimation (or fusion plus evaporation) of smoke dyes. AMCP 706-187[32] gives 30.85 kcal/mole or 130 cal/g for 1-methylamino anthraquinone, a frequently used red smoke dye, and 28.9 kcal/mole or 128 cal/g for 4-dimethylamino azobenzene (Sudan Yellow GGA). Shidlovsky[3] gives 26.8 and 110 for auramine, 21.2 and 80 for indigo, and 25.3 and 180 for paratoner. Combining the data for heat output from the chlorate and sugar or sulfur, and the heat requirements for the evaporation of the dye (disregarding other heat consumption factors), it is obvious that a large excess of heat is provided. Unfortunately, because of the flammability of the vaporized dye, this heat may suffice under unfavorable conditions to convert the bright, billowing colored cloud into a miserable, thin, greyish plume of burning dye. However, in most cases this fault occurs only as a temporary interruption of the smoke emission. In the case of grenades, it may be caused by dry grass catching fire from the hot fumes and igniting the dye, or otherwise from burning tape at the vent holes, perhaps even from an overheated metal burr at the orifice. In designs where an empty space exists between the top layer of the composition (plus first fire) and the orifice, the danger of ignition is reduced.

Colored smoke mixtures are compacted by the usual pressing techniques. This adds to the unpleasantness of processing because of the unavoidable dust and consequent staining of the skin, which is not cherished by female workers in particular.

More serious and apparently not realized until fairly recently is the toxic and potential or actual carcinogenic character of several smoke dyes—Indanthrene Golden Yellow, MIL-D-50029B; the popular red dye methylamino anthraquinone, MIL-D-3284A; and the much used auramine hydrochloride.[335] An earlier study on the subject has been issued by the Army Medical Center, Maryland.[336]

In order to avoid this nuisance and hazard and possibly gain other advantages such as better dye recovery, experimental smokes have been formulated with solutions of plastic binders. In one type, intended for a *Drill Mine Signal* (Formula 138), a small amount of binder in a large amount of highly volatile solvent forms together with the standard ingredients a pourable slurry that solidifies to a spongy mass after evaporation of the solvent. In another more conventional formula (No. 139) for a little spotting charge, the very firm small pellets, compacted at only 5000 lb/in.^{2} from the nearly dry granules of the mixture, give an excellent smoke. Future improvements in colored smokes, based on the conventional reaction mechanisms, may well lie in this direction. However, work being undertaken by the author's company during recent years is concerned with preserving a much larger percentage of dye by separating a hot gas source and the dye. This is an obvious and by no means original idea but the proper arrangement of the dye and the selection of propellant present numerous problems. Earlier efforts along similar lines are found in a U. S. patent[337] in which a compressed heat source containing iron powder and an oxidizer salt is used, segregated from the dye.

Examples of munitions containing colored smoke mixtures are the *M18 Colored Smoke Hand Grenade* with 11 1/2 oz of either red, green, violet, or yellow mixture;[38] the *M130 Yellow Streamer (hand-held) Ground Smoke Signal* in which (as in certain light signals) a propellant charge is ignited by manual actuation of a primer and the undelayed actuation begins to produce a smoke streamer during the ascent from about 50 ft above the ground. The burning time is 10 sec and the altitude reached, up to 700 ft.[33]

Certain Coast Guard floating orange smoke signals have been mentioned above and the *AN-Mk3 Mod 0 Marine Smoke and Illumination Signal* in Chapter 16. The latter contains separately a flare and an orange smoke composition of about equal duration (18 sec). However, only one signal must be initiated at one time. Canisters with colored smokes (also HC) can be expelled from *Howitzer Shells M2A1 and M4*, preferably close to the end of flight, whereupon the canisters follow essentially the trajectory of the shell.[41] Obviously the functioning of the pyrotechnic part of such munitions is imperiled by the severe conditions of delivery, such as an impact on rocky ground.

While there are a number of uses of colored smokes, the varieties are fewer than of light signals. A hand grenade with a clay body mixed with dyestuff to mark persons in maneuvers is a color "signal" of sorts.[338] As mentioned by the author,[4] this is a type of device often suggested by military personnel as a desirable addition to *Booby Trap Simulators*. More recently,[339] similar ideas have been expressed for the marking and identification of rioters. The dispersion of invisible marking material that appears under ultraviolet light has been suggested in addition to the use of dyestuffs. Colored and white smokes have also been tried for artillery tracers and lately for the tracking of missiles.

A final word on the visibility of a smoke signal on the ground. Red and yellow are generally regarded as the best-visible colors, green as one of the poorest; against a greenish background of vegetation the smoke may be nearly useless.

Great stress is often laid on duration of the smoke emission—a tentative specification the author reviewed not so long ago proposed that a signal for nominally 15-min emission should be penalized as 100% defective if functioning for 10 1/2 min and 25% defective at 13 1/2 min! Since in any wind exceeding, say, 3—6 m/sec the smoke is promptly dissipated, visibility and color recogQition near the orifice is all that can be expected and such visibility is better (other conditions being equal) the higher the rate of emission. From this viewpoint, a shorter lasting, brightly colored smoke should be preferable to a prolonged but barely recognizable trail. High volume and long burnin timeg are mutually exclusive at a constant given weight and volume of composition, hence it appears preferable to aim at a high rate of emission with a consequent relatively high "duration,"—i.e. persistence of the smoke cloud—rather than on insistence on a long duration of smoke emission.

Since the purpose of signaling smokes is quite different from that of screening smokes, the formerly-described test method designed to measure obscuring power is probably not, or only conditionally, applicable to colored smokes. On the other hand, an accurate measurement of color is possible, as evident from a U. S. patent[340] for a smoke colorimeter.

chapter 19

Dispersion of Chemical Agents

In a manner that closely parallels the creation of aerosols for obscuration or signaling, one can also form aerosols of definite chemical entities. As a rule, the substance that is to be evaporated and recondensed is mixed with the heat- and gas-producing pyrotechnic mixture. Exceptionally, the evaporated substance derives from a chemical change within the heat-producing system. Arrangements where a gaseous heat source or gas itself furnishes a physically separate dispersing force fall under a subject treated in the next chapter on nonspecific gases.

Dispersion of chemical agents by pyrotechnic means has advantages and disadvantages. Where the tactics of dissemination require a multitude of relatively small, discrete units, and in particular where the area is not directly accessible to the acting party, the self-contained, compact and reliably functioning simple pyrotechnic canister or grenade is preferable. Even the fact that a large amount of the agent may be destroyed in the process can often be tolerated. In this respect, the techniques and means of achieving them parallel the dispersion of signal smoke. Dissemination over large areas, quite similar to that of screening smokes, cannot be performed economically by such chemical mixtures. Purely physical dispersion is resorted to, as in crop dusting from small airplanes.

The agents considered for pyrotechnic dissemination are *pesticides* and *fungicides*, *irritants* for riot control and warfare, and highly *toxic substances* such as nerve gases. To these must be added certain crystalline *cloud-seeding substances* that promote precipitation of moisture from the air and are useful for *rain-making*, hail prevention, fog dispersal, and possibly the breaking up of hurricane formations. The production of disinfectant gases such as sulfur dioxide will be treated in Chapter 29.

As to the means of dispersion, the principles and practices of cool-burning compositions for smokes apply to the dispersion of agents with only slight variations from the formulations for colored smokes. Sometimes, especially in the older patent literature, one finds unusual combinations. The dispersion of pyrethrum or DDT appears in such patents.[341] These claim the active material, on a substratum, to which carbohydrates or other cool-burning fuel and an oxidizing salt are added. Green and Lane[342] report that DDT or gammexane with a standard smoke mixture create particles below 3 μ in diameter.

A remarkable "monopropellant" of cool and often even flameless exothermic decomposition slightly above its melting point of 203°C is dinitrosopentamethylenetetramine, with the even more formidable chemical name 3, 7 dinitroso-1, 3, 5, 7 tetra-aza bicyclo [3,3,1] nonane. It was developed in Germany originally as a blowing agent for rubber, [343] but has been used in mixture with pyrethrum, lindane or DDT, [344,345] (Formula 140).

Shidlovsky, citing Russian references,[346,347,348] and one British,[349] divides the agriculturally useful thermal mixtures into *insecticides* in which gammexane, HC, DDT, and others are combined with chlorate and anthracene, dicyandiamide, or hexamine; *fungicides* with dichloronaphthoquinone as the active agent; and *acaricides* (mite control agents) using a compound named tedion with chlorate/fuel mixture.

Irritants, which affect the mucous membranes of the eyes and the respiratory system, are used in war and police work to force people out of hide-outs or to disperse crowds. Chloroacetophenone (CN) has long been an effective tear gas (lacrimator) and lends itself to dissemination by heat when mixed with sugar, potassium chlorate, potassium bicarbonate, and diatomaceous earth, as in the *M7A1 Tear (CN) Hand Grenade*. The effect of CN (this designation has nothing to do with the chemical symbol CN for the cyanide radical) can be enhanced and prolonged by mixing with the vomiting and sneezing agent (sternutator) diphenylamino chloroarsine or Adamsite (DM), as used in the *M6 Irritant Hand Grenade, CN-DM*. This one employs smokeless powder as the heat and propellant source and magnesium oxide as a diluent.[38] DM is also used in a recent U. S. patent.[350] The particles are coated with boiled linseed oil and further coated with precipitated chalk and nitrocellulose; the granules are then highly compacted.

Recently, these agents have been replaced by the compound o-chlorobenzal malononitrile, coded CS, as in the *M7A3 Tear Hand Grenade CS*. Because of its sensitivity to decomposition, this material is mixed with sugar and a little wax and water, compressed in pellet form, coated with sugar syrup, and dried. The pellets are then embedded in a modified chlorate/sugar composition (Formula 141). In another procedure, the powdered CS is enclosed in gelatin capsules, which are admixed to the same or a similar heat powder combination.

Irritants can also be dispersed by explosive force, as in the plastic (to avoid lethal fragments) baseball-sized *M25 Riot Hand Grenades*, which are filled with CN or with CS.

The above-mentioned agents, while unpleasant even in minute concentration and extremely distressing in larger doses, generally leave no lasting after-effects. Soon after removal from the contaminated atmosphere recovery from the effects of CS is prompt and complete. But as little as 1 mg of CS per cubic meter of air will irritate the eye. This is less than one-thousandth and perhaps as little as one ten-thousandth of the threshold value for chlorine or sulfur dioxide. The psychological after-effect of exposure to CS is remarkable: a person who has been exposed to it, even though the exposure was not extreme, may relive the burning sensation on lips, tongue, and nose merely by thinking about it—as in composing or typing of these passages.

Lethal agents, such as the "poison gases" of World War I, have always been dispersed by explosive action, with the exception of the original chlorine gas attack of 1915 at Ypres, where commercial cylinders were used.

Turning from the dirtier aspects of pyrotechnic applications to something more constructive, we enter the field of creating aerosols to cause condensation of rain clouds and fog.

The crystals of silver iodide (AgI) resemble ice crystals quite closely, both belonging to the hexagonal crystallographic system. As such, they can form the nuclei for the condensation and crystallization of water in a super-cooled and hence super-saturated moisture-laden atmosphere.[351,352] Lead iodide (PbI_2) behaves similarly and the organic compounds metaldehyde[353] and phloroglucinol[354] have in the laboratory proved their activity. In fact, metaldehyde forms ice crystals from water vapor up to 31°F while silver iodide does this, according to the article, only up to 25°F.

Two variations of dispersion are possible: the iodides may be present as such and are dispersed by a mixture of the composite-propellant type (Formula 142); or the iodide may be produced chemically, with several ways available to reach that goal. Shidlovsky cites mixtures of lead powder and iodine donors such as ammonium iodide, iodoform

* If the reader is baffled by the use of the letter J in lieu of I used in Shidlovsky's original text and copied faithfully in the translation to the third edition, this undoubtedly derives from the adoptation of the German word *Jod* for iodine and also from the casual use of the J in German where it is not regarded as a letter at all but rather as a variant of the letter I.

(CHI_3*), or "iodanil" ($C_6O_2I_4$) mixed with propellant material (Formula 143). In the project "Storm Fury" (NOTS), the iodates of silver or lead were used as oxidizers with gas-forming binder and heat-producing fuel metal (Formulas 144 and 145).

While interest in the United States has focused on the aborting of hurricanes and to a lesser extent on rain-making, it appears that the major Russian interest is in hail prevention. A secondary effect of causing precipitation by cloud-seeding is the removal of clouds and of fog. It has been possible to create enough of a "hole" in a cloud and fog layer over an airport to enable planes to resume operations (Salt Lake City, Utah, December 1962).[24]

Recently reported studies by R. M. Stillman and E. R. LaChapelle[355] demonstrated that a type of snow avalanche prevalent in the Rocky Mountains is caused in part by sublimation of the lowest snow layers with formation of weak fragile layers of crystals. This "depth hoar" may cause the whole mass of snow above it to break off in big slabs. Traces (a few grams per cubic meter) of long-chain organic molecules such as 2-octanol, the compounds benzaldehyde, n-heptaldehyde, and (less effectively) ethylene glycol inhibit ice-crystal growth by sublimation and thus promote formation of a strong bottom layer of snow. The effectiveness of avalanche control by dissemination of those chemical agents has not yet been conclusively proven. However, it is obvious that pyrochemical dispersion of such chemicals would be an ideal way of spreading out these materials in finely distributed form in not easily accessible mountainsides just prior to the first snowfall by dropping canisters or shooting rifle grenades or artillery shells in areas where potentially hazardous snowbanks form.

A recent concrete development is the dispersion of metals and compounds at high altitude (80—100 km and more). While 1 cubic kilometer of air at atmospheric pressure weighs 1 million tons, the same volume at 150 km altitude weighs no more than 2 kg: In such a rarefied medium, a metal vapor such as sodium dispersed from a rocket will create a large brilliant yellow train visible at twilight, useful for the study of ionospheric wind and of temperature. The sodium release has been performed by the heat from 8 kg of thermite evaporating (at a temperature of 900°C) 2 kg of sodium powder mixed with the thermite. Oxidation of the sodium takes place slowly.

An important modification of this technique is the replacement of sodium by barium. This allows the creation of a barium ion-electron plasma suitable for the study of magnetic field lines. Beside these, potassium and cesium have also been employed for certain explorations. As mentioned earlier (Chapter 13), ion formation of these alkalis also takes place in high-energy flare mixtures and in rocket propellants when the alkali is in the form of oxidizer salts.

The explosive dispersion of aluminum or the release of trimethyl aluminum at high altitude leads to formation of the lower oxide AlO that emits a blue glow useful for temperature measurements. Other studies with different goals involve the release of ammonia or nitric oxide. The subject is lucidly treated, with individual references, in an article by Rosenberg.[355a]

Part V

Kinetic Energy

chapter 20

Creation and Uses of Nonspecific Gases

To Arundell House, where first the Royall Society meet.... Here was an experiment shown about improving the use of powder for creating of force in winding up of springs, and other uses of great worth. And here was a great meeting of worthy noble persons. Thence to the office and then home to supper....

Samuel Pepys, Eighth Year (1667)[356]

Gas: (A word invented by the chymists). It is used by Van Helmont, and seems designed to signify, in general, a spirit not capable of being coagulated: but he uses it loosely in many senses and inconsistently (Harris).

Johnson's Dictionary[357]

The science and the techniques of gun propellants, rocketry, and explosives are separate disciplines, though together with pyrotechnics they have a common pyrochemical basis and involve overlapping pyrophysical phenomena. As a matter of practice rather than of classificatory nicety, the following uses of pyrochemically-produced gases are generally regarded as being in the area of pyrotechnics.

Moderately-sized and simply-built rockets, cardboard mortars, or integral parts of fireworks items that propel such fireworks or military signals into the air are always regarded as pyrotechnic components. Subminiature rocket projectiles and very small control rockets for steering missiles and spacecraft in flight may or may not be considered in this category.

*C*artridge-*A*ctuated *D*evices (CAD), now preferentially called *P*ropellant-*A*ctuated *D*evices (PAD), are used for ejection of stores from dispensers of airplanes, for the complex mechanisms of ejection of pilots from fast-flying aircraft as well as astronauts from spacecraft (such as Gemini) in cases of low-altitude aborts. The major devices are classed as *Removers* (*Retractors*), *Thrusters*, *Catapults*, *Guillotine-Type Cutters*, and certain *Initiators* (for operation of other PAD's).

Gas generators for pressurization operate hydraulic accumulators;

activate galvanic cells by forcing alkali solutions into the actual battery; operate fire extinguishers, turbo-blowers, and electric generators; and inflate balloon-like flexible cells such as emergency flotation bags for military helicopters. They may also act as dispersants for finely powdered solid agents (of the type described in the preceding chapter) or of solutions. In such applications it often appears to be desirable to throttle strongly the gas flow from the cartridge. Here, the orifice is kept so small that delivery of gas proceeds essentially independently of the geometry and resistance of the secondary device. This is called a "high-low" system. While it has the advantage that the gas-flow can be predetermined and tested independently, the heat losses are large so that more than two and a half times as much propellant is required than in standard PAD's. A further disadvantage is the possible obstruction of the small orifice by unburned propellant slivers or debris from the initiation system.[358]

Guillotine-Type Cutters are part of escape and stage separation systems, since they sever wires or cables preparatory to implementation of the procedure. Such cutters function in recovery systems to cut parachute shroud lines, load lines suspended from a high altitude balloon, or anchoring cables of a submerged moored mine.

Explosive Switches are devices in which the electrieally initiated explosive prime ignition charge, reinforced by some propellant powder, moves a piston, opening and closing electric contacts. Similar action is produced by *dimple motors*, *pistons*, or *bellows*. Dimple motors work by inversion of a curved flexible disk in the manner of a Belleville spring. The movement effected may be as small as 1/10 in. in a dimple motor against an 8-lb load, 5/16 in. in a piston motor against 20 lb, or 1 in. in a bellows motor against 10 lb, but the latter may work against as much as 100 lb.[178] In the same group belong *explosive valves*, which allow nearly instantaneous opening or closing of a channel through which gases or liquids can flow. Perhaps the word "explosive" in such items is a misnomer since the small amount of primary explosive acts essentially without destructive effect and in the manner of a fast propellant. Added propellant powder may do little more than maintain pressurization and prevent reversal of the piston. Explosive valves are part of automatic inflation devices on certain life-vests, such as those for astronauts.

Actual explosive force is employed in explosive bolts and nuts. Here, efforts are made to literally channel the explosive force inside a cavity toward a predetermined weak spot in order to effect a clean fracture. These items serve the purpose of stage separation of multistage missiles. Some act by "brute force" but others by mere over-pressurization. Choice of type may depend on the toleration or exclusion of fragmentation.

The development of PAD started prior to World War I, and research and development between the two wars included both civilian uses, such as the design of a humane cattle killer by a cartridge-operated captive bolt, as well as military applications.[6] The greatest boost to the modern PAD came, however, from the need for systems that ejected pilots from fast-flying aircraft. This problem was given to the Ordnance Corps in 1945 and led, together with other parallel efforts, to the various systems and designs of actuating devices. They have been covered in a Frankford Arsenal manual[359] in loose-leaf form with inserts that bring the collection up to date. Specific Navy items are found in two loose-leaf handbooks.[47,47a] The one out of NWL furnishes advice in Chapter 5 on amounts of various propellants for achieving a certain pressure in a fixed volume; in Chapter 7, formulas for the theoretical optimization of power cartridges are given.

More detailed basic information can be obtained from the ample literature on rocketry and rocket propellants, examples of which are the books by Sutton,[360] Wimpress,[361] and Warren,[362] while the *Rocket Propellant Handbook* by Kit and Evered[363] is merely a convenient reference for properties of some ingredients.

A comprehensive report useful for reference, on the use of PAD for escape systems from advanced aerospace vehicles has been prepared by McDonnell Aircraft Corporation for the Air Force.[364]

Taylor's book[6] is quite detailed on a great variety of applications of PAD other than for ejection. It deals, of course, mainly with the materials and applications in the United Kingdom where apparently a greater use is made of mixtures and compounds of high nitrogen content.

Private companies in this country have issued catalogs, pamphlets, and data sheets on the items that they have produced in the past and are able to furnish on short notice. The nature of these devices and the great variety of existing types on the one hand, and the small individual requirements and irregular demands on the other, as a rule, keep the manufacturer from having these things as "shelf items." This applies to all pyrotechnics, excepting only such merchandise as railroad fusees and torpedoes that are regular products of commerce.

Our concern here is mainly with the pyrochemical means of obtaining the gas from solid compounds or mixtures. The person not dealing specifically with the problems of internal ballistics must at least be conversant with the gas laws of physics as they are found in general texts such as Semat.[365] They will suffice for understanding the relation of chemical composition and physical conditions to performance of gases, but it may be necessary for actual calculations to take cognizance of the fact that the gas laws are idealized concepts and that under high pressures and near the point of condensation corrections must be applied.

Since one gram mole of any compound in gaseous form at STP (standard conditions of temperature and pressure—0°C and 760 mm Hg) occupies a volume of 22.4 liter, the optimal ratio of weight of gas former to obtained gas volume demands the use of elements or compounds of lowest molecular weight. Therefore, the ideal substance would be hydrogen (H_2), but this element is difficult to obtain pyrotechnically. Next, in order of ascending molecular weights, are water (H_2O), mostly to be considered a useful gas as long as higher temperatures are maintained after actuation; nitrogen (N_2); nitric oxide (NO); and carbon monoxide (CO). Least desirable are hydrogen chloride (HCl), carbon dioxide (CO_2), and chlorine (Cl_2), as well as any products of higher molecular weight.

If the average molecular weight of the mixture of the reaction products is near 22, the gas output obtained should be about 1000 cm^3 of gas per gram of reacted propellant. This figure can be approached with nitrocellulose powders and slightly exceeded with certain high nitrogen-gas formers. Mixtures such as black powder that leave large amounts of "fixed" non volatile residue furnish much less. Of course in actual operation the volume obtained, or at fixed volume the pressure reached, will be higher in proportion to the temperature during and following reaction. In general, a mechanical action is irreversibly completed at the end of maximal extension or before a decline in pressure can set in. Sometimes, however, the pressurization must be maintained at a certain level and for a certain time after peak.

Some details follow on the most important gas-forming substances and combinations, i.e. *black powder*, *nitrocellulose powders*, *high nitrogen propellants*, *composite propellants*, and *high-temperature propellants*. All of these are in practical use and are either specified for standardized cartridges or are in the process of replacing older propellants.

Black powder (gun powder), the jack-of-all-trades of the older fireworks art, is characterized by versatility, low cost, easy ignitibility, excellent fire transfer properties at a wide range of temperatures and pressures, and good stability at elevated temperatures. It is, however, difficult to adapt to orderly slow burning progression, is low in gas output, high in corrosive residue, and uncertain as to very long-range surveillance properties. Its manufacture is relatively hazardous and specialized. Smaller amounts under good housekeeping conditions are fairly safe.

The specification MIL-P-223B (1962 with Amendment 1—1963) is mandatory for all Department of Defense uses. Its formula is given under No. 146. Black powder is not strictly a mechanical mixture of potassium nitrate, sulfur, and charcoal. During the ball-milling phase of mixing the charcoal and sulfur, the latter is forced into the porous structure of the carbon. This makes the individual grain very dense. Hard compressed larger pellets or columns are relatively nonporous, so that they burn in orderly fashion.

In the manufacturing process black powder is granulated or corned. After finishing and polishing, which involves a slight coating with graphite (omitted on the finer particles and sometimes also on coarse powder), it is graded—separated into various grain sizes. Of the nine classes of fineness, Class 1 is the coarsest and Class 8 the finest powder, while Class 9 is called sphero-hexagonal with a complex shape of 0.6 in. diameter, each grain weighing 1/128 lb or 3.54 g (55 grains).

Older stores of black powder may have been labeled in accordance with the superseded Specification JAN-P-223A, which in turn is separated into Army and Navy application. The Army numbers are from A1 for the coarsest to A6 for the finest; the Navy names are Cannon, Musket, FFG, Shell, and FFFG to about 50 mesh, and Class 6; while the grades Fuze, FFFFG, and Meal correspond to Classes 7 and 8.

The specific gravity of the coarser grains must be 1.69—1.76 for unglazed, 1.72—1.80 for glazed powder.

Charcoal, JAN-C-178A, is not truly pure carbon but rather presents an undefined intermediary state between its organic origin and the element carbon, retaining both hydrogen and oxygen. Because of this composition and its high adsorption capability, it contains as much as 6% water; black powder specifications allow, therefore, a maximum of 0.70% moisture. The equilibrium water content varies with the relative humidity of the ambient air. At 90% RH, a weight gain approaching 2% may be measured but at normal RH (20—60%), the equilibrium water content has been given as 0.2—0.6% and between 80 and 90% RH as 1.0—1.5%. This moisture problem accounts for the deterioration of the powder and sometimes of its container under unfavorable conditions. Replacement of charcoal by a purer, non cellular carbon made from sucrose (sugar) is the subject of a U. S. patent.[366] Powders made with such carbon are slower burning than regular black powders.

Much has been theorized about the reactions that take place when black powder "burns." Undoubtedly, the products vary with conditions of the reaction, i.e. temperature and pressure. In one older investigation, it was found that 56% solid residue, 43% gas, and about 1% water were obtained. The solids were predominantly potassium carbonate (K_2CO_3)—61%, potassium sulfate (K_2SO_4)—15%, and potassium sulfide (K_2S)—14½%. The latter might actually be K_2S_2, since the analysis gave also 9% sulfur and several minor products. The gases were carbon dioxide 49%, carbon monoxide 12½%, and nitrogen 33%, with some hydrogen sulfide, methane, and hydrogen.

The variability of its reactions (aside from variations in composition) is evident from the wide range of the volume of permanent gases measured, i.e. from 198—360 ml/g.[367] The same can be said about the range of heats of explosion: 562—837 cal/g.

The heat of combustion, i.e. the complementation of the internal reactions in the presence of sufficient ambient air to convert all oxidation products in the highest normally-stable oxides, has been given as 1425 cal/g.

By older standards, black powder has a very good heat stability. Its explosion or self-ignition temperature has been reported as varying from 180°C (356°F) to 266°C (510°F), but some lower and higher data exist, depending on the method of testing. These figures do not apply to sodium nitrate powder with higher self-ignition temperature, described below.

The sensitivity to impact is low, friction sensitivity very small, but sensitivity to ignition by flame, high. Being a "low" explosive, its brisance as measured by the Sand Test is only 8 g as crushed sand *vs.* 48 g for TNT.

Under Specification X-PA-PD-623 (superseding JAN-P-540), a slow burning and fine (Type 8) powder, specifically for use in fuzes, is described (Formula 147). It differs from the standard military powder in the replacement of charcoal by a semibituminous coal and by a higher sulfur content.

By substituting sodium nitrate for potassium nitrate, a slower burning, somewhat more hygroscopic, less powerful black powder, introduced by DuPont in 1857, is obtained. Called "Blasting Powder B," it is used mostly in the form of large compressed pellets for shooting coal in underground mining, clay, and shale[222] (Formula 148). A government specification for such powder covering various grain sizes for uses in saluting charges, practice bombs, etc. is JAN-P-362.

The foregoing may already transgress the specific interest in black powders for PAD. For further details, one may refer to the *Encyclopedia of Explosives*, Vol. 2;[49] Taylor;[6] Davis;[8] and, historically, Biringuccio[13] and Partington.[14] Interesting variants of black powder are disclosed in British patents by Bowden and Blackwood who claim control of performance variations by extraction treatment of the charcoal;[368] improvement by replacing sulfur with alkali formate;[369] an increase in burning rate as high as 1000% by use of a low-melting mixture of potassium, sodium, and lithium nitrate;[370] and an increase in reaction rate by the addition of lead acetate to standard black powder.[371]

German efforts during World War II to eliminate the ill-defined and water-adsorbing charcoal by tetranitrocarbazole (TNC) led to mixtures with potassium nitrate and aluminum, useful as intermediary fires for illuminants, [315] though hardly resembling actual black powder.

Douillet[372] reports experiments with modified black powders of high sulfur (25%) and charcoal (20%) content, the latter declared to be best when high in volatiles. The advantages of this mixtures are claimed to be burning at "lowest pressures" at a rate governed by the often cited formula

$$V=KP^n$$

where V is the burning rate in mm/sec, K a constant between 2.9 and 3.8, P the atmospheric pressure expressed in *deci*meters of mercury (1 dm $=100$ mm), and the coefficient n is 0.62 for powders high in sulfur and charcoal. According to Shidlovsky,[3] ordinary black powder will burn at only 0.1 lb/in.^{2} ambient pressure.

Historically, the next types of propellant for PAD are the *Nitrocellulose* Powders*. They are high in gas, low in residue, and adaptable by formula and geometry to a wide range of burning rates and predeterminable burning-rate variations in the course of their deflagration. In a special group, certain additives make the burning rate nearly independent of the pressure (plateau or mesa burning), but this property

* The word "nitrocellulose" (NC) is generally used notwithstanding the fact that the so-named products are cellulose nitrates (esters) and not nitro compounds.

pertains only to a limited pressure range. Heat resistance is only fair, and all NC powders tend to gradual, self-accelerating decomposition, especially at elevated temperatures. This, however, can be counteracted up to a point by stabilizing additives.

When cellulose, such as cotton fiber, is nitrated, between two and three nitrate groups enter the basic $C_6H_{10}O_5$ group causing between 11 and 14% nitrogen (N) to enter the molecule. The lowest nitration with 11% N leads to products that are soluble in numerous solvents. As pyroxylin or collodion cotton, they are part of lacquers and plastics, the best-known being celluloid (Formula 149). Not suitable as a propellant, such material is sometimes of pyrotechnic interest as a binder or as a consumable structural component that requires very little oxygen for combustion.

Pyrocellulose (MIL-P-231A—1955) and Guncotton (JAN-N-244—1945) are higher nitrated, pyrochemically valuable types of nitrocellulose with 12.6 and 13.35% N, respectively. Pyrocellulose is still soluble in ether/alcohol mixtures, while guncotton has more limited solubility. The two are often blended with each other. Wet extruded, with small additions of nitrates and stabilizers, they form the so-called *single base powders*. Examples of such powders used in PAD are Type M6 (JAN-P-309) and Type M10 (JAN-P-715) smokeless, and IMR ("Improved Military Rifle").

By colloiding guncotton with nitroglycerine (actually glyceryl trinitrate), the so-called *double base powders* are obtained. It is possible to process and extrude these without the use of volatile solvent. This permits the manufacture of large grains without the need for the time-consuming removal of solvent and the disadvantage of ensuing shrinkage. Such a propellant is first converted into "carpet roll"—0.080-in. thick rolled-up sheets resembling thin rubber mats, which then are extruded in large, powerful hydraulic presses located in armored bays. A quite different mode of propellant preparation is the manufacture of "ball powder" starting from a lacquer-type solution and converting it into tiny spheres by agitation in an aqueous solution of a colloid. The process is applicable to both single- and double-base propellant.

The ratio of NC to NG in double-base powders varies widely and with it the energy and behavior of the propellant. The double-base propellants M5, M2, M26 and M9—all found in PAD—contain 15, 20, 25 and 40% NG, respectively. They are solvent extruded with the exception of M9, which consists of small flakes.

Any solid body of a propellant will furnish a decreasing amount of gas per unit time when the surface area becomes smaller during combustion at constant burning rate. In order to keep gas evolution constant or even increasing, geometrical, physical, or chemical means are employed. Perforations in cylindrical grains with simultaneous external and internal burning cause the inner surface or surfaces to increase while the outer decreases. The result is a more-or-less constant volume of gas produced. If an external surface is completely prevented from ignition, then the increase of the internal one causes *progressive* burning, i.e. gas development at an increasing rate. The physical means of such restriction of burning is called inhibition and is done by potting, wrapping, or plastic coating of the surface. If the conditions of removal of the gases are such that the pressure in the chamber or "motor" rises throughout the burning phase, then conditions for acceleration of gas-formation rates may occur independently from the geometry of the powder grain or grains. However, the addition of 2-5% of organic lead salts such as salicylic or acetyl salicylic acid to double-base propellants has the curious effect of making the burning rate independent of pressure in a limited range, a phenomenon called *mesa* or *plateau* burning[373] because of the appearance of the pressure/burning-rate curve. Ball powder can be surface treated in order to counteract the *regressive* burning of the spheres.

The burning rates of NC propellants are somewhat difficult to compare with ordinary pyrotechnic rates, since NC powders perform poorly at atmospheric pressures and their BR is generally represented in a graph of BR *vs.* pressures—about 10,000 lb/in.^{2} for regular NC powders and about 1000 lb/in.^{2} for specially phlegmatized, slow burning mixtures. The latter have inverse burning rates ("burning times," given here for easier comparison) of 2—5 sec/in. at 1000 lb/in.^{2}, while the regular NC powder, both single and double base, are below 1 sec/in.

Nitrocellulose varies in *heat of explosion* from 734 cal/g for collodion cotton to 1160 cal/g for a high-nitrogen NC with 14.14% N, but its *heat of combustion* declines with increasing nitration from 2606 cal/g for dinitro- to 2179 cal/g for trinitrocellulose, calculated from the formula of Jessup and Prosen[374]

$$\text{Heat of Combustion (const. press. 30°C)} \\ = -4176.7 + 141.26\, N \text{ cal/g}$$

where N is the percentage of nitrogen and the minus sign expresses exothermic reaction.

While all NC's are thermochemically underbalanced in oxygen, furnishing CO, CO_2, H_2O, H_2, and N_2 as products, NG is overbalanced and yields CO_2, H_2O, N_2, and O_2. Since the heat of explosion of NG is 1486 cal/g, the double-base propellants profit both from the higher heat of explosion of NG and its excess oxygen effecting a more complete reaction of the NC. Gas volume (water as gas) for both types varies from 900—1000 ml/g at STP.

Both single and double base propellants will survive temperatures of 200°F (129°C), but such exposure is only recommended for very short periods. The self-ignition temperatures (within 5 sec of exposure) for pyrocellulose, guncotton, and nitroglycerine have been given as 170—230°C.

Specific information on NC, NG, and their combinations, in addition to the literature quoted earlier, is found in books by Ott and Spurlin[375] Miles,[376] Naoúm,[377] and the classified CPIA (formerly SPIA) Abstracts.

The *High-Nitrogen Propellants* comprise the mixtures based mainly on ammonium nitrate, guanidine nitrate, nitroguanidine, and some more unusual gas-formers such as nonexplosive azides, dicyandiamide, 5-amino-tetrazole, and others. This group has the highest gas output, little or no residue, and often furnishes "cool" gas at slow rates.

The decomposition of ammonium nitrate by itself is an only moderately exothermic reaction according to the equation.

$$NH_4NO_3 \longrightarrow N_2O + 2H_2O + 8.8 \text{ kcal (exothermic)}$$

With 110 cal/g, it is not initiable from ambient temperature and barely self-sustaining at elevated temperature; therefore, pyrotechnically it is of no interest. While it produces a physiologically interesting and useful gas ("laughing gas"), the reaction is not free from side reactions in which not only harmless nitrogen but also the physiologically most undesirable higher oxides of nitrogen appear. Eight modes of decomposition have been postulated.[6] These are mostly much more exothermic than the one leading to N_2O, furnishing from 219—292 cal/g. The same reactions prevail when certain catalysts are added, and in the presence of fuels considerably more heat is evolved. Taylor and Sillitto[378] describe mixtures of ammonium nitrate with 2% charcoal or 12% starch that yield 878 and 860 cal/g and 909 and 938 ml gas/g respectively, calculated at constant pressure and with water considered as a gas. By using cool-burning or hotter fuels, a wide range of flame temperatures can be achieved—from 360—2000°C and gas volumes up to 1350 ml/g. Among fuels mentioned that are also high gas formers are ammonium oxalate monohydrate, $(NH_4)_2 C_2O_4.H_2O$, and nitroguanidine.

British patents by Taylor[379] show the use of ammonium dichromate as catalyst for cool-burning mixtures, which can be cast because they melt below 120°C. Burgwald *et al.*[380] claim additionally the catalyst Prussian Blue and in excess of 20% of plastic binder and plasticizers. Similar combinations are claimed by Eiszner and Stanley[381] and Marti.[382] These three rightly belong in the class of composite propellants because of the sizable amount of a fuel-type binder.

A disadvantage of ammonium nitrate is its hygroscopicity and a crystal-structure change that increases its volume by 3.8% at the unfortunate transition temperature of 32°C, which is also the "popular" summer temperature of 90°F over most of the United States. This transition point can be shifted, i.e. depressed by about 20°, but not abolished, by admixture of 8% of potassium nitrate.

While ammonium nitrate furnishes the slowest burning propellant-type mixtures, the use of catalysts and fuels decreases the burning time (inverse burning rate, used here for easier comparison with strictly pyrotechnical performance) to 2—5 sec/in. at a pressure of 1000 lb/in.^{2}

Guanidine nitrate and nitroguanidine are both exothermically decomposing materials of similar reactivity and similarly catalyzed decomposition reactions but without some of the disadvantages of ammonium nitrate, though they are much less available and of much higher cost. Gas volumes on decomposition by themselves or with admixture of ammonium nitrate or various fuels are also in the same bracket —about 1000 ml/g. Disclosures by Taylor and Hutchison[383] and by Hutchison[384] give examples of these types of propellant. Burning-rate catalysts or "sensitizers" claimed are vanadium oxide (V_2O_5), cuprous chloride (Cu_2Cl_2), molybdic acid, ceric oxide, and others. The burning times achieved are generally long—14—25 sec/in. Nitroguanidine produces hotter gases and presumably higher burning rates (under comparable conditions) than guanidine nitrate. But in general these compositions are chosen because of their nonviolent reactions and relatively cool gas formation, which makes them useful for turbo starters for aircraft and for dissemination of pesticides as discussed in Chapter 19. Formulas 150—154, all taken from the above-quoted British sources, pertain to the high nitrogen gas-formers cited above.

It remains to have a look at *Composite and High Temperature Stable*

Propellants. They will be needed more and more for supersonic aircraft and space missiles, which develop high skin temperatures or encounter extreme environmental conditions.

Composite propellants, i.e. mixtures of oxidizer salts and various plastic binders (generally about 25% of the latter), have been touched upon above in connection with ammonium nitrate. Plastic binders are pyrochemically quite inert until pyrogenic breakdown occurs at very high temperatures; the most temperature-stable should be the fluorinated polymers. The stability and temperature resistance will then hinge on the properties of the oxidant. From this viewpoint, composite propellants using potassium perchlorate should have the best properties. The conventional ammonium perchlorate composite will in general withstand 300°F for days and 400°F for hours while the corresponding figures for potassium perchlorate are 100°F higher. A mixture of sodium nitrate, sulfur, and rubber, vulcanized by heating, has been patented[385] as a "composite explosive" suitable to replace black powder, and might possibly be a heat-resistant gas former.

Mixtures of potassium perchlorate and Hycar (acrylonitrile copolymers) have not only good temperature stability but seem to be free from forming toxic gases on prolonged storage.[386]

Epoxy resins have also suitable physical properties, but silicone resins proved disappointing. One must always consider that PAD's are often small devices and their propellants should be available as easily handled extruded shapes if at all possible.

In connection with high-temperature resistance, attention has been given to actual explosives. As a rule, the so-called high explosives will deflagrate without destructive brisance on ordinary ignition if not extremely confined or burning in large amounts. They are, however—a strange and seemingly perverse fact—generally much less heat-resistant than azides and styphnates. Exceptions are HMX, trinitrobenzene (TNB), and Hi-Temp, a proprietary material based on RDX. Other explosives, possibly suitable as gas-formers, are still classified.

Commercially available is the explosive Tacot (du Pont)[188] of m.p. 378°C (710°F), at which point it decomposes rapidly, while claimed to be stable below that temperature. Under development in government laboratories are compounds that in combination with certain plastic binders have cook-off temperatures up to 1000°F.

Part VI

Noise

chapter 21

Explosive and Whistling Sound

Pyrotechnic means of creating sound include the "joyful noise" of explosive fireworks items described in Chapter 17 and whistling devices. The latter are found in fireworks for pleasure and in a few minor military items. The strictly utilitarian explosive sound producers are sometimes identical with fireworks items. They are parts of weapon-simulators that make maneuvers more realistic or are actually used for deception of real enemies; some provide sound signals for troop training to show that the trainee "succumbed" to a concealed weapon; a few have been tried as crop-protecting bird deterrents and for clearing airport landing strips from hovering birds; others represent a sizable commercial item in the form of warning signals of the railroad torpedo type. Various combinations are used for the simulation of noises and other effects on battlefields and for various catastrophes in the motion picture industry. Military salutes might also be included, though their arrangement differs from that of the other devices.

Military training devices employ flash and (explosive) sound as in the earlier-described *M*110 *Gunflash Simulator* (Chapter 15 and Formula 43) and the *M*115 *Projectile Ground Burst Simulator* (Formula 44), in which a whistling sound precedes the flash and explosion, and the *M*117 *Booby Trap Flash Simulator* (Formula 45). In fact, all flash signals function also as auditory signals if operating at a distance not much exceeding 2000 yards, such as the *M*27*A*1*B*1 *Projectile Airburst Simulator.*[33] The *M*80 *Firecracker* (and the apparently quite similar Navy *Mk1 Mod 0 One-Inch Salute*) are the military counterparts of the civilian "salute"—a small cardboard cylinder with a piece of firecracker fuse of 3 sec.-delay or longer and 3 of a flash cracker mixture of aluminum, sulfur, antimony sulfide, and potassium perchlorate. It can be attached to a variety of pull or pressure release mechanisms[39,387] for the improvisation of booby-trap simulators. In unconventional "conventional" warfare or guerrilla activities, such noisemakers can be used to divert attention or even simulate a skirmish, disturb patrols, etc. Blasting caps have been used in such noisemaking devices. Electric detonators are especially suited for simulation of the repetitious sound of machine-gun fire but require for this purpose a battery, a clockwork, and a system of electric contacts, not to forget the shielding of the individual caps from each other—all of which makes the device cumbersome.

For the purpose of military protocol, as well as for simulated heavy gunfire, various sizes of primed gun cartridges are partially filled with black powder either in loose form or as compressed pellets. A felt wad and a closing cup held in place by Pettman Cement (JAN-C-99) confine the powder charge and by obturation of the developed gas and the subsequent sudden expansion help in producing a loud noise. Such a powder charge may be quite substantial—in the *Blank Ammunition Double Pellet Charge for 75 mm Guns or Howitzers*, the two pellets combined weigh 0.87 lb.[41]

Since the advancing motion of a railroad engine on a steel rail is in effect comparable to a glancing sledge hammer blow, a tablet of an interposed impact-sensitive pyrotechnic mixture will explode with a a report loud enough to be heard in the engine cab above the other noises. In the domestic form of the well-known *railroad torpedo*, a chlorate-containing mixture (Formula 155) compressed in the form of a rather substantial pellet with flat bottom and rounded top is wrapped in moisture-resistant paper. Sand particles glued to the bottom and a pair of lead or aluminum straps hold the item firmly onto the rail—an important feature—since the torpedo must not be dislodged in a severe rain and windstorm.

Because these torpedos are handled rather casually, it is imperative that their impact-sensitivity not be excessive. In testing, a 25 lb drop-weight of a certain profile is used, which must not set off the explosive mixture when dropped from a height of 8 in. but must function (at least on the second try) from 16 in. The sound level at a distance of 20 ft should be no less than 105 dB. The torpedo must not disperse any damaging fragments (as evidenced by its effect on a denim cloth at 3 ft), and conditions of water and heat resistance are spelled out by commercial convention.[388]

A number of formulas appear in U.S. patents,[389,390] but whether these would be adequate by modern standard or pass the testing conditions described would have to be ascertained by trial and error.

Certain foreign railroad torpedos are actuated by a primer and a black powder charge confined in a metal capsule.

Explosive sound rarely evokes a question about the actual mechanics of its propagation as an unwanted by-product of the sudden expansion of propellant gases in firearms or of explosive action in blasting. The subject seems to be generally ignored in books on sound. It only appears, and rarely, in the special literature when means for attenuation are sought where the sound is carried from the test site through the air and through the ground to a populated area. While the subject is of interest to the harassed pyrotechnics manufacturer either in connection with photoflash-cartridge testing or while he is engaged, additionally, in some explosives work, it cannot be more than merely touched upon here. The principle in a gun silencer is to stuff the area through which the sound waves travel with material of the highest possible density, such as steel wool. By the same token, heavy steel curtains or sand- and water-filled enclosures act similarly on explosives test sites. A report emanating from Fort Belvoir [391] is instructive since it contains the replies to an inquiry on noise abatement by a number of institutions on the problem of avoiding air-blast effects to the neighborhood when up to 25 lb of explosives are set off. It is apparently possible to confine charges of this size within a steel sphere.

The noise from flares merely burning depends on the speed of combustion and on the emitted gas and is sometimes a strong hiss or as much as a "gentle roar" as in the "waterfall" spectacle (Chapter 17). However, the decrepitation of the crystals of certain acids of aromatic structure in combination with oxidizer salts may under proper conditions proceed with a shrill *whistling sound.*

The active substance most often used in pyrotechnic whistles is gallic acid (3, 4, 5 trihydroxybenzoic acid). The potassium salts of benzoic acid; of 2, 4, dinitrophenol; and of picric acid (2, 4, 6 trinitrophenol) and the sodium salt of salicylic acid (o-hydroxy benzoic acid) are also effective. They are combined with potassium chlorate, perchlorate, or nitrate (Formulas 156—160).

In World War II, the Germans used picrate whistles as a psychological adjunct to bombs[7] and they also used pyrotechnic whistles as gas alarms. In this country, the *M*119 *Whistling Booby Trap Simulator*, ignited by a pull cord, produces a shrill whistling sound for 3 1/2 sec from a 2-in. column of Formula 156. The *M*117 *Booby Trap Flash Simulator*, pull-cord ignited, starts with a whistling sound followed by an explosive flash.

The only study known to the author that treats the subject of pyrotechnic whistles in a scientific manner is by Maxwell.[392] His experience confirms that of the manufacturers of these items—that the compositions used are capricious, and even products from the same batch but not pressed at the same time yield widely varying results, not to speak about variations in grain sizes, compaction, or percentages. The decomposition of the aromatic acids seems to progress in an oscillating manner whereby the crystals "explode" in a manner of speaking—the word "decrepitate" used earlier may perhaps be more to the point. Thus, a rhythmic acceleration and stoppage of the reaction takes place, leading to alternation of pressure and rarefaction in the tube on whose length the pitch of the sound depends.

If a mixture of this kind does not whistle, its flame characteristics, as shown by Maxwell, are quite different from those during whistling; the nonwhistling condition produces a much larger and more brilliant flame and for a much shorter time.

Caution is advisable with these highly active and (even in moderate quantities) potentially explosive mixtures. While Maxwell mentions a tube diameter as high as 3 in. in one of his experimental series, it is the author's experience that a whistling mixture in a steel tube of 1-in. diameter performed properly at first but ended with a deafening report.

Maxwell's article mentions also the interesting fact that a pyrotechnic whistle not only burns well under water, which is to be expected, but also that the sound can be heard above the water, though at greatly reduced intensity.

A PAD could undoubtedly be used to produce a whistling noise or the noise of a siren by acting strictly as a compressed gas source on a suitable acoustical device, but the author does not know of any such item proposed or actually designed.

Part VII

Heat Production Per Se

chapter 22

Uses of Heat

Creation of flame or glow from pyrochemical reactions proceeding from ambient temperatures was the subject of Part II, Primary Flame and Glow. In all subsequently described applications, heat was the source of useful secondary phenomena, viz. radiation (light), aerosolization (smoke), kinetic energy (PAD, noise), but the heat was not utilized as such, e.g. for heat transfer plain and simple, except in an accessory capacity.

The following five chapters are concerned with heat, pyrochemically produced after initiation by a prime ignition source and employed for heat transfer exclusively, though other functions may be connected with it, such as timing.

The specific functions that fall into this category are *first fires* and *igniters* for pyrotechnic mixtures and solid propellants; *fire transfer lines*, *time fuses* and *delay columns*; *fire-starting devices* for nondestructive uses; *incendiaries* and related *destruct units*; *heat cartridges* and other means for *"mild" heat transfer*; and finally, special *torch-flame* combinations.

Some of these uses are served by discrete units, such as fire starters, fuse trains, and heat cartridges, but the fire transfer systems of pyrotechnic items are generally integral components of a flare, smoke candle, etc., and are formulated and adapted to the specific function of bringing the main item to self-sustaining reaction after having been initiated themselves by a prime ignition system or a fire transfer line.

It would appear that such heat transfer would be amenable to theoretical treatment, so that ignition and heat transfer systems might be calculated from the amount of heat, temperatures, and the laws of heat flow. Unfortunately, because of the complexity of the conditions, efforts along these lines and the successful applications of mathematical formulas, such as the calculations for the progression of the glow front in delay columns have been limited. However, the increasing variety of available materials, especially of "modern" metal-powder fuels, continues to advance the state-of-the-art along empirical lines and to increase the reliability of pyrotechnic devices. Since every point of fire transfer may be a weak link, the importance of these ancillary and subordinate formulations is obvious.

chapter 23

First Fires and Igniters

The name "first fire" should semantically and logically apply to prime ignition, but in pyrotechnics it refers to the intermediary ignition source between primary initiation and the main item. The terms "starter mixture" or "ignition mixture" are occasionally used in the same sense. In flare candles and other highly compacted items, the first fire is always put in intimate contact with the major reactant body by pressing it on top or painting it in slurry or paste form on the surfaces to be ignited. Sometimes, additional loose powder or small granules of an ignition mixture help in fire transfer, especially when flash from a prime igniter has to bridge a somewhat larger gap. A separately confined mass of first fire in a tube or capsule that ejects its contents and spreads hot gases and particles over a larger surface area is called an *igniter*, and if it is in the form of a miniature rocket, a *pyrogen igniter*. These devices are used for solid-propellant ignition.

The theory of initiation has mainly been expounded in the field of solid-propellant ignition. A recent effort for pyrotechnics has been published by Johnson.[393] The calculations are difficult since they involve calories transferred to the surface of the initiated column at certain temperatures and over a time interval; heat absorption and flow in the main item, heat developed in the main item from the incipient final reaction, and, of course, heat losses. Johnson advocates on the basis of theoretical considerations a small area on which fire transfer is concentrated; low thermal conductivity, density, and specific heat of the first fire; and use of high-energy, flare-type first fire mixtures in such a way that radiant heat transfer is optimal.

The need for first fires arises from the fact that many fuel-oxidizer combinations are difficult to initiate and that if there is a choice the least sensitive main item is the most desirable one from the viewpoint of safety in manufacturing and handling. The first fire, on the other hand, is often a much more hazardous mixture. However, only a small fraction of the weight of the main item has to be processed, and this quantity can sometimes be handled and applied in a liquid medium that greatly reduces the accidental ignition hazard.

While it may seem advantageous to keep the number of components of a pyrotechnic system as small as possible, great discretion must be displayed when it is a question of omitting an intermediary charge in a chain of ignition. A prime igniter should not be increased in size or potency in order to make a first fire unnecessary, and the latter should similarly not be enlarged or upgraded for the purpose of eliminating a second intermediary mixture. The rule should be to keep the more hazardous compositions at as low a sensitivity as practical and at minimum size.

A number of mechanical tricks can be applied in order to facilitate fire transfer between two solid layers. They consist of increasing the area of contact between the two compositions. Mixing part of the first fire with candle composition could be regarded as such a step, but mere consolidation of the last (top) increment with the layer of first fire powder or granules in one pressing operation also improves intimacy of contact. "Nesting" the increments by use of a stepped ram, i.e. a ram with a truncated conical protrusion, is often resorted to—not only between first fire and candle, but between increments of the main items themselves. In fireworks-type items, coarse grains of black powder sprinkled on top of the mixture before consolidation cause embedment of these grains in the surface and may suffice as a first fire.

Failures of fire transfer from first fire to main object occur frequently. The cause is transfer of too few calories or transfer at too low a temperature to bring the main object to a self-sustaining exothermic reaction. In these cases, the first fire either does not reach a sufficiently high temperature because of its composition and nature of reaction, or it may be present in too small a quantity or dissipate its heat because of faults of design; it may further react too fast. In some cases, the first fire, though of seemingly excellent quality, may be blown off and away from its contact surface, leaving the dudded surface without any evidence of fire transfer. Such failures are more likely to occur on very small surfaces such as delay columns, where the ratio of metallic enclosure material to pyrotechnic mixture is large and detrimental. In these cases, the behavior of the first fire may also be influenced by the action of the primary ignition system such as a percussion primer or an electric igniter. In rare cases where there is a choice of alternate prime igni tion sources, one may be good, another one marginal.

Since initiation may take place at high altitude, fire transfer may be impeded by the changes in burning rate or flame temperature caused by diminished atmospheric pressure. This can be counteracted if the fire transfer takes place in full confinement or under adequately obturated conditions. Also, certain first fire mixtures, as shown below, are less influenced by pressure changes than others.

Transfer lines between prime ignition and first fire must be regarded as a part of the secondary ignition system. Quickmatch, a type of black powder fuse that has a more-or-less well-defined burning time and thus acts as a delaying element, will be described in the next chapter. It becomes a nearly-instantaneous transfer line if enclosed in a narrow paper tube and is thus used as "piped match" in fireworks for simultaneous initiation of several effects.[7]

Where the beginning of a fuse train is not part of a self-contained system, as is the case in fireworks ignition and certain blasting operations, the fuse must be lit by hand. Ordinary matches are difficult to handle in the open, and some black-powder fuse trains, especially safety fuse, are very hard to start with ordinary matches except by employing some time-consuming tricks. They are more reliably set off by a sharp tongue of flame, which can be furnished by a *Lead Spitter Fuse Lighter*. This is a coil of thin lead tubing filled with black powder and burning at a speed of about 36 ft/sec. Another device is a sparkler-like coated wire called a *Hot Wire Fuse Lighter* of 7, 9, or 12-in. length and burning 1, 1 1/2, or 2 1/4 min, respectively, by means of which a whole series of fuses can be lit in sequence. These items, including the *Pull-Wire Fuse Lighter* described in Chapter 12, are found in the *Blasters' Handbook*.[222]

A new and vastly different fire transfer and method of ignition is by means of explosive line charges, which contain relatively small amounts of high explosives. *Pyrocore*[180] is the tradename for such an item. It consists of metal tubing 1/8 in. or less in diameter and a combined explosive and ignition mixture, 4—40 grains per foot, which can be contained with little additional shielding so that the line may be threaded through other components of a system without damaging them. The terminal explosive effect will cause the ignition of numerous metal/oxidant mixtures or of propellants. It is claimed that even coarse commercial thermite can be initiated from Pyrocore if Alclo pellets (see Formula 182) are used as a starter. All this seems to contradict the concept of nonviolent fire transfer in order to preserve the relative position of the compositions during burning and heat transfer. Undoubtedly, the extreme speed of the detonating wave (12,000—21,000 ft/sec) and the high instantaneous explosive pressure exerted on the substratum (in the absence of damaging brisance) effects the fire transfer in this novel manner. Special "mild" explosive initiators and transfer connections are available for the Pyrocore systems.

Before going into the subject of specific compositions, it should be mentioned that the same mixtures that are useful as first fires may also have certain uses in their own right, as in heat-producing cartridges, or be applicable for delay trains. Conversely, a delay or a "heat powder" mixture may be also an ignition mixture or become one by the addition of a binder or an ash-forming, heat-retaining, inert additive such as a siliceous material. Furthermore, mixtures, (judging from their use of hazardous ingredients) that seem to be suitable only for prime ignition may be needed in special cases for fire transfer purposes, confronting the manufacturer, if he has any choice in the selection, with a dilemma between the desire for the most reliable fire transfer and for safety in operations.

First fires without metallic fuel* are nowadays mainly found in fireworks items. The most important one of this class is black powder. It is fairly safe to handle, often serves as a combined ejection and ignition material, and is itself easily ignited over a wide range of temperatures and at diminished atmospheric pressure down to about 0.1 atm.[3] Its property of "fouling" metal parts because of the corrosiveness of its reaction products need rarely be considered since pyrotechnic devices are one-shot items and if ejected from reusable dispensers their own expendable case is the actual "gun barrel." Black powder has been discussed in some detail in Chapter 20.

Formulas 161, 162, and 163 are of the same type as black powder formulated for a variety of applications where extreme heat is neither needed nor desirable. Here, the evolution of gaseous reaction products is either innocuous or actually beneficial, be it for creating a certain pressurization above the initiated area or for coincidental expulsion or separation of the ignited item and container. It might be mentioned that most of the formulas of this type demonstrate an effort at creating a black-powder substitute, but no single formula or family of formulas has yet achieved this goal, even though the newer

* It is customary to group boron and silicon with the metal fuels because of similarity in pyrochemical behavior.

formulations may have specific advantages in the applications for which they have been created.

Formula 164 is a somewhat unusual specialty item for paper destruction. Formula 165 was developed by the author as FIC-2 for the *XA2A* parachute flare, the developmental predecessor of the *Mk 24*. The mixture is applied to several layers of surgical gauze. Originally, a solution of nylon was used as binder for this, as well as for a second expulsion and fire transfer disk of a somewhat different formulation containing a sizable percentage of tetranitrocarbazole (TNC).

More conventional hot first fires are represented by Formulas 166—172 based on silicon or silicon and other fuels of high heat output and high heat retention. Of these, Formulas 166 and 167 are suitable for HC smokes that are fairly hard to ignite. They produce a certain amount of gaseous reaction products. While this tends to carry some of the heat generated away from the substratum, it may actually be beneficial by creating temporarily a slight overpressure during the fire transfer in grenades, canisters, or smoke pots. For Formula 168, the same considerations apply, since here the illuminating flare candles for which this formula is especially created are in transient confinement during the ignition phase.

The next group of formulas (No. 169—172) are of the "gasless" type and even though relatively low in calorific output have good fire transfer properties where the conditions are not too exacting. Formula 171 has, curiously, led a semi-anonymous existence as an ignition powder for the phosphorus mixture in *Drift and Float Signals* (Formula 41) and its history, including the strange selection of cuprous oxide (Cu_2O) as one of the oxidizers, is obscure. Appearing on Navy drawings, there is no identifying specification, but it has been given the nickname "six-six-six."

Much better fares the closely related and probably derivative Formula 172, extensively characterized as the *Mk* 25 *Mod* 2 *Starter Composition*. It exceeds "six-six-six" moderately in calorific output (380 cal/g against 344), has a very high ignition point of about 500°C, and is insensitive to accidental initiation. A considerable amount of attention has been given to this item in reports emanating from NAD, Crane, Indiana.[394,395,396]

An extremely potent, but seemingly very little known mixture, is the one of flake aluminum and sulfur in approximately stoichiometric ratio of about one to two parts. It can be ignited with an ordinary match and reacts slowly with brilliant white glow, forming beads of aluminum sulfide. It is hot enough to initiate mixtures that are quite difficult to start. Because of the nature of its ingredients, its practical application is, however, limited. Some degree of consolidation is possible by adding small amounts of binder and thoroughly moistening the low-density mass. Remy[60] mentions the reaction and its application as a "booster" initiator for thermite, but the pyrotechnic literature seems to have by-passed this remarkable reaction.

A somewhat specialized group of first fire compositions falls under the heading of *tracer ignition.* Tracers for artillery ammunition have been described in Chapter 16. Because of the stresses on the tracer composition at the moment of firing, the tracer mixtures are extremely heavily compacted, which makes ignition difficult. On the other hand, the use of very active oxidizers (and fuels) in the igniter is to be shunned because under the conditions of ignition by the highly-compressed propellant gases explosive action might occur. The solution has been the employment of alkaline-earth peroxides, mostly barium peroxide (BaO_2) with 12—25% magnesium. Binders are asphaltum or calcium resinate, and zinc stearate with "red toner" added as a distinguishing coloring agent. Some formulas also include Parlon, a chlorinated rubber product (Formulas 173—176).

It is desirable that the first fire from the tracer emit little light so that the gunner is not blinded by it while, with the light trail appearing at some distance from the gun, the purpose of the tracer is fulfilled.

On the other hand, by filling the tracer cavity with a strontium-peroxide-type igniter mixture only, a purposely misleading effect can be achieved, since the tracer seems to pertain to a projectile of much larger caliber. A projectile for this kind of "psychological warfare" is called a *headlite bullet.*[257]

"Dark" igniters for tracers have been the subject of several patents by Heiskell[397] and by Clay and Sahlin.[398]

Mixtures similar to the tracer igniters are Formulas 177 and 178, useful for ignition of thermite. For the same purpose, a mixture of magnesium and barium peroxide, formed into a small ball with the help of some collodion and provided with a length of magnesium ribbon for ignition, has been used in Germany under the name of *Zündkirsche* (ignition cherry.)[61]

Other first fires of highest heat concentration, though not of highest calorific output, comprise combinations of the elements boron, titanium, and zirconium in the finest powdered state with the oxides of iron, manganese, and of other nonvolatilizing metals. The oxidizer may be an alkaline-earth chromate. All these oxidizers must also be in finely powdered form. Such mixtures are all of the "gasless" type, hence free of any binder. Because they may be quite hazardous and tend to accidental ignition, they are often mixed under water or under a volatile organic liquid. Mixtures that are not used in complete confinement may contain a binder. This promotes good compaction, produces more stable granules, or permits use as a surface coating. Addition of glass powder or other siliceous material may dramatically improve fire transfer by converting a disintegrating residue into a sintered well-adhering coating. On the other hand, when the ignition is effected by blowing the glowing particles onto the ignited surface, the need for persistence of heat retention may require admixture of coarser fuel particles. Representative types of this class are Formulas 179, 180 and 181. It should be noted that aside from the indicated additions, Formula 180 may be varied in proportions and choice of chromate to yield a family of formulas with a wide range of heat outputs; the same effect can be achieved when boron is replaced by zirconium. These mixtures are also described in Chapter 34 on calorific output.

Compositions of highest heat output using boron, aluminum, and titanium or zirconium, with oxidizers such as potassium nitrate or perchlorate, are not normally needed in strictly pyrotechnic fire transfer, but they are important in the initiation of solid propellants, be it for small items such as gas cartridges or for larger grains. Formulas 182, 183, and 184 are some of the better-known examples.

The theory of fire transfer from a pyrochemical ignition mixture to a powder grain, such as the surface of a double-base propellant, has been the subject of much speculation and falls outside the scope of this book. Total heat output, pressurization by gaseous reaction products, glowing solid particles, and condensing vapors will contribute in varying degree to the creation of a stable flame-front on the propellant surface.

First fire formulas appear sporadically in the patent literature occasionally claiming unusual compounds and alloys as fuels but generally being in line with formulas cited. Of historical interest is the original work of Hale[399] at Picatinny Arsenal, which was the start of the low-gassing mixtures. The formulas of Magram,[400] Zenftman,[401] Magram and Blissel,[402] McLain and Ruble,[403] Clay and Sahlin,[398] Heiskell,[397] and Hart[404] are examples of patented ignition mixtures.

The most modern trends include the use of fluorine-containing polymers playing a part both as oxidizer and as binder, as shown in a patent by Williams and Gey.[405]

chapter 24

Delay Trains

The mediaeval method of measuring time by prayers (Paternosters) or the recitation of creeds, very common in Europe, especially in Catholic countries, was used by gunners in timing their fuses.

J. R. Partington (1960).[14]

Introduction and History

If a pyrotechnic powder mixture that is neither extremely violent nor too sluggish in its reaction is spread out in the form of an elongated loose-powder train or tamped or compressed in a narrow column and ignited at one end, the reaction generally proceeds in an orderly fashion as shown by a glow or flame-front that can be followed visually. By coupling prime ignition and the final pyrotechnic event with interposition of such a train or column, certain time delays are engendered. They comprise an important phase of modern pyrotechnics.

The history of pyrotechnic delay timing until about the year 1929 can be summarized in one word—black powder. This all-purpose material of the older fire-art furnished not only propulsion, fireworks effects, and explosive force, but in proper arrangement and with modifications in composition formed compressed powder columns or elongated fuse lines that are (especially the latter) still used in fireworks items and some military devices. However, the modern requirements of extension of the timing range, high stability, and complete confinement (obturation) led to a search for low-gassing delay formulations. The earliest efforts by G. C. Hale,[399] starting around 1929, involved silicon and red lead (Pb_3O_4) or lead chromate ($PbCrO_4$). Glycerine was added, probably since it has binding power with lead oxide. These early mixtures were of course gas-forming to some extent. They were followed by truly low-gassing mixtures such as of barium peroxide and the chalcogens patented as early as 1934.[406] Iron or antimony and potassium permanganate were tried and used extensively in Germany, but never became popular in this country. World War II provided a new impetus to research and development of the "gasless' delay formulations and the work on improving them has continued ever since—by no means steadily but rather by sporadic efforts.

Delay Ranges and Uses

Pyrotechnic delay timing comprehends an approximate range of $^1/_{100}$ to 40 seconds per linear inch, a ratio commonly referred to as "burning time" (BT), or occasionally as "inverse burning rate," the burning rate (BR)—expressed as in./per sec or centimeters or meters per second—being more popular for characterization of propellants and explosives. For the indicated range, the figures would be a burning rate of 100—0.025 in./sec.

Since there are practical limits to the length of a powder column as well as to the shortness at which reproducible functioning can be expected, pyrotechnic delaying action of an event can be said to be practical for a time interval of about 5 msec to 5 min. Where this range is insufficient, it can be extended to the microsecond range (one microsecond equals one millionth of a second) by using prime explosives, sometimes in combination with highly active pyrotechnic mixtures; in the long time-delay zone of minutes, hours, days, or weeks, certain chemical or physicochemical effects of corrosion or of gradual softening of high molecular solids can be utilized. These means for long delay times are not pyrochemical in nature, but they are treated in some detail since they are associated with incendiarism and explosive demolition.

Delay action below 1 msec and in the lower millisecond range is used in projectiles and bombs that explode on impact on the target but which may do greater damage if a time delay is interposed in order to detonate the missile after penetration of the target.

In the millisecond range, say from 25—1000 msec, the delay action provides for effective blasting in quarries, salt mines, and numerous other commercial applications of explosive charges. Staggering the explosions in series of boreholes reduces vibration and improves fragmentation. It has special advantages in grade construction work.[222]

The range of 1—6 sec is used in ground chemical munitions such as smoke pots, tear gas and smoke grenades, and fragmentation hand grenades, as well as in photoflash cartridges and similar air or space-launched signals. With such items, it is imperative that the exploding or otherwise hazardous object is safely removed from the point of initial activation—the person throwing the grenade in one class, the dispenser in the skin of the vehicle in the other. Hand-thrown grenades, which emit smokes, avail themselves of a 1.2—2.0 sec delay effected by the well-known "bouchon"-type fuzes *M*200*A*1, *M*201*A*1, *M*207*A*1, and *M*210. Certain riot grenades allow 1.4—3 sec delay action in an integral fuze that terminates in a detonator. The bouchon-type detonating fuze *M*206*A*1 is used in a white-phosphorus-smoke hand grenade with a delay of 4—5 sec.

Longer delay times, up to 60 sec, are found as part of so-called delay switches. Following an electric initiation, these devices undergo a pyrotechnic delay action that is followed by a mild explosive effect from a small charge of a prime explosive. The latter causes a piston to move, opening or closing one or more electrical connections. Timers of this kind are used in missiles and nuclear devices to take part in sequential actions when there is no need for high accuracy of timing but where there is a stringent requirement that the device function properly with highest reliability and not before a certain minimum time interval has elapsed.

In space technology, sequential timing may well go beyond 60 sec. It is possible to arrange delay trains of slow-burning mixtures to allow up to 5-min delay time, but for longer periods the items become rather cumbersome. At the present state of the art, the limit of reliably burning mixtures is 40 sec/in. at atmospheric pressure and considerably less in confinement. Thus, about 10 in. of column length have to be accommodated in a 5-min delay unit. This has been achieved by Ellern and Olander in an internally vented arrangement known under the trademark *Pyroclok*.[407]

We come now to the devices for longest delay timing, which are nonpyrotechnic. They are used as bomb fuzes and mainly constitute a harassing factor—greater still if they refuse to function properly and remain a hazard even after the end of hostilities! Similar long delay timing is used in connection with the clandestine emplacement of explosive or incendiary charges or in leaving concealed charges behind during a retreating action. The units that permit such delaying action for hours, days, weeks, and in extreme cases for months, will be described below. They are quite small and compact, while clockworks for longer time intervals become quite large and heavy and have other disadvantages; they are indispensable when the required time interval must be accurate to minutes or seconds.

The events outlined for the various timing ranges are only a few examples out of a large number of delay-timing sequences. There are many others such as the bursting of a fireworks-display subassembly at the apex of a rocket flight into the air, or the similar display of an emergency signal after being shot out of a special pistol for signals. Parachute flares may use consecutive delay times for free fall, opening of the chute, and ignition of the flare candle. A few gun-projectile fuzes employ powder trains rather than mechanical timers.

We shall see later that all chemical delaying devices (pyrotechnic or otherwise) have large temperature coefficients and only fair accuracy at any one time. The reason for their use is economical and tactical, since no expensive component parts or assembly jobs are involved and millions of delay columns can be literally "knocked out" by unskilled labor on easily-setup assembly lines for millions of hand grenade fuzes or photoflash-cartridge delay inserts. In addition, those small, highly compacted and heavily enclosed devices are rugged, stable, and have no moving parts that might need servicing after prolonged storage. In short as well as long delay timing, the added advantage of smallness is a factor, though in the minute range, clockworks may be lighter and more compact. Then, however, the problem of initiation of clock action would have to be considered.

Factors in Formulating and Design

In discussing the field of delay timing by pyrochemical means, the first question to arise might be the calculation of the burning time (rate) from the reactions of the components and their chemical and physical properties. Japanese scientists have attempted to arrive at complex formulas,[408] but it is doubtful whether such formulas are of practical help, partly because of insufficient data on the factors involved and partly because of the numerous external factors other than those residing in the chemical components themselves.

Burning time depends on the following factors, enumerated roughly in order of significance:

1. Nature of the reactants and products;
2. Particle sizes (especially of fuels);
3. Relative amounts of ingredients (especially of metal fuels);
4. Ambient pressure;
5. Ambient temperature;
6. Compaction;
7. Diameter of column;
8. Nature of envelope.

Without going into detailed analyses of this seemingly inexhaustible subject, a few random notes will be added on the various parameters of delay performance.

As far as the *nature of reactants and products* is concerned, one may cite[3] as one of the slowest mixtures (though not likely to be used as a delay formula) the one of 96% ammonium nitrate and 4% charcoal of burning time 318 sec/in. at a compaction of d=0.94 g/cm^3. This corresponds in slowness of progression to the item called "slow match," a type of string fuse that, according to a dictionary definition,[174] "now consists of loosely twisted hempen cord steeped in a solution of saltpetre and lime-water and burns at the rate of one yard in three hours," i.e. 300 sec/in.* Davis[8] defines slow match as containing lead nitrate but gives no further data. Rope or wick-like, extremely slow-smoldering fuses made of cellulosic fibers and perhaps very lightly impregnated with a nitrate are occasionally used in clandestine work, but they require ambient air. A study on the burning of cellulose[409] may be of significance in this context: the impregnation of alpha-cellulose paper with 1$^1/_2$% of potassium bicarbonate ($KHCO_3$) completely suppresses flaming but it increases the sensitivity to glowing ignition and presumably to continued burning.

Another interesting topic is the relation of the burning rate and temperature of the tobacco leaf and its potassium salt or chloride content, which exert opposite influences. The burning temperatures are roughly inversely porportional to the potassium content. The glow time of a tobacco leaf was more than twice as long at the tip than at the base.[409a] Cigarettes, as everyone knows, have been used as delay fuses in fire-setting since they do not go out (except the Russian *papirosi*); with some precautions as to support and positioning, the rate of glow progression for American cigarettes is of the order of 6-8 minutes per linear inch in still air.

The slowest delay mixtures that function properly under a great variety of conditions are those based on very finely powdered and chemically active metals of highest density, especially those developed in the author's laboratory by D. E. Olander.[410] Of these, the ones with tungsten (wolfram) metal and up to 40 sec/in. burning time are the subject of specification MIL-T-23132 (1961).

At the other end of the timing scale, considering the influence of specific materials on delay time, are mixtures containing zirconium

* The author wishes to apologize to the pyrotechnic fraternity for quoting the OED as a source of pyrotechnic information, but sometimes the most commonplace items are hard to find without a special study of the more obscure literature.

of subsieve fineness. The previously described first fire mixture A1A and similar combinations have a BT of the order of 100 msec/in. Still faster are mixtures of an excess of finest molybdenum powder combined with potassium perchlorate.

Because of the limitations of choice of oxidizers in delay mixtures, especially of the low-gassing variety, data on changes in burning time with change of oxidizer are scarce. Hale and Hart[411] have paired red phosphorus with a number of oxides and chromates and not surprisingly the relatively most active oxidizers such as silver oxide (Ag_2O) and barium peroxide (BaO_2) produce very fast compositions, highly stable oxidizers such as barium chromate ($BaCrO_4$) and the oxides of iron (Fe_2O_3 and Fe_3O_4) yield very much slower burning times, while cuprous oxide (Cu_2O) stands in between. The latter oxide appears also in combination with zinc powder in an article by one L. B. Johnson—Jr![412]

In the intermediate pyrotechnic timing range of about $\frac{1}{2}$—10 sec/in., modern practice employs a variety of metal powders and other elements that behave like metals such as silicon, boron, selenium, and tellurium. These must be pyrochemically quite active for reliable fire transfer to them but not so active as to yield only very short reaction times. They may be used with two oxidizers, one of them often being the rather inert barium chromate, while the other is much more reactive and heat-forming, such as potassium perchlorate. This system permits a wide variety of formula changes and with it of burning times, though the type and amount of the fuel is generally the superseding factor. Suitable fuels have been found by trial and error: Besides silicon and boron, the most popular are certain alloys of zirconium and nickel, and the metals manganese, tungsten (wolfram), molybdenum, and chromium. All the delay formulas, no matter how seemingly well standardized and specified as to properties of ingredients, mixing, and pressing procedures, are capricious and a constant source of difficulties to the manufacturer—sometimes more so than the main item. If one can choose, the so-called D16 series, with a range of about 3-13 sec/in. and using manganese, barium chromate, and lead chromate, seems to give less trouble than the others.

Formulas 185-195 are examples of typical low-gassing delay combinations used in numerous military items and mostly taken from the military literature such as drawings and specifications. Quite a few combinations have appeared in the patent literature, examples of which are the claims by Hale and Hart,[411,413] Bennett and Dubin,[414]

Hale,[415] Hart,[416] Patterson,[417] and a large number of patents specifically referring to delay blasting caps and their ignition mixtures by inventors connected with Hercules Powder Co.[406,418–422]

In German World War II ordnance, the combinations of antimony metal and potassium permanganate were favorites.[48]

In the United Kingdom, work on slow-burning mixtures has continued to follow the trend of employing high-gassing mixtures using such fuels as tetranitro oxanilide or tetranitrocarbazole with barium nitrate.

The author has explored somewhat casually the qualifications of certain tantalum and niobium (columbium) powder samples and found that they reacted with barium chromate without the addition of a more active oxidizer. This is an encouraging feature pointing toward the potential usefulness of more of the very heavy and pyrochemically active metals, if only scientific research can provide us with means for defining the surface properties of these metals so that they can be made reproducibly to pyrotechnic requirements.

Particle size of the ingredients of delay mixtures is of great influence as far as metal fuels are concerned. And not only the measurable factors of particle surface area or average diameter but some quite elusive properties of reactivity are factors. It is often necessary, in order to match an existing composition, to proceed by trial and error to procure samples of various production batches of the fuel metal. One may find in some cases, in particular with tungsten, that the pyrochemical behavior of such metal powders may vary enormously even when particle sizes seem to be little at variance. On the other hand, the grain size of an active oxidizer such as potassium perchlorate has little effect on the burning time, though it may influence ignitibility.

The *percentage* of metal fuel in a formula has a decisive influence on burning time, mainly insofar as burning times become faster and faster with excess of metal. This appears to be a function of the increased heat conductivity of the mixture rather than of the increased surface area of the active fuel, since the admixture of silver powder in lieu of part of the excess fuel has been observed not to alter the burning rate. It is certainly remarkable that whatever the mechanism of action is, high metal powder content supersedes the calorific output as a factor in the burning rate up to a point where as much as 95% metal powder is present.

As to the percentage of oxidizer, if two competing compounds, such as potassium perchlorate and barium chromate are present it is important that the metal is in stoichiometric excess in relation to the perchlorate, at least in obturated fuzes, since indications are that free oxygen is otherwise released. Beyond this condition, it seems to have little influence on burning time if, say, 5 or 10% perchlorate is present.

Even though the greatest effects of gas *pressure* on burning time are found where high pressurization is encountered, as in completely enclosed systems, the rarefaction of the atmosphere due to geographically encountered variations of altitude must be considered for open-burning fuse trains. Safety fuse (see below) of a burning time of 120 sec/yard at or near sea level will be 5% slower at 5000 ft and burn for 126 sec/yard.[222] Extreme conditions would be encountered on high flying airplanes or on missiles but delay columns for proper functioning in these cases are designed to be independent of external pressures. Such enclosed systems do not react, however, at the same burning time as open ones since no "gasless" mixture is free from developing a certain amount of volatile and pressurizing substances derived from traces of adsorbed or chemically-bound water or hydrogen, or other volatile or gas-forming substances. In addition, self-heating during the reaction, interstitial air contributing to pressurization with rising temperature and the gaseous products of primer fire all take a share in speeding up the chemical reaction of the column.

Theoretically it might be possible to design sufficient internal vent space to annul the influence of such pressurization and also provide a large "heat sink," but this is hardly practical. A compromise is possible, however, and now often resorted to, as found in the U.S. patent by I. Kabik *et al.*[423] It features essentially a delay column concentrically surrounded by an "empty" (i.e. air-filled) volume. One might think that evacuation would further improve the conditions, but it is easily seen that an overpressure of many atmospheres (corresponding to as many volumes of gas at atmospheric pressure) will be little altered by "taking away" one atmosphere of pressure due to evacuation—not to speak of technical problems and the increased difficulties of fire transfer.

If the confinement of low-gassing mixtures includes only a minimal "vent" volume, the burning time of the column is considerably speeded up. In one case where a very slow-burning tungsten mixture was thus employed, the burning time was about halved. But even the internally better vented devices are rarely provided with adequate vent volume to approach open burning conditions and the burning time is shortened the skimpier the alloted vent volume.

The *temperature coefficients* of low-gassing delay columns have been studied quite extensively. Within the most often tested interval, —65° to +160°F, a temperature coefficient of 0.08%/°F is a reasonable figure. This means that the variability of delay time over the indicated range of 225°F would amount to ±9% attributable to temperature change only. Exceptionally, the range may be less, but more often is much higher. Crowded design, because of size limitations, often provides inadequate vent space, causing a lack of dissipation of heat that in turn leads to excessive pressure build-up. If we add a variability due to uncontrollable factors of manufacture and of burning behavior itself, we may have to allow a burning-time range approaching ±20%.

Significant design features of the delay column and its envelope comprise the degree of *compaction* of the mixture, the *diameter* of the delay column, the *thickness* of the tube or block into which the powder is pressed, and the *material* from which the enveloping support is made.

The burning time and, in some respects, the reliability of the item itself vary with these factors; increased compaction means slower reaction speed, the variation being considerable at first, but when a certain degree of compaction has been reached, an increase in pressure changes neither density nor burning rate any more than by trifling amounts. In other words, the compaction pressure *vs.* burning time can be represented by a curve steeply ascending and then asymptotically tapering off, the turning point for the more common delay compositions being at 30,000-50,000 lb/in.^{2}

The *diameter* of the delay column may become critical when the composition is slow burning and the heat output is small, as is often the case. A column diameter of 0.20 in. is usually adequate and if ambient temperature is never expected to be very low, successful burning is possible at 0.11 in. diameter.[45] Otherwise, a composition may perform adequately at normal and at higher temperature but at extremely low temperatures enough heat may be "soaked away" in the surrounding metal that the glow front becomes narrower and literally "tapers off," so that burning may cease before the terminal fire-transfer point is reached. By the same token, decrease in the tube-wall thickness will tend to concentrate the heat in the column (or, better, will dissipate less heat per unit of time). The result may be a faster burning time with decrease of wall thickness. Studies on the influence of the *nature* of the container—i.e. of its heat conductivity—have been less conclusive, but undoubtedly the selection of ceramic bodies *vs.* metallic would have a favorable influence on marginal situations and especially where multiple columns occur in close proximity.[424] It is obvious that deficiencies caused by heat losses are mainly encountered with slow burning columns; very fast ones are not affected at all.

One effect of heat conductivity along the column wall may be a premature initiation of whatever terminal charge follows the delay column. This "anticipatory effect" has been attributed by Picatinny Arsenal researchers to permeation of the unburned column by hot gases. Whatever the actual reason, some possibility exists that the terminal effect by which the delay time is measured is not caused by the glow front of the column actually reaching its end. In some unhappy circumstances, the delay column may be by-passed in its entirety when—by reverse loading "bottoms up," for example—a sensitive first fire is pressed first and some of it is retained along the tube wall during subsequent filling of the tube with delay mixture. Mechanical breaking up of the column would have the same result. There is of course hardly a more unfortunate effect than instantaneous action in lieu of delayed action because it may cause the user's death, as in the case of hand grenades.

As can readily be seen, the actual design and assembly of delays, where high reproducibility is required, involve numerous technical tricks, and even then unexpected deviations from proper behavior baffle the pyrotechnician, Highly polished internal surfaces, use of dead-load presses or exertion of excellent pressure control, automatically preset advance and dwell times of the pressing ram, pressing in small increments, perfect leak-proof sealing of the unit, and use of the most uniform prime ignition sources are some of the conditions that will improve pyrochemical delays.

Commercial Fuse Lines

Ignition lines and tubular fuse trains can be used for delay timing in fireworks items and in some military signals, but in general do not burn consistently enough and are not physically suited to withstand the rigors of military surveillance tests and actual exposure in use.

The *Blasters' Handbook*[222] describes *safety fuse* as a tightly wrapped train of potassium-nitrate-type black powder of 120 or 90 sec yard burning time (3$^1/_3$ or 2$^1/_2$ sec/in. burning time—0.4 or 0.3 in./sec burning rate) with a normal variability of $\pm 10\%$. The coating of such a fuse is most often an impregnated fabric, with more and more effort being made to improve the moisture-sealing effect by use of a wax coating or a plastic tubing, as in the polyethylene-covered M-700 safety fuse.

Quickmatch (MIL-Q-378B) consists of cotton strands impregnated with black powder (Formula 146) mixed with starch paste. "Mealed" (Class A) quickmatch contains additional fine black powder as a coating, which increases the rate of burning. Of the three types specified, Type III is intertwined with annealed copper wire. The specified burning times lie between 8 and 17 sec/ft. When the rather rigid and brittle pieces of quickmatch are bent, threaded through holes, and thus partially broken up, the burning rate increases considerably, hence quickmatch is more suitable for fire transfer as such rather than for timing.

Another more modern type of delay ignition line is known under the tradename of Thermalite Ignitacord. It burns with an external, short, very hot flame and comes in two types: 0.75 and 1.5 sec/in. average BT. This type of fuse is used in commercial blasting work. In the same applications, but for multiple hole detonations, the very fast (1 sec/ft) Quarrycord is used.

The most recent filiform delay element is the material known under the tradename of Pyrofuze.[425,426] It consists of an aluminum wire or braided strands of such wire coated or, more correctly, clad with palladium metal. On heating to 660°C, the two component metals combine in prodigously exothermic reaction, which progresses at 0.8-1.8 sec/ft, depending on the type of wire. By winding the material on a core of low heat-conductive supports, better than fractional seconds of delay action can be effected and lower time tolerances than for ordinary delays are claimed.[427] Because of its very high temperature stability and the fact that at high temperatures the temperature coefficient becomes very small, this unusual item has some promise, but its invariable and high speed of "burning" as well as its costliness limit its uses.

Table 14 shows the rated delay times of commercial fuses expressed in several ways and gives nominal figures for compressed columns for comparison.

Safety fuse, MIL-F-20412, is used for delaying action in the *Aircraft Smoke and Illumination Signal AN-Mk* 5 *Mod* 1 ("drift signal") where ignition is effected on impact on the water and a 9-sec delay permits time for submersion and resurfacing prior to smoke and flame emission. Similarly in the *AN-Mk* 6 *Mod* 2 ("float light"), a much longer delay time (90 sec) is provided by safety fuse since this

Table 14
Fuse Trains

	Inverse Burning Rate				*Burning Rate*			
Type	*sec/in.*	*sec/ft*	*sec/yd*	*sec/cm*	*in/sec*	*ft/min*	*in/min*	*cm/sec*
Safety Fuse	3⅓	40	120	1.3	0.3	1.5	18	0.8
	2½	30	90	1.0	0.4	2.0	24	1.0
Ignitacord								
Type A Green	¾	9±1	27	0.3	1⅓	6⅔	80	3.4
Type B Red	1½	18±2	54	0.6	⅔	3⅓	40	1.7
Qarrycord			ca.3			ca. 60		
Pyrofuze			2.4—5.4			ca. 60		
Delay Columns								
very fast	0.1	1.2		0.04	10	50	600	25.4
fast	1.0	12		0.4	1	5	60	2.5
intermediate	5.0	60		2.0	0.2	1	12	0.5
slow	10.0	120		3.9	0.1	0.5	6	0.25
very slow	30.0	360		11.8	0.03		2	0.085

signal is initiated at the moment of release from an aircraft.[33]

A different function is added to the delaying action of safety fuse in the selective fuze for the *Mk* 5, 6, and 10 aircraft parachute flares as well as the new fuze for the *Mk* 24 flare. The delay time is preselected by the position of a clockhand-type plunger pointed toward a length of peripherally placed safety fuse. The ignition of a transfer fire charge and simultaneous propulsion and penetration of the hollow plunger tip into the delay fuse initiates the fuse and depending on the length of fuse between plunger and transfer fire to the flare (the other part of the fuse dead-ends) the sequencing of parachute display and flare candle ignition is matched with the height of desired drop from the aircraft prior to activation.[258]

In certain artillery ammunition fuzes, such as the *M*54, *M*55*A*3, and *M*77 *TSQ Fuzes* and the *M*84 *Time Fuze,* two circular channels with compressed black powder, one above the other, can be moved so as to lengthen or shorten the useful path of the powder train; thus the delay time can be preset. In the first three fuses mentioned, the train can also be by-passed for "super-quick" action (hence the SQ designation) aside from a dead-ending safe position for all four.

With the exception of these two schemes, selective timing by pyrotechnic means is hardly ever heard of. Rigid columns of pressed delay mixtures are little suited to such variability, either continuous or stepwise. The idea of tapping delay columns or sensing the glow front through the envelope at selected points has no doubt occurred to everyone in the field, but its effective execution is another matter.

The principle of fabricating delay columns in "endless" lines and cutting these to size rather than resorting to the tedious individual production of single items seems enticing. If soft lead tubes of $\frac{1}{2}$-in. internal diameter and 11/16-in. outside diameter are filled with delay powder, closed on both ends, and swaged to a $\frac{1}{4}$-in. outside diameter, a compacted delay fuse line results, which can be cut in short lengths or bent and formed into a spiral if desired. The development of this type of delay has been pursued in the United Kingdom, while in this country similar processes are employed for explosive line charges. Considering that a delay unit must be provided with ignition and terminal fire transfer material and fitted in a larger assembly of some kind, it seems doubtful that this type of fuse train is of much use except under conditions where existing safety fuses are useful. One exception is the insertion of delay trains into detonators. Mixtures of barium peroxide and selenium, or the slower ones with tellurium

(Formulas 193, 194, and 195) have been tamped into lead tubes and consolidated by swaging, starting from 3-ft length and ending up with 4 ft of fuse, a process used by the Hercules Powder Co. for over thirty years.

As a final note to pyrochemical delay timing, a U.S. patent[428] will be mentioned in which an electrically ignited heat source transmits its heat to a metallic rod or cup that on reaching a certain temperature sets off a heat sensitive charge such as a blasting cap. Whether or not the scheme has merits for the claimed purpose, i.e. as a delay, it is interesting because it effects pyrochemical heat transfer into an enclosed space, which remains sealed off. This may have significance in certain design problems.

Nonpyrotechnic Long Delays

Since there is a limit to the length of any fuse train, delaying action by pyrotechnic means must yield to other progressive action for long duration. Clockworks for periods of days and weeks are even more cumbersome and expensive than for short duration, but chemical action or swelling and dissolving action of a liquid on a solid are simple means for long-delay timing.

The *M1 Delay Firing Device*, MIL-F-13893B (ORD) (1962), commonly called the delay *pencil*, is a World War II invention of the Allies but captured by the Germans and used by both sides[15] and is based on the corrosion of a tinned steel wire by solutions of cupric chloride in water/glycerine mixtures. The wire is held under 16-lb spring tension and its corrosion and eventual break releases a firing pin acting either on a percussion primer (M27) or on a miniature incendiary containing the tip of an SAW match as described in Chapters 11 and 12. Because the glass ampoule with the corrosive liquid is an integral part of the device, each delay time requires a separate pencil marked by a lacquered metal band of a different color, which also serves as a safety detent.

The pencils are made for the minutes as well as the hours range, assuming an ambient temperature of about 70°F. The temperature coefficients are very high and increase with the length of the nominal delay time.

The device is used for incendiary or demolition action in clandestine activities. Two pencils attached to one item increase reliability of firing and reduce the probability of a "delayed delay action" or of failure due to a hung-up firing pin, which occasionally occurs. A hazard of the device is corrosion of the wire in storage, be it because of breakage of an ampoule or through external influences. While a peephole allows a check on a "fired" but retained striker, it is not possible to discern if a partial corrosion has left the device at the brink of firing.

A different principle is involed in fuzes, of which the *Long Delay Tail Fuzes AN-M*123 are examples.[44] Omitting the preliminary arming mechanism necessary to assure that a bomb can be safely handled and released, the delay action is engendered by the softening action of solvent on a celluloid collar that retains a firing pin. Vials containing solvents of varying activity toward nitrocellulose (acetone, being the most active, esters least) allow a nominal delay range of 1-144 hours. Again, the temperature coefficient is high—for up to 12 hours nominal, a timing ratio varying from 1:9 to 1:24 applies within the relatively small temperature range of 115°F down to 25°F. The slow-acting solvents become quite unreliable at low temperatures because they may not dissolve but only swell the disk, which then may remain strong enough to hold back the striker. As a result, this type of fuze may stay suspended for many months or years, and either never go off or if at all, only due to unpredictable gradual shear.

Different in design but based on the same principle are the *Fuzes AN-M*132, 133, and 134, all of nominally 16-min delay time (6-80 min range between 120°F and 10°F).

Devices of this class that in size and application correspond to the above-described delay pencil, except that they permit much longer delaying action, have been used since World War II[429] but enjoy a more anonymous existence. They are suitable for underwater demolition. Because their glass ampoules are not part of the device but are only inserted prior to emplacement and activation by crushing, one item plus a set of ampoules serves the whole delay range.

It has been shown earlier that pyrochemical action between sulfuric acid and chlorate mixtures can be used in combined delaying and ignition devices (Chapter 10). This can be done by use of the corrosive or hygroscopic actions of the acid, or the active components can be brought together by tilting or spilling, an effect that may be produced from some gradual, external process of a simple nature.

A short glance must be given to long-delay timing based on electronic phenomena. A DOFL (now HDL) report[430] describes an electronic delay timer that can be set in 5-min increments to up to 49 hours 55 minutes.

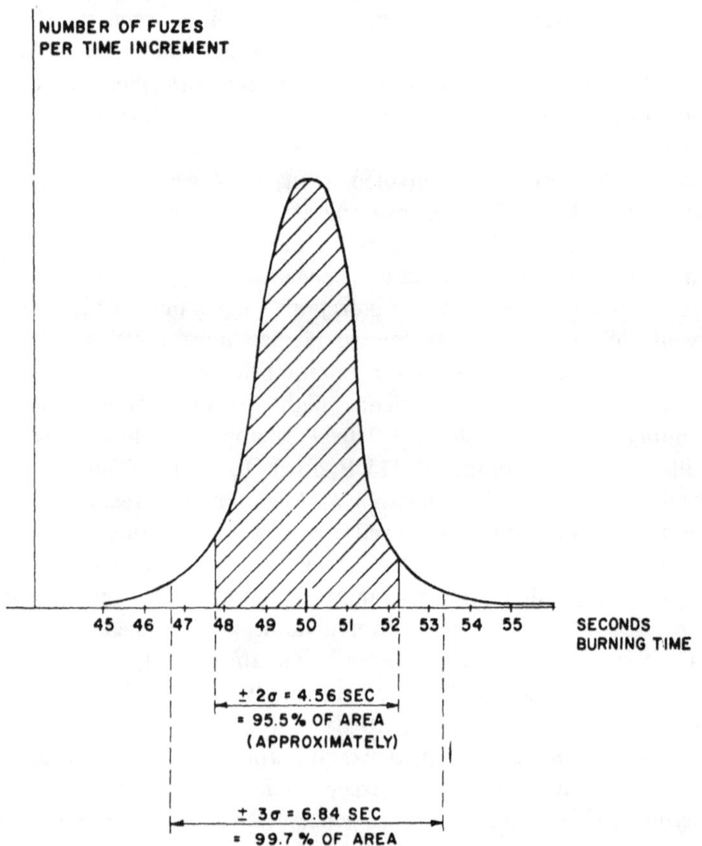

Fig. 3 Nearly Normal Distribution of Delay Times for a Fuze Production

An electronic delay element for certain propellant-actuated devices is described in a Frankford Arsenal report.[431] About 3 in.^{3} in volume and 4 oz in weight, it is claimed to permit delay times of 0.1-15 sec with a $\pm 3\%$ accuracy over an interval of —65 to +200°F. By comparison with pyrotechnic standards superior performance, this is surpassed by another device,[432] for which incremental delay times of 0.1 sec up to 1000 sec are claimed with an accuracy of $\pm 0.1\%$. The future will show if the above-described advantages of pyrotechnic delays (economy, ruggedness, producibility) will outweigh the performance characteristics of electronic or mechanical delays.

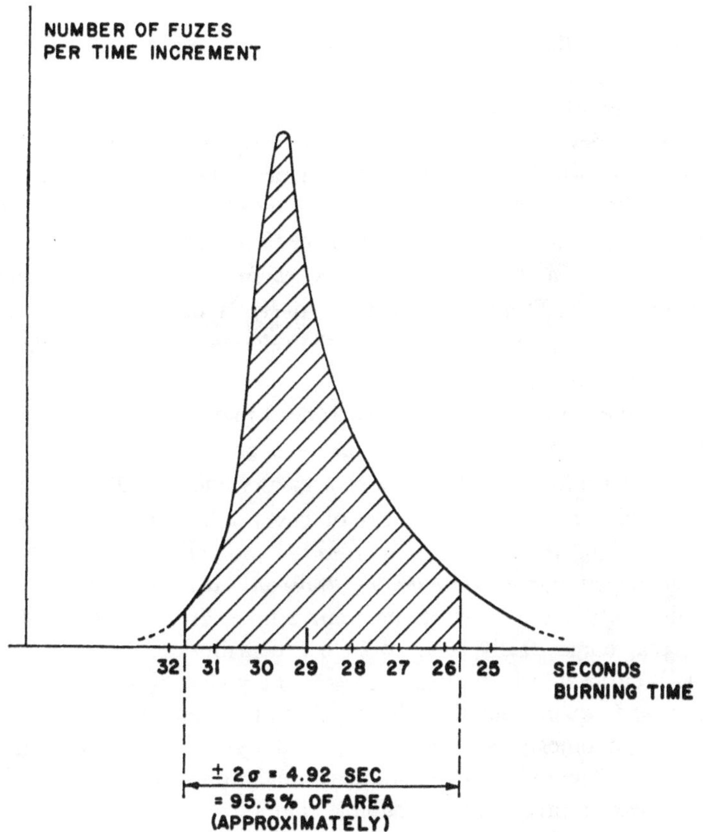

Fig. 4 Moderately Skewed Distribution of Delay Times for a Fuze Production

Note to Chapter 24

In order to evaluate the delay-time data of a production lot, it is necessary to delve into the methods and the terminology of statistics, of which a few highlights will be given here.

Mean value and *range* are the simplest expressions and the most common. They are generally understood, but do not suffice at all in the characterization of a *population* (complete set) of values, or of a sample out of such a population. For variations in measurable quantities, such as delay time, the most useful measure of dispersion of a frequency distribution of these variations is the *standard deviation*, with which the reader must familiarize himself. The term standard deviation (represented by the Greek letter σ) is the root-mean-square deviation of the observed numbers from their mean. Expressed algebraically, σ is obtained by taking the square root of the sum of the squares of all such deviations from the mean, divided by the number of deviations. A second term of significance is the *coefficient of variation*, which is σ expressed as a percentage of the mean value. This coefficient permits a comparison of the dispersions of two or more series. If a number of sets of values are to be compared—for example, the quality of a 2-sec fuze and a 4-sec fuze—and have a standard deviation of 0.02 and 0.04 sec respectively, a coefficient of variation of 1% for each would signify that the relative variation in delay time is the same for both fuzes.

If all the values measured are arranged in order of length of timing from the lowest to the highest, using a conveniently small interval of say 0.2 sec for a delay unit with a mean value of 50.0 sec, a graphic presentation of time *vs.* number of cases will form a characteristic *histogram*, and the frequency distribution will form a line approaching a bell-shaped curve (*Gaussian or "normal" distribution*) under ideal conditions. Figure 3 represents a curve taken from actual production data that comes fairly close to such a distribution.

This curve shows that delay times that are close to the average value are most frequent and that the number of extreme values, either low or high, becomes smaller and smaller, the greater the deviation from the mean. The value of the standard deviation lies in the fact that it can be used to predict the probability that a certain value will fall within specified limits from the mean. In the example of Figure 3, a standard deviation of 1.14 sec and a coefficient of variation of 2.28% were calculated. Taking $\pm 2\ \sigma = 4\ \sigma = 4.56$ sec, the statistician will then calculate that the probability of all values falling within the range of 4.56 sec is 95.5%; this is the ratio of the shaded area of the curve to the total. Similarly, with a $\pm 3\ \sigma$ spread, or 6.84 sec, the probability of all values falling within this range is 99.7%.

Figure 4 shows a curve approaching a normal distribution for a similar item, bur for complex reasons the distribution of values shows a definite deviation from the ideal curve, in other words, the curve is *skewed* to the right (the direction in which the excess tail appears). However, in either case, the major portion of the area under the curve is covered within the $\pm 2\ \sigma$ or $\pm 3\ \sigma$ limits, so that any probability relations will still be fairly close to the theoretical. A comparison of the coefficients of variation of the two curves shows that the relative variation in delay time is less in the population represented by Figure 3 than that of Figure 4.

Sometimes the harried manufacturer may find that the distribution of values shows two distinct peaks. This is called a *bimodal* distribution and may mean that two influences are at work, such as the use of two strongly deviating powder lots or differences in the fabrication methods on two production lines.

Since pyrotechnic time delays are "one-shot" items, the quality of a fuze must be determined from samples. Other mathematical relations govern the probability that a sample will approach the true value for a given population. These and many other statistical considerations are for the specialist. Books on statistical analysis, some of fair readability and others quite difficult, are found in the references.[433–436]

Selection and evaluation of samples themselves for acceptance of ordnance materials that can be tested by measurable properties is made according to MIL-STD-105D (1963) *Sampling Procedures and Tables for Inspection by Attributes.*

chapter 25

Fire Starting and Fire Setting

Everyone vicariously enjoys witnessing a devastating fire, and its appeal is sufficiently elemental to surmount intellectual and cultural differences.
Nolan D. C. Lewis and Helen Yarnell (1951).[437]

This chapter will deal with starting fires for comfort under difficult conditions, as well as with incendiarism and the destruction of documents and related items by burning or application of intense heat.

The means of achieving these purposes are not always pyrotechnic but the presence of special initiation systems, the choice of unusual fuels, or conversely the supply of solid oxidizers to the common fuel make this subject one of pyrotechnic interest.

As an aid in quickly starting a fire in the open for strictly civilian purposes, mixtures of fuels and oxidizers are avoided because of their hazards in storage, but occasionally one is found in the patent literature.[438] Easily ignitible mixtures of sawdust and pitch, or rosin compressed into pellets or cakes, serve adequately; so do waxed cartons or a dash of kerosine. Solid oxidizer material is sometimes marketed in small paper cups and is relatively harmless.

Firemaking for comfort and for warming food under adverse weather conditions may be a problem for military personnel such as downed flyers, for whom the carrying of liquid hydrocarbon fuel would be impractical. The *M*1 and *M*2 *Fire Starters*[38] are handy gadgets for starting nearly instantaneously a prolonged (13 and 4 min, respectively) high-caloric quietly-flaming combustion that will cause sustained burning of even wet wood and other solid fuels. Both items consist of congealed kerosine in a celluloid container, a safety match igniter and striker, and are waterproofed. They are too small for effective incendiary action except under very favorable conditions.

The *PI* or *PTI* (for pocket incendiary) consists of a celluloid case that just fits an average-sized pocket. It contains several ounces of gelled kerosine but has no ignition system. Two clips are provided for attaching two *M*1 *Delay Firing Devices*, each holding a "matchhead"—originally a small celluloid cylinder into which the tip of an SAW kitchen match was cemented and which was filled with a mixture of antimony sulfide and potassium chlorate.

The development of the *M*1 "*Candle*," the *PI Case*, and the "*pencil*" and its *igniter*, have been described by Fieser.[18] The author's company produced the original celluloid matchheads for the "pencils" that later became the *Magnesium Matchheads* designed and formulated by him and his co-workers.

The development in World War II of stable gels of various consistencies from liquid fuels for use in flame throwers, incendiary bombs, and the above-named small devices was the work of scientists at the Chemical Warfare Services Laboratory at Massachusetts Institute of Technology and of others. The most satisfactory of these was *Napalm* by Fieser,[18,439] the name derived from its main ingredients—aluminum NAphthenate and PALMitate. Napalm is now standardized as *M*1 *and M*2 *Thickener* [MIL-T-589A (1953), MIL-T-13025A (1955), and MIL-T-0013025B (1964)]. Related types are *M*4 *Thickener*, (ISO), MIL-T-50009A (1960), and one consisting of natural rubber and isobutyl methacrylate (polymer AE) developed by du Pont.[15]

Concurrently with these efforts at the start of World War II, Standard Oil Development Company worked on a sodium soap thickener; indeed, as a field expedient, ordinary laundry soap does quite well where long-lasting stability is no object. The most recent developments omit the soaps altogether and use as a basis a large amount of synthetic plastic compound. The mixture, called *Napalm B*, while solidifying at —30°F, will remain a viscous liquid even on exposure to lower temperatures because of slow heat transfer. The idea of a thus-thickened hydrocarbon fuel is also found in a Belgian patent.[440]

The thickening principle is also used in *Composition PTI*, a complex mixture of magnesium oxide and carbon, with hydrocarbon in paste form ("goop") compounded with hydrocarbon thinners, magnesium, and sodium nitrate. Another mixture of plastic thickener—gasoline, magnesium, and sodium nitrate—is *Composition PTV*. All are described in TM3-215.[35]

Congealing a liquid fuel has the advantage in incendiary practice that the burning gel does not run out even if unsupported, e.g. if spattered from a bomb or after a celluloid casing burns away. Thus consumption is slowed down and the flame remains in one place. Since with flames of relatively low temperature (where heat transfer is essentially convective and conductive) duration is of the greatest importance for ignition of heavier wooden structures, Napalm—a term now popularly standing for the congealed liquid rather than for the thickener—is one of the most effective incendiary agents.

Other approaches to solidification of liquids are the gelling of nitroparaffins with dextran nitrate esters[441] or of methyl nitrate* with nitrocellulose or kieselguhr.[48] An older study of theoretical interest and not specifically pertaining to incendiaries is the article by Thorne and Smith[442] concerned with solidified alcohol.

Sterno is a well-known commercial solidified liquid fuel. It burns with a nearly invisible flame and little residue, and these properties as well as its odor proclaim it to be mainly ethyl alcohol. It is a handy heat source for campers, etc. According to a U.S. patent,[443] it consists of an alcoholic solution of nitrocellulose, solidified by addition of a miscible nonsolvent for the nitrocellulose. The latter is thus "thrown out" of solution and forms a voluminous mass that traps most of the liquid portion.

Since the appearance of pyrogenic, extremely fine silica (0.015-0.020 μ particle size), commercially known as Cab-O-Sil[444] and other tradenames, it has been possible to gel many liquids instantaneously by the addition of several percent of the powder. A stable, firm incendiary gel can be made from Stoddard Solvent with 8% of Cab-O-Sil. The finest type (7 mμ) will even congeal liquid hydrogen.[445]

In the patent literature, we find in addition to the quoted Fieser patent[439] one for a more stable incendiary fuel by Bauer and Broughton,[446] and various methods by Schaad,[447] Vannucci,[448] and Long,[449] the latter actually pertaining to a gelled propellant. A surprisingly large number of additives for improvising gelled gasoline are described in the technical manual on incendiaries.[47c]

A relatively cool-burning incendiary is white phosphorus. Some of its igniting and incendiary properties have been discussed in Chapter 6. Phosphorus by itself is a poor incendiary, effective only against easily ignited objects. Thus the first air attacks in World War I by hydrogen-filled dirigibles (zeppelins) were soon abandoned because phosphorus-filled bullets spelled death to lighter-than-air craft.[450] In the late summer of 1940, British airplanes carried nitrocellulose sheets, two inches square and covered with yellow phosphorus (kept in water prior to release), and dropped them in the region of the Black Forest

* According to Stettbacher, methyl nitrate was the agent used in the infamous Reichstag burning (1933).

in Germany. These fire leaves or "visiting cards" did little damage because of cold weather, which often led to slow oxidation rather than to ignition. Similarly, Japanese efforts in World War II to do damage by means of small balloons carrying phosphorus, which drifted toward the northwestern U.S. coastal region, were ineffectual.

Notwithstanding its deficiencies as an incendiary, phosphorus has a certain tactical value because of the power of obscuration caused by the volatility of its final combustion product, phosphorus pentoxide (P_2O_5), the spontaneous re-ignitibility of doused, unburned residues, and the severity of the wounds caused by burning phosphorus when it comes in contact with human skin. Its burning properties can be improved by mixing it with synthetic (GRS) rubber, a mixture called *plasticized white phosphorus* (PWP).[35]

After what has been previously said about high-energy flares, one might wonder why the magnesium/salt-type oxidizer mixtures should not be useful as incendiaries. The heat from flare candles is rapidly dissipated in the cloud of volatile or aerosolized reaction products. The duration of any hot-burning flare of reasonable size for an incendiary unit is short. In addition, the deposited salt and metal-oxide particles interfere with the combustion of structural parts. This is not to say that certain slow-burning magnesium/sodium nitrate/binder mixtures cannot be specially formulated to serve as incendiaries, but a simple experiment will show that light-producing pyrotechnic candles with flame impinging on a substantial piece of wood will mostly have a disappointing ignition effect.

Burning magnesium in bulk will be more-or-less effective while acting mainly on the substratum on direct contact. A quietly resting, fully ignited piece of magnesium will form a molten pool that becomes covered with loosely adhering oxide while the ambient air gradually diffuses to the unburnt metal beneath. The combustion becomes more lively if there is a strong movement of air or if the quiescent burning is disturbed and fresh metal is exposed through shifting of the solid surface. Thus the effect on a wooden floor may consist mainly of deep charring, but incendiary action can be good against difficult-to-ignite substances such as rubber and plastics, and of course against all fairly easily ignited solids or liquids—at least as long as there is no overwhelming quenching effect by the liquid.

In order to make magnesium work properly, a hollow magnesium body is filled with a mixture that creates temperatures high enough to ignite and melt the magnesium body. Actually, in the magnesium bombs of World War II an alloy of magnesium with a small percentage of aluminum, some zinc, and manganese was used.[451] The use of such magnesium alloys goes back to World War I, when the Germans developed an alloy they called "Elektron" containing 13% aluminum and some copper. These additions increased the mechanical strength of the magnesium without affecting its ignitibility. Sometimes, a much larger percentage of aluminum is mentioned in sources of information, but actually such mixtures—the 50:50 alloy for instance—are very brittle, more useful for conversion into fuel powder.

The filler for magnesium bombs such as the 2-lb *AN-M52A*1 and the 4-lb *AN-M*50*A*2 is a modified thermite called *thermate*—TH3, formerly Therm 64-C[35] (see Formula 196, and 197, a similar mixture with a synthetic resin binder, called TH4). TH3 thermate is also the filler in the *AN-M*14 *Incendiary Hand Grenade*[38] [MIL-G-12297C (MU) (1965)]. Even though by design this is a grenade for throwing, its main function is as an equipment-destroying item that might function more effectively if carefully emplaced.

In line with the development of thermates are the patent by Thompson,[452] filed in 1944 but awarded in 1957,* and the similar one by Woodberry *et al.*[453]

The small hexagonal—cross-section magnesium bombs were dropped in clusters, and since they are fairly easily made harmless by covering with dry sand or graphite powder, an explosive charge with variously delayed action was added to a design variant, the *M*50*XA*3, to discourage firefighting.[257] As an incendiary, such an item that scatters the burning magnesium becomes much less effective. In the mass raids of World War II,[454] such as those on Hamburg and Dresden, the 4-lb magnesium bombs were dropped by the hundreds of thousands together with other bombs and incendiaries, causing so many fires simultaneously that the updraft produced firestorms of hurricane force. The loss of life caused by these firestorms may have been of the order of that caused by the atomic bombs at Hiroshima and Nagasaki, though recently these estimates have been doubted.

Very little can be found in the quoted manuals and other publica-

* There is no such thing as a "secret patent" under U.S. patent law. A patentable invention that the federal government regards as undesirable for publication is processed like any other patent application (though in a special department) and kept under a *Secrecy Order* until the time when the order is rescinded and the patent can be awarded.

tions about incendiary projectiles. *Naval Airborne Ordnance*[258] speaks of 20-mm aircraft ammunition with incendiary or high explosive *and* incendiary filling for use against other aircraft. White-phosphorus-filled smoke shells, such as the M302 for 60 mm mortars and other WP shells,[41] while normally employed for spotting and laying down smoke screens, have a certain incendiary effect. A patent by Stevenson[455] claims mixtures of powdered magnesium/aluminum alloy powder (50:50) with oxidizer salts with improved burst duration by the addition of zirconium of 20-60 μ particle size for use in projectiles.

Thermite itself (Formula 198) causes an extremely high heat concentration and is able to penetrate metal by means of one of its two reactions product, iron, in molten state high above its melting point at approximately 2500°C. Under some conditions, solid steel plates as thick as one inch can be perforated by a properly directed stream of the white-hot mass. While thermite can act as an incendiary because of its combination of convective or conductive and radiative heat, it is more valuable because of these penetrating properties, such as for demolition of machinery. Small amounts are too easily quenched to exert the proper persistent heat flow necessary for ignition of combustible structures.

Thermite is most efficient in its typical manifestations when used without a binder, though two patents[456,457] claim improvements by binders and other additives. Because of their cooling effect, such binders, make ignition more difficult and cause heat-flow away from the substratum. Counteracting this by the addition of a stronger oxidizer can partially remedy these faults, but in the end the character of the thermite reaction is lost.

An interesting combination of incendiary actions is the subject of a German patent[458] which claims thermitic high-temperature reactions between aluminum ("Pyroschliff") or calcium silicide, with phosphates such as tertiary calcium phosphate as the oxidizer. Since phosphides are the products of reaction, the application of water during fire fighting will cause formation of self-flammable phosphine. It would appear that such secondary action might be more of a nuisance than tactically efficient.

A combination that resembles thermite in its character of destructive effects, though it is quite gas-forming, is the one claimed by Linzell[459] for mixtures of aluminum powder and anhydrous calcium sulfate, Sulfates as oxidizer substitutes for nitrates have been mentioned earlier. While the patent is listed as "Incendiary Composition," the author speaks of it also as a heating composition for refractory materials, and it is obvious that what constitutes destructive or constructive action depends on the desired result rather than on the nature of the mixture, the ability to control the result being one criterion.

Plain thermite (TH1) in relatively substantial amount is the ingredient in the *M1A1*, *M1A2*, and the *M2A1* cryptographic-equipment-destroying incendiary units. Their purpose is the complete breakdown of electronic equipment by melting, burning and severing of connections of certain classified devices. This task is not as easy as it appears, especially on a small scale and when the heat has to be confined, e.g. under indoor conditions. Where the purpose of annihilation can be performed by combustion, as with security-sensitive information written on paper, celluloid boxes containing sodium nitrate and interspersed between the documents in heavy filing cabinets are successful, and the cabinet confines the sizable conflagration. The *M4 Incendiary File Destroyer* of this class and the preceding destructors are described in some detail in TM3-300.[38]

Larger quantities of documents can be speedily disposed of by mixing them in an open pit, drum, or the like with about 140% of their weight of sodium nitrate. Provisions must be made to prevent the powerful flame and updraft of this spectacular reaction from carrying away any unburnt paper. The destruction of the paper and also of admixed plastic items is very thorough and any somewhat more resistant combustibles are likely to be gradually consumed in the large molten residue after the main reaction is over. Any substantial admixture of combustible metals such as magnesium or aluminum must be avoided, since a disastrous explosion may occur.

Alkali chlorates can be used in the same manner and are valuable when the destruction is on a small scale and where the reaction of the paper, etc. with a nitrate might be too sluggish. However, chlorates do not form a substantial molten residue, since the alkali chloride is partly aerosolized. Admixture of some nitrate may correct this deficiency,[460] and perchlorate may also be added to increase oxygen supply.

The destruction of paper in the presence of oxidizer salts is of great importance where speed is necessary; otherwise, plain burning under proper conditions of loosening up and disintegration of the residue or mechanical communition is preferred. Tactically, the former situation may arise during a hasty withdrawal action or a surprise infiltration into an area where classified or compromising documents are stored. The principle can be extended to "push-button" destruction of papers in a briefcase or a notebook by interspersing the sheets with a film composed of equal parts of celluloid and sodium nitrate, as reported by Fieser.[18] Unfortunately, these and similar devices act with strong flaming. A quick, gasless and gas-scavenging, cold or cool-smoldering, fast-acting chemical destruct system has not yet been invented.

A few remarks should be added about testing the effectiveness of incendiaries. It is quite appropriate to compare small incendiary units with the help of standardized wooden structures of moderate dimensions. They imitate actual conditions by creating some updraft between vertical parallel boards or by addition of horizontal and vertical members of wood of standardized size, nature, and moisture content. Similarly, the incendiary may be applied to other materials and the procession of the fire transfer observed. As a rule, however, any miniaturization is likely to give a false picture as to actual effectiveness.

In rare instances, permission can be obtained to destroy or attempt to destroy a building prior to its being razed and under controlled conditions of fire setting. A book with the disingenuous title "Operation School Burning"[461] describes such tests, which were performed following a disastrous school fire in Los Angeles in 1958.

Some common staple foods resist sustained burning in moderately large amounts no matter what incendiary effort is applied. This is specifically true of carbohydrates or substances rich in carbohydrates, such as sugar or corn. The subject is rather obscure and one might be permitted to speak of a "critical mass" below which sustained burning will not take place.

chapter 26

Heating Devices

In civilian life, in technology, and in a few specialized military applications, other than the aforementioned ignition or destruct systems, it is sometimes convenient to have an intense, self-contained, instant heat source available. This type of pyrotechnic gadgetry has been boosted by the discovery of the "gasless" heat powders. They can be confined in fully sealed cartridges, whereby not only the heating takes place without flame or sparks but more heat can be transferred in orderly fashion without losses through vent holes. But even where such efficiency is not possible or not needed, and the heating process is rather uncontrolled, as in some applications in welding or foundry practice, the capacity to bring heat to any spot, without power lines, gas burners, or welding equipment can be a great convenience.

Stones heated in a fire and dumped into a shallow depression in the ground served as one of the oldest means of warming food; a bag of roasted chestnuts fresh from the roasting pan is a fine hand-warmer while the contents are being consumed; and the hot water bottle, while obsolescent, can still be found in the home as a bed-and body-warming device. All these items have in common with pyrotechnic heat sources that they contain just so much heat and for useful heat transfer must be at a high temperature, without causing damage or discomfort.

By insulation, heat transfer can be slowed down and attenuated, and in a pyrotechnic device, the heat itself can be gradually released though rarely slowly enough to make much difference. Also, these schemes are of little help when the heat transfer must be fast and still not excessive as to temperature. If we add the requirement that certain devices must be functional and effective at extremes of ambient temperatures, at least between —65 and +160°F as in certain power sources described below, the problem only becomes solvable by compromise. One could devise auxiliary heat sources activated through thermal monitors at the lower temperatures, but the smallness and innate simplicity of pyrotechnic arrangements allows such complexity only exceptionally. Another expedient could be a built-in heat sink, i.e. a material such as solid zinc, which absorbs much more heat by melting than by mere heating in solid state and which returns the latent heat of fusion on subsequent solidification. This scheme is also rarely practical.

Examples of nondestructive applications of pyrotechnic heat are the heating of *canned foodstuffs*, or of *water* for the preparation of food by sealed heat cartridges or in double-walled cans; the activation of *galvanic cells* with solid, fusible electrolyte; auxiliary heat sources for some other types of galvanic cells; heat cartridges for *soldering irons*; heat sources for *comfort* in extremely cold weather; means for *welding wires*; and heat applications at various levels of intensity in *foundry practice* for prevention of cooling of top layers of melts or of so-called risers.

A self-heating soup can is shown in Taylor's book[6] together with the results of the investigation by Caldwell and Gillies[462] of suitable heat sources. One such a source consisted of equal parts of calcium silicide ($CaSi_2$) and "hammerschlag" (Fe_3O_4) with 0.58 kcal/g. Overheating was reduced by lining the cartridge with asbestos paper. However, instead of a self-contained initiation, the first fire for the heat source had to be ignited by means of quickmatch; this means that the cartridge was vented to the atmosphere. Even with a "gasless" mixture, this should cause a considerable loss of heat.

Brock[2] mentions the use of burnt lime and water as a heat source for food cans. This reaction furnishes 0.26 kcal/g of calcium oxide (CaO), and while not exactly classifiable as pyrochemical is said to be able to produce a temperature as high as 450°C—enough to set fire to guncotton, sulfur, wood, and "sometimes straw."[67]

In patents (rarely to be taken too seriously as to usefulness, practicability, or economic potential), one finds mixtures containing iron filings, aluminum powder, ammonium chloride, potassium chlorate, mercuric and cupric salts, and alkali permanganate in various combinations, all activated by water and purportedly useful for heating foodstuffs, to assist in hair-waving or to act as heat pads on the human body.[463-467]

Serious attention has been given to modern approaches to chemical body heating systems, though in these reactions we move farther away from pyrotechnics proper. One of these systems,[468] investigated by the Ethyl Corporation, channels exhaled air over sodium aluminum hydride ($NaAlH_4$), producing heat as well as hydrogen gas that at the most exposed body parts in shoes and gloves can be catalytically and highly exothermically reacted. For anyone trying his wits on the problem, the following figures should be of interest: Under severe arctic conditions, each glove requires 3 and each boot 7 watt-hours for reasonable comfort but after long exposure (7 to 8 hours) these figures must be multiplied by a factor of about six. (1 watt-hour equals 860 calories.)

It is possible without resorting to the use of highly refractory metal as an envelope to design a heat cartridge that will heat water in seconds to near-boiling, provided that the greater part of the tube remains submerged in the water. The strong quenching effect prevents the steel body from melting and since no consideration of scorching arises, such a concentrated heat source is (exceptionally) permissible.

Direct creation of steam within a pyrochemical system seems to belong in the category of unsolved problems because of the enormous cooling effect of the evaporating water. The term *direct* creation out of the pyrotechnic-system proper is used, since the application of encased high heat producers to create steam for driving underwater projectiles, for instance, is quite another matter and has been proposed and patented.[469]

A commercially available, practical pyrotechnic heat source is the fully enclosed, primer-activated cartridge, a type of which is described in patents by Bennett.[470] Mixtures of zinc, barium chromate, and manganese dioxide are claimed as the heat-producing mixture. A soldering iron is marketed in the United States that uses such a cartridge activated by the pull and release of a spring-loaded pin. The inexpensive cartridge furnishes 10 kcal and is of merit for use in isolated spots where electrical current is unavailable or use of a torch flame for heating is impractical or hazardous.[471]

Worthy of mention also is the burning portion of self-vulcanizing tire patches consisting of a cool and slow-burning mixture of potassium nitrate and a carbonaceous material such as sugar with some binder materials.[7]

The activation of solid-electrolyte cells by gasless heat powders has received a large amount of attention—all buried in the classified literature. What can be disclosed in this book is taken from a commercial brochure:[472] The most common heat source is Z-2, a mixture of zirconium metal and barium chromate converted by means of glass and silica fibers into so-called heat paper. The battery itself consists of the metal pair Ni/Mg, while the solid electrolyte is a eutectic mixture of potassium chloride and lithium chloride with certain additives acting as depolarizer and thickener. The value of a solid-electrolyte battery derives from the fact that the ordinary galvanic cell undergoes chemical reactions whether in use or in storage, while the heat battery is completely dormant until the moment the electrolyte is melted. This advantage is of course somewhat balanced by the transient nature of the current and the fact that one deals with "one shot" devices with all the inherent difficulties of establishing reliability data that that term implies. Heat output data on the zirconium/barium chromate family of heat sources are found in Chapter 34.

Pyrotechnics plays a more subordinate role in other galvanic cells of high storage life where the inactivity is achieved by separation of battery proper and of electrolyte. In ammonia activated batteries, pyrochemical heat aids in transfer of the ammonia, and in certain alkaline batteries, a concentrated potassium hydroxide solution is projected into the battery cavity with the help of a propellant-gas cartridge.

At low ambient temperatures, dry cells and other galvanic cells[473] lose their faculty to give their nominal current output. Pyrotechnic heat sources can be employed to remedy this deficiency. Experimentally, one such system consists of cheap, easily molded, relatively cool burning pellets, inserted into metal tubes, which are immersed in the electrolyte. The pellets consist of fine iron powder and sulfur in the proportion 70/30 (theory 64/36; calc. 250 cal/g). They are easily ignited and while expanding in a semiliquid state, glow at bright red heat.

A variety of other mixtures will be described in Part IX and specifically in Chapter 34 where their calorific output is given.

In metallurgy it may be desirable to prolong the molten state of a metal before and after pouring. Contamination may be tolerable if the additives float as a slag on top of the metal. This type of pyrochemical heating is the subject of a British patent[474] that claims mixtures of aluminum and silicon with salt-type oxidizers and sand. These are molded into slabs and put into packages to be laid on the surface of the molten metal.

In United States foundry practice,[475] less exothermic varieties called "hot topping compounds" are found more often than strongly exothermic heat sources. These are used on the surface of risers where by mere glowing they add heat to the riser and at the same time act as heat insulators. The mixtures are water-wet molded in any desired shape and "baked." After initiation, they retain their shape. Such a mixture is also effective in the form of a hollow cylinder surrounding the wall of the riser.

A patent by Camp and MacKenzie,[476] remarkable for its claims of unusual combinations in unusual form, concerns fluid mixtures of metal alloys such as Li/Mg/Al with fluorocarbon oils, which, dropped on a hot surface, react with evolution of 1.2 to 2.3 kcal/g mixture "without appreciable gaseous by-products." The reactions proceed with formation of the fluorides and chlorides of the fuel metals and of carbon. The reactivity of fluorinated organic compounds is something rather new in pyrochemistry and extends also to the explosives field, as evident from a patent by Gey and Van Dolah.[477]

Because of its erosively destructive effects, the thermite reaction previously described, must be modified in such a way that instead of a separation into liquefied layers, the whole reacting mass sinters and stays in place. This is achieved by replacing the aluminum, all or in part, by magnesium, whose highly refractory oxide permeates the liquefied mass. Such a composition is said to be useful for the welding of aluminum or aluminum-alloy wire cables, as explained in a Swiss patent.[478] Formula 199, quoted from this patent, is actually not part of the patent claims, which are concerned with means of consolidating mixtures for use as heat cartridges.

Along exactly the same lines, Shidlovsky[3] emphasizes, as shown by several drawings in both the 2nd and 3rd editions, the usefulness of "magnesium thermite" for welding telephone and telegraph wires. He describes a cylindrical pyrocartridge, longitudinally perforated and slipped over the abutting wire ends that are temporarily conjoined by a kind of plier. He also quotes the composition (Formula 200) of a special thermite said to be used to weld ground wires to metallic structures.

While the author has not worked along these lines, he found from some incidental experiments that the addition of coarse atomized magnesium to commerical thermite did produce the sintered ash but, strangely, the mixture became very hard to ignite, even though one would have expected just the opposite. Furthermore, the reaction was much slowed down.

In the future, this type of operation, i.e. welding or brazing by means of a small, self-contained highly concentrated heat source, will probably be used for emergency repair operations in outer space.[479]

Welding on a larger scale, whereby both the heat from the reaction and the produced metal play a part, will be described in Chapter 30, Solid Products.

chapter 27

High-Temperature Flames

For welding or cutting, torches function by means of a constant supply of gaseous fuel and oxidizer such as acetylene and oxygen. The resulting flame is the hottest of all those using common fuel gases, better than with hydrogen or natural gas. It is hotter than the flames of pyrotechnic mixtures, which are unsuitable for most of the specific uses of torches because of their short duration and the larger amounts of solids in most of them. When pyrotechnic mixtures display a perforating power comparable to that of a cutting torch, this is caused by the erosive force of extremely hot molten material such as from a thermite reaction.

If the subject of high-temperature flames has any place in the context of pyrotechnics, it is because attempts have been made to increase the temperature and with it the potential efficiency of torch flames by the use of unorthodox fuels, some of which are found in pyrotechnics and, to a lesser extent, of unusual oxidizers.

The combustion of hydrogen in oxygen is superior in heat output to all pyrochemical combinations on a calories-per-gram basis. Hence the greater energy concentration in other flames must either derive from the presence of more stable combustion products or from feeding fuel and oxidizer at an increased rate. The latter of course is only useful if the speed of reaction is adequate to cope with the amounts available. An example is the comparison of the hydrogen/oxygen and hydrogen/fluorine reactions: The heat output for both reactions is identical—3.2 kcal/g—but by volume the hydrogen/fluorine reaction is superior—2.9 against 1.7 kcal/liter of reactants. The decisive point is, however, that under the same conditions the speed of the hydrogen/fluorine reaction is ten times as fast as that of the hydrogen/oxygen. As a result, the flame temperature is over 1000° higher, namely 4300°K against 2930°K.[480]

Heat output can also be boosted by the use of *endothermic fuels*, i.e. those that decompose into the elements with release of energy. Two of these are cyanogen [$(CN)_2$] and carbon subnitride (C_4N_2), the latter actually the dinitrile of acetylene dicarboxylic acid; thus, NC. C≡C. CN. They furnish, with oxygen, flames measured at 4800 and 5300°K, respectively—higher than any other measured. Acetylene itself (C_2H_2) is an endothermic compound as is diacetylene (C_4H_2), said to be better for gas welding than acetylene, with a flame temperature with oxygen of about 3900°K.

A recent member of this group of endothermic, hot flame-formers is the compound tetracyanoethylene, $(NC)_2C{=}C(CN)_2$, of M.p. 199°C and B.p. 223°C, synthesized in 1957, highly endothermic, and furnishing a flame of 4000°K or higher, as reported in the chapter on cyanocarbons in Kirk-Othmer's Encyclopedia.[24]

Still another way to increase energy output is the use of endothermic compounds as *oxidizers*. The oxides of nitrogen (N_2O, NO, N_2O_4) are such endothermic compounds. Ozone (O_3) has been considered and tried, though it is hazardous to handle in undiluted state. On the other hand, even a stable exothermic compound such as nitrogen trifluoride (NF_3) can furnish with hydrogen an extremely hot flame (Kirk-Othmer,[24] Volume, 9, p. 629).

Some interest was displayed in the fifties in the use of metal powders, especially flake aluminum with oxygen, for torch flames. A very thorough report on the subject has come out of the Research Institute of Temple University,[481] whence other articles and reports on the subject of high temperature flames have also emanated.[480,482-488] The subject is, however, not so new, as shown by an article by Baker and Strong from the year 1930.[489]

While some spectacular results can be shown by the action of the aluminum/oxygen torch,[490] the use of a solid fuel and the fact that the flame is not self-sustaining and requires the presence of a "pilot gas" complicates the operation of such a torch. Actually, the aluminum/oxygen flame probably owes its effectiveness to the heat concentration in the product, since at 3800°K peak temperature it is not spectacularly hot. One of the reasons may be that the stability of aluminum oxide (Al_2O_3) at very high temperature is relatively poor and the reaction

$$Al_2O_3\ (l) \longrightarrow 2AlO\ (g) + O$$

highly endothermic.[491] On the other hand, the most stable molecules known, "stable enogh to exist at temperatures of the sun," are nitrogen and carbon monoxide. Since between 5500 and 6000°K one reaches "the limits of chemistry,"[490] this means also that pyrochemically-created heat cannot rise above this limit.

Part VIII

Chemical Production

chapter 28

Introduction

The products of pyrotechnic devices are as a rule an unextricable mixture of worthless solid residues and of various gases and dispersed solids. Even in the case of aerosols, where one species such as the dye in a colored smoke or the phosphorus pentoxide in a white smoke may appear in more-or-less pure state, one is not interested in isolating or producing the substance as a chemical but only in observing it as a physical phenomenon.

A number of pyrochemical exothermic reactions produce pure metals or compounds, alloys, or specific gases as chemical entities. Most of these processes are part of common production methods in metallurgy and chemical technology, such as the roasting of sulfidic ores or the synthesis of chalcogenides or halides from the elements, but a few lend themselves to performance in the manner of pyrotechnics. A condition for this is, of course, that the reaction be self-contained, because it is self-sustaining owing to the amount of heat evolved and the level of temperature reached. If the desired product is volatile, separation from a residue is a minor problem. In the case of solids, it is necessary that the product appears first as a liquid and that it is of much higher (or lower) specific gravity than the remainder. Exceptionally, the total residue may be the product, as in the well-known conversion of a mixture of iron powder and sulfur into iron sulfide (FeS).

Pyrotechnic production of chemical substances will be practiced where one wants to bring the material during its liberation or synthesis to the place of use and be able to achieve such production without any of the devices of chemical or metallurgical technology, such as furnaces, feeding mechanisms, ducts, etc. The reason for such procedures and the means of achieving them will become evident in the following two chapters.

chapter 29

Gas Production

In order to have a supply of a specific gas handy for occasional use, such as an emergency need for oxygen, one would think that keeping the pure gas in highly compressed or compressed and liquefied form would be the only sensible means of storage. For most applications this is undoubtedly the case, especially when greatest purity is demanded and limitations of space and weight are not necessary. There are, however, excellent reasons for preferring a chemical, and especially a pyrochemical and pyrotechnic, preparation of a gas in situ: Nonliquefiable gases require strong and heavy containers for relatively small amounts of compressed gas and the operation of valves, perhaps by remote control, adds to the weight, size, and complicacy of the system. The situation is not much better when the gas is liquefied. Then we have to consider, beside the low density of such a liquid, the fact that at different ambient temperatures the rate of release of the gas may vary so much that auxiliary heating devices must be employed. Third, and this may come as somewhat of a surprise to many, the volume of the ingredients of a chemical mixture may be lower and the weight of the gas released from it higher than that of the gas itself in condensed form in the same space, notwithstanding the fact that the residue is considerable—in the case of hydrogen even much larger than the weight of the released gas.

Pyrochemical gas production has, however, some serious disadvantages: The gas is delivered at a high temperature and the canister from which it emanates may become heated to the point of being a hazard; salt fog and miscellaneous debris may be admixed to the gas stream, requiring filtration; worst of all, the chemical reactions as presented by textbook equations are hardly ever realized without side reactions taking place. This may mean that the gas contains undesirable and harmful contaminants, as will be shown below. The best known pyrotechnic gas generator is the oxygen candle.[492,493,494] It is based on the thermal decomposition of alkali chlorates according to the equations.

$$2NaClO_3 \longrightarrow 2NaCl + 3O_2 + 25.0 \text{ kcal (exothermic)}$$

$$2KClO_3 \longrightarrow 2KCl + 3O_2 + 21.4 \text{ kcal (exothermic)}$$

These exothermic reactions are easily promoted by catalytic agents that depress the temperature of decomposition, but the evolved heat is insufficient for self-sustaining decomposition. A small amount of a fuel that furnishes no gas and is nonhazardous in mixture with chlorates—such as powdered iron or steel wool—is added, as well as barium peroxide, glass powder, and glass fibers. For respirable oxygen, the formation of chlorine and carbon monoxide must be completely suppressed. Compositions of this kind deliver about 40% of the chemical mixture as oxygen, or 280 ml/g at STP (Formula 201).

If such a mixture is carefully heated and kept only a few degrees above the melting point of the chlorate, it can be cast rather than compacted by pressing. Another method, which takes advantage of the low melting point of sodium chlorate (255°C), consists of hot pressing at high compaction pressure (120,000 lb/in.^{2}) and 225°C, whereby the theoretical density is approached. Higher temperature (245°C) produces still better compaction at much lower compression but the mass becomes sticky or even liquefies at only 7000 lb/in.^{2}, according to an NRL report.[494]

Jackson and Bovard[495] claim to avoid the dangerous heating up of the canister to red heat during use by means of a layer of potassium perchlorate. The low heat conductivity and the near zero heat of decomposition of the perchlorate into chloride and oxygen is the reason for the protective action—not a purported endothermic decomposition.

Water is sometimes recommended as an additive during the mixing and pressing of chlorate formulas but, as pointed out in the quoted NRL report, any residual moisture contributes to formation of sodium chlorite ($NaClO_2$) and hypochlorite ($NaOCl$) and hence to chlorine formation.

A somewhat different approach to the oxygen candle is described in a technical report,[496] which also includes an extensive bibliography. This candle consists of lithium perchlorate with the addition of manganese powder and lithium peroxide (Formula 202). The yield of oxygen is 49% of the candle weight compared to about 40% for the chlorate candle. At a compacted density of 2.32 g/cm^3, it contains as much oxygen as an equal volume of liquid oxygen at -183°C. A problem with otherwise desirable components such as lithum perchlorate is their hygroscopicity, though small amounts of moisture, according to the report, neither caused chlorine formation nor interfered with the decomposition reaction itself.

While not strictly pyrotechnic in nature, two closely related sources of oxygen should be mentioned. The first is a German method during World War II to supply oxygen to the diesel engines of submerged submarines.[48] The reaction between calcium permanganate and highly concentrated hydrogen peroxide not only produced oxygen, but the reaction also furnished 552 cal/g, using 80-85% hydrogen peroxide. Water was released as superheated steam and utilized first as a power source in a steam turbine and then as a source of "pure" water for the submarine crew. A practically identical system was described more recently for the production of respirable oxygen and electrical energy through decomposition of 90% hydrogen peroxide.[497]

The other system, described in an article by C. B. Jackson and R. C. Werner,[498] consists essentially of potassium superoxide, (KO_2) (formerly regarded as K_2O_4), which is used in a self-contained breathing apparatus. It appears to be ideal for the purpose since it evolves oxygen by breathing on it (i.e. with water), controls humidity by absorbing water, and removes the excess carbon dioxide in the exhaled air by chemical absorption according to the equations:

$$2KO_2 + H_2O \longrightarrow 2KOH + O_2$$

$$2KOH + CO_2 \longrightarrow K_2CO_3 + H_2O$$

Certain metal oxides catalyze these reactions, but in a canister as part of a breathing apparatus, only a small portion of the superoxide needs a catalyst. Potassium superoxide contains theoretically one-third of its weight of oxygen in the form that can be liberated as a gas, but in actual practice the amount is smaller, amounting to 190 cm^3/g. In a pamphlet,[499] the use of sodium superoxide—NaO_2 (not to be mistaken for sodium peroxide, Na_2O_2)—has also been mentioned. This should furnish about 30% more oxygen than the potassium compound.

More recently, the use of alkali ozonides such as KO_3 has been mentioned for air revitalization.[500] These compounds were formerly regarded as quite explosive and excessively reactive.[61] Such introduction of substances once regarded as "impossible" for practical use shows not only the trend of the times but also that what appears in the literature and is faithfully copied by later compilators often needs a reappraisal.

The production of sulfur dioxide by reacting sulfur with oxidizer salts is rather simple, though in practice two difficulties are encountered: If for safety's sake a rather unreactive oxidizer is used, the sulfur may tend to burn off without initiating the decomposition of the oxidizer salt, which in a closed space will of course end the gas formation right at the start; however, as soon as the reaction is well initiated by means of a hot starter mixture, the heat of reaction may be excessive. This problem was solved in the writer's laboratory by Kowarsky and Olander[501] by coupling an exothermic and an endothermic sulfur-dioxide-producing reaction as follows:

$$2S + KClO_4 \longrightarrow 2SO_2 + KCl + 71.5 \text{ kcal (exothermic)}$$

$$2S + CuSO_4 \longrightarrow CuS + 2SO_2 - 30.5 \text{ kcal (endothermic)}$$

This reaction can thus be kept at a brisk pace without excessive heat output. The SO_2 candle or canister, at one time considered in a certain tactical biological application, could be used as a quick-acting fungicide, disinfectant, or animal repellent. Small cardboard cartridges for insertion in the runs of moles or gophers also contain mixtures with sulfur, perhaps similar to those recommended by Izzo[9] for destruction of wasp nests by means of a modified black powder (58 parts potassium nitrate, 10 parts sulfur, 32 parts charcoal) that combines an incendiary and a repellent effect by its emission of noxious gas.

It is not possible to produce nitrous oxide (N_2O) in the pyrotechnic manner. The heat output of the reaction

$$NH_4NO_3 \longrightarrow N_2O + 2H_2O + 8.8 \text{ kcal (exothermic)}$$

is insufficient for a self-sustained process and the gas is not pure enough for inhalation as an anesthetic ("laughing gas"). Very toxic nitric gases are formed in side reactions. Catalysts such as ammonium dichromate make the decomposition of the ammonium nitrate self-sustaining, but instead of obtaining nitrous oxide, a mixture of nitrous oxide, nitrogen, nitric oxide, and nitrogen dioxide is produced.

Nitrogen can be obtained by the metathetical reaction of sodium nitrite and ammonium chloride and the subsequent decomposition of unstable ammonium nitrite:

$$NaNO_2 + NH_4Cl \longrightarrow NH_4NO_2 + NaCl \text{ (0 kcal)}$$

$$NH_4NO_2 \longrightarrow N_2 + 2H_2O + 52.5 \text{ kcal (exothermic)}$$

In solution the reaction yields pure nitrogen, and the comparison of the nitrogen thus evolved with that obtained from air led Rayleigh to the discovery of argon in air because of the difference in density of the two "nitrogens." The dry reaction proceeds briskly with 0.42 kcal heat developed per gram of mixture and has been used with a stabilizing additive in a mixture called "Hydrox" for safe blasting in coal mines, the explosive effect being caused only by pressure from expansion of the gases.[6] Such nitrogen is, however, not pure and contains some ammonia and oxides of nitrogen.

It is more difficult to make hydrogen by pyrochemical means. This was done in France prior to World War I for filling balloons. The so-called *Silicol* process, described in Mellor,[67] by Remy,[60] and also in a French patent,[502] combines silicon in the form of ferrosilicon and soda lime into a mixture called *Hydrogenite*. The reaction proceeds exothermically, perhaps according to the equation

$$Si + Ca(OH)_2 + 2NaOH \longrightarrow Na_2SiO_3 + CaO + 2H_2$$

and furnishes 270-370 liters of hydrogen per kilogram of mixture. The activation of the cartridge has been described as done by "forcing a glowing ball or wire into the mixture." From the very limited experience of the writer, it appears that the mixture may be unignitible in small amounts even with the hottest first fires, though theoretically its heat of reaction is 0.41 kcal/g. However, pyrotechnic compositions or single substances that contain water in any form are notoriously difficult to ignite or to initiate. It is possible that the reaction between silicon and hydroxides can only become self-sustaining when large amounts are initiated with an ample heat source.

The newer hydrides, such as sodium borohydride ($NaBH_4$) or sodium aluminum hydride ($NaAlH_4$) and the corresponding lithium salts, are prodigious sources of hydrogen, which is released by addition of water.[152,153] Such a system could be called pyrotechnic only if one considers the introduction of the water by some driving or breaking force, which in turn would be engendered by a power source.

The reaction

$$LiAlH_4 + 2H_2O \longrightarrow LiAlO_2 + 4H_2$$

furnishes 2.36 liter of hydrogen per gram of hydride or about half as much if the weight of the water is also considered.

Undoubtedly, one can produce other gases by pyrochemical reaction or by combining a heat source with a pyrogenic decomposition reaction. These include the halogens and pseudo halogens, such as cyanogen. The latter is formed from heavy-metal cyanides:

$$2AgCN \longrightarrow 2Ag + (CN)_2$$

$$Hg(CN)_2 \longrightarrow Hg + (CN)_2$$

$$Hg(CN)_2 + HgCl_2 \longrightarrow Hg_2Cl_2 + (CN)_2$$

At this time, no requirements appear to exist for such pyrochemical production.

Note to Chapter 29

The behavior of the alkali chlorates on heating is complex and warrants special attention. Many points regarding the decomposition reactions of the dry salts at elevated temperatures, in the pure state, and in the presence of catalytically acting oxides are still not fully cleared up. A Navy report[508] and Mellor, Supplement I to Volume II,[67] summarize the studies concerning the divers reaction mechanisms postulated. From these, one special point should be mentioned because it is not usually cited, namely, the finding of chlorite ($Me^{I}ClO_2$) formation as an intermediate in uncatalyzed decomposition of the alkali chlorates.

In perhaps simplified form it can be stated that the pure alkali chlorates melt on careful heating without decomposition, but slightly above their melting points at least three major reactions begin to take place:

$$2MeClO_3 \longrightarrow 2MeCl + 3O_2 \qquad (a)$$

$$4MeClO_3 \longrightarrow 3MeClO_4 + MeCl \quad \text{(b)}$$

$$MeClO_4 \longrightarrow MeCl + 2O_2 \quad \text{(c)}$$

With increase of atomic weight of the alkali (Me), reaction (b) becomes predominant over (a), observed as follows[503]:

Li 40% Na 67% K, Rb, Cs 87%

It is well known that the decomposition reaction of the chlorates is greatly facilitated by the addition of various heavy-metal oxides and by silica. The catalytic effect is one of lowering the temperature at which oxygen appears first and of eliminating the perchlorate formation. Different researchers have given different sequences of oxides arranged in order of diminishing catalytic effectiveness. Contradictory statements have been made about the relative position of Fe_2O_3 and MnO_2 in such a series. However, it is agreed that these two, as well as Co_2O_3 and Cr_2O_3, head the list of effective agents.

A mutual reinforcement of activity of the oxides of iron and manganese has been assumed in commercial manganese dioxide. It has furthermore been shown that manganese dioxide actually loses and regains oxygen during the reaction. The amount of the additive is important and 10-20% are necessary for optimal results. Under these circumstances, complete decomposition seems to take place above 275 and below 400°C. An undesirable side reaction of chlorate decomposition is the formation of free chlorine gas. Proper catalysis can keep this to 0.25% or less, but for respirable oxygen this amount is still excessive and special additives (Formula 201) are required to reduce it. A statement that potassium chlorate catalyzed with manganese dioxide furnishes oxygen already between 70 and 100°C is obviously erroneous and would seem to derive from a misread statement that the decomposition temperature of the chlorate is *lowered* by 70—100°.

The involved decomposition mechanism of chlorates is not paralleled by the behavior of bromates or iodates, which decompose with simple evolution of oxygen, but bromine is found in oxygen from bromates.

While all this is important for chlorates as an oxygen source, it probably has little bearing on fuel/chlorate mixtures of approximately stoichiometric proportions or with excess fuel where either the strongly exothermic reaction takes precedence over any perchlorate formation or makes it irrelevant because of the high temperature reached in the reaction.

The decomposition of potassium perchlorate by itself or in the presence of catalysts or fuels has been explored mainly in connection with propellant chemistry and the newer information has often been obtained under security restrictions. Some of the complexities of the reactions are outlined in studies by Marvin and Woolaver,[504] Patai and Hoffmann,[505] Bircumshaw and Phillips,[506] and Harvey.[507] A great amount of information on perchlorates in general has been collected in the book by Schumacher,[170] and on both chlorates and perchlorates in the above-quoted three volume NavOrd report.[508]

Are chlorates and perchlorates hazardous in storage if we speak only of the alkali salts, excepting ammonium perchlorate? Stettbacher[17] cites a report on the destruction of a whole factory of potassium chlorate when 156,000 kg in *Fässern* blew up during a fire. Assuming that the containers were wooden barrels, such as the ones in which chlorate was shipped from Europe before World War II, it would by no means be hard to understand that during a conflagration the mass of burning wood and paper liners, melting chlorate, and structural timber of a burning building constitutes altogether a veritable giant bomb. Thus the hazard of this and other active oxidizer salts is conditional rather than implied in the nature of the oxidizer. Senze *et al.*[509] state that pure sodium or potassium chlorate cannot be exploded by intense blows or friction, but that such explosions take place if the chlorate is in contact with flammable organic matter, not necessarily mixed. It has even been claimed[510] that the dusts of chlorates and perchlorates extinguish "cool" hydrogen or illuminating gas flames while potassium nitrate is a better dry-powder fire extinguishing agent than sodium bicarbonate. In industrial practice, these oxidizer salts are probably safe if kept carefully separated from fuel-type materials, especially sulfur, phosphorus, and active metal powders, and if they are stored in fireproof and relatively small buildings while the materials themselves are contained in metal drums rather than in wooden barrels.

chapter 30

Solid Products

The thermite process described earlier for destructive purposes and as a mere heat source was originally destined for the production of metals, alloys, and certain compounds, and the uses of the physical effects of the reactions were an afterthought. Tissier in 1856[67] and Beketov in the 1860's[3] first observed the strong reducing power of aluminum powder in dry reaction, but aluminum, as a new material (Wöhler 1828), was then much too expensive for commercial uses. Dr. Hans Goldschmidt, in numerous patents from 1895 on,[511] claimed production of the majority of metals and their alloys, and of phosphides, arsenides, silicides, and borides by reduction of the oxides or respective salts with aluminum, both in more-or-less finely dispersed state. He also coined the original tradename *Thermit**, which eventually became the common term "thermite" (sometimes, and of dubious legality, "thermit"). The technique itself is important enough to employ the special term *Aluminothermics*, and many details on the subject are found under this heading in the *Encyclopedia Americana*.[25] A similarly extensive description of the thermite process is found in Volume 3 of Ullmann's encyclopedia[22] under *Aluminothermie*. Information on the present uses of thermite in the United States was given to the author privately.[475]

Pyrotechnic production delivers the metal at extremely high temperature so that it can be used in situ for welding of rails, repairing heavy machine parts, or producing small castings. Some metals that are difficult to liberate by conventional methods are easily obtained, pure or as definite alloys, for convenient admixture in the manufacture of specialty steels. Metal produced by the thermite process is entirely free from carbon, which is always a contaminant in metallurgical processes with coke or charcoal. Such carbon is a detriment to certain usages. Finally, nothing could be simpler and

* Thermit is still a registered trademark owned in the U.S. by Thermex Metallurgical, Inc., Lakehurst, N. J.

more convenient than preparing a limited amount of any one of a large variety of rare or rarely produced metals and their alloys by what may be called a "push-button method." For these reasons, pyrotechnic production of solids by the Goldschmidt process or one of its variants is an important technological method.

In a general way, the thermite reaction is the reduction of a metal oxide by a reactive metal leading to the oxide of the reductant and to the metal from the reduced oxide. The prototype reaction, also the commercially and technically most important, is the following:

$$8Al + 3Fe_3O_4 \longrightarrow 4Al_2O_3 + 9Fe + 795 \text{ kcal (exothermic)}$$

The heat output of this reaction per gram of reactants is 0.87 kcal/g or 3.7 kcal/cm^3 (theoretical) and must be called moderate, both on a weight or actual volume basis (the density of the unconsolidated mixture, the form in which the material is used, is about 2 g/cm^3). The heat output is a little higher for the reaction

$$2Al + Fe_2O_3 \longrightarrow Al_2O_3 + 2Fe + 203 \text{ kcal (exothermic)}$$

which yields 0.95 kcal/g. The preference for the use of "black" iron oxide, sometimes called ferrosoferric oxide, may derive from the fact that a coarse but reactive, scaly product of relatively large surface area is available as *hammerschlag* (German for hammer-beaten, meaning blacksmith's scales)—the result of surface oxidation of bulk iron during heating in air. Together with a coarse "grained" aluminum, this yields just the proper reaction speed and heat concentration for optimal temperature and promotion of coalescence of the metal below and the slag above it.

The temperature reached by the reacting mass has been given as 2000-3000°C (one may encounter similar figures given in °F that are erroneous), but is probably limited by the boiling point of aluminum, which is below 2500°C. If the reaction is too fast because of small particle size of the components, or especially because oxides are used that have low heats of formation, such as the oxides of copper, lead, or bismuth, the process may exhibit explosive violence.

On the other hand, when the heat of formation of the aluminum oxide is only moderately higher than that of an equivalent amount of the oxide that is to be reduced, the reaction may not take place, or only sluggishly, or lead to alloys of unused aluminum and the liberated metal. A typical case is that of titanium, where the reaction between aluminum and titanium (IV) oxide (TiO_2) comes to an equilibrium leading to alloys of the two metals and also to formation of titanium (II) oxide (TiO), which collects in the slag.[22] For zirconium, the alloy $ZrAl_3$ has been named as the aluminothermic product.[512] In such cases, some modifications of the basic process can achieve the goal of making a pure metal. There are at least four such modifications, all of which are either used or have been proposed or patented in the period since 1894.

One of these variants is the addition of a more active, more heat-forming but chemically related oxide. Thus in the case of chromium a few percent of a Cr(VI) compound ($K_2Cr_2O_7$ or CrO_3) will increase the yield of a consolidated "regulus" of chromium on the bottom of the crucible from the main reaction:[67]

$$2Al + Cr_2O_3 \longrightarrow Al_2O_3 + 2Cr + 129 \text{ kcal (exothermic)}$$

which on an equivalent basis amounts only to 21 kcal compared with 33 kcal for the Al/Fe_3O_4 reaction. (Equivalence means the amounts of metal and oxides able to react with or containing a gram equivalent of oxygen, i.e. 8 g.) Similarly for manganese where the dioxide reacts too violently and the monoxide is unsuitable, a mixture of the two is effective, as quoted from a German patent by Ullmann,[22] or the compound Mn_3O_4 can be used.

The second method involves the addition of high heat producing extraneous oxidizers. Frequently mentioned are barium peroxide and other strong oxidizers such as lead dioxide, sodium persulfate (Na_2SO_5) and even chlorate. Kühne[513] claims such additions for the aluminothermic production of numerous metals, and Cueilleron and Pascand for aluminum/titanium alloys.[514]

A third modification and one that already had been claimed in the original Goldschmidt patent[511] and by many successors is the replacement, all or in part, of the aluminum by other reductants. Among the replacements or additives are calcium, silicon, magnesium, calcium carbide (CaC_2), misch metal, and boron. Misch metal has been used by Weiss and Aichel[515] for producing vanadium and niobium (columbium), the latter, however, forming aluminum alloys rather than pure metal. Vanadium is now made from the vanadium oxides (V_2O_3 or V_2O_5) with calcium in the presence of calcium chloride, "thermally initiated" in an argon atmosphere.[516]

The last modification to be mentioned here consists of starting the reaction at a previously elevated temperature, thus gaining the advantage of promoting it without introduction of foreign materials or expensive reductants. It must however be noted that in such a scheme we move away from a simple pyrotechnic production and are back at a more-or-less conventional metallurgical operation; in fact, this is precisely the case in metallurgical processes where the thermite reaction is performed in furnaces. Thus Ullmann[22] cites the production of 250 kg of ferrotitanium alloy from 600 kg of the mineral Ilmenit, aluminum "griess," and lime, the reaction time being only 7 min. Other examples of the high-temperature thermite process are the formation of titanium and zirconium from their dioxides with magnesium powder and fluxes after heating the mixture under argon above 1000°C.[517] Under hydrogen, zirconium hydride is obtained in a quite similar operation.[518] Iron/titanium alloys are obtained, according to a Japanese reference,[519] with varying titanium content from 10—50%, the amount increasing with increased starting temperature.

As the above-mentioned formation of zirconium hydride shows, not only metals but also various compounds can be the end product of a thermitic process. Silicides, such as the compound $BaSi_2$ from the peroxide and oxide with silicon, have been quoted by Ullmann[22] from German patents. The lower oxide of titanium (TiO) has been described as obtainable from the dioxide and magnesium.[67] A very hard form of artificial corundum (crystalline aluminum oxide) of superior quality for grinding and polishing operations is the by-product of the above-described production of chromium metal.[22]

Because the sulfides of metals have a lower heat of formation than the oxides, some difficult-to-obtain metals can be made from their sulfides with aluminum. This process has been described by Gardner [520] for niobium (columbium) and tantalum from their disulfides (NbS_2 and TaS_2), whereby the relatively volatile by-product Al_2S_3 distills off above 1550°C. It is an interesting coincidence that the production of the same two metals by reduction of the pentoxides Nb_2O_5 and Ta_2O_5 with silicon can be performed under formation of silicon (II) oxide (SiO), which volatilizes in vacuum.

Solid SiO, reacted with the oxides of zinc, magnesium or manganese in a vacuum, forms the metals, which in turn distill off.[61]

The welding or the repairing of heavy machinery with *thermite* is somewhat different from the formerly described welding and brazing operations by pyrotechnic heat (Chapter 26), because here the metal produced becomes part of the joint or crack, or replaces a worn or broken-off piece. For these purposes, a forging type (with manganese), a cast-iron type (with added ferrosilicon), or a so-called wabbler thermite of great mechanical strength for building up worn machine parts have been formulated.

While other methods of reparing machine parts, such as electroslag and submerged-arc welding, have become popular, the thermite process has remained competitive because it requires no capital investment and is easily performed as a one-at-a-time operation. Normally, a certain amount of preheating is desirable, but recently a "Self-Preheat" process has dispensed with this.[521]

In the field of rail welding, the thermite process supplements and sometimes replaces flash or gas-pressure welding. The latter methods require very costly factory installations. They furnish rails about one-quarter of a mile long, which are hauled to the point of field installation. Here one can resort to the old-fashioned mechanical joining (a source of most maintenance on a railroad) or use field welds. The best of these is the thermite weld. It may be used exclusively with specialty items such as crossings, for haulage tracks in coal mines, and for crane rails.[475]

The thermite process is also suitable for the butt-welding of reinforcing bars of large diameter. Here the thermite process has remained competitive with other methods such as arc welding because it produces a true fusion weld of high tensile strength. Economically, it requires a lower labor cost, thus offsetting the higher expense for material.[475]

A relatively new development is the combination of metals and ceramic in compacts called *cermets*. The purpose is to combine the high refractoriness, resistance to osidation, electrical insulation, and retention of compression strength on heating—all properties of ceramic bodies—with the ductibility and thermal shock resistance of metals.[522]

Obviously the thermitic processes can serve this purpose when, for instance, the oxide of the finished cermet is produced from aluminum powder, which liberates the metals in reaction with an oxide of a suitable metal such as chromium or cobalt. An excess of aluminum oxide and addition of other refractory oxides or clay to the thermitic formulation is needed for the desired cermet.

The reaction can be extended to the formation of complex mixtures with heavy-metal silicides, borides, or carbides in lieu of metal.

When zirconium silicate ($ZrSiO_4$) or a mixture of ZrO_2 and SiO_2 is reacted with aluminum in the presence of aluminum oxide and then reheated, zirconium silicide ($ZrSi_2$) becomes the major product. Titanium dioxide (TiO_2) and boron (III) oxide (B_2O_3) with aluminum similarly form titanium boride (TiB_2). If the reduction of the oxides such as TiO_2 or SiO_2 with aluminum is performed in the presence of carbon black, the carbides TiC and SiC are formed embedded in aluminum oxide.[523] This subject is also treated in a British patent titled "Autothermic Fired Ceramics."[524]

In all these cases it appears that the exothermic process must be initiated by heating the preformed pellets, etc. in a furnace. This is again the above-described high-temperature process in which the need for promoting reactivity by an initial high temperature derives either from the dilution of the reactants with oxides, the low heat output of the main reaction, or both.

In Chapter 19, it was indicated that the dispersion of alkali metals at high altitudes is a new tool in space exploration. Instead of merely evaporating the metals by pyrochemical heat sources, one might think of producing them by chemical reaction and dispersing the resulting metal vapors in the rarefied atmosphere.

Lithium, sodium, potassium, rubidium, and cesium have been produced by reacting their chromates, dichromates, sulfates, molybdates, or tungstates with an excess of zirconium powder. The reactions, in general, are violent or even explosive and yield impure metals, but if the zirconium is in larger excess, the process is said to take place smoothly and gently and the metals are obtained in high yield and nearly free from oxide. The processes are cited in Supplements II and III to Volume II of Mellor's Treatise[67] as performed in a vacuum and by heating the mixtures to temperatures of 250—1000°C —mostly below 500°C. Since these reactions are strongly exothermic, there is not the slightest doubt that pyrotechnic initiation will lead to satisfactory performance. The dissemination of the evaporating metals should take place under the high-altitude conditions just as well as the distilling off "in vacuo" in the laboratory.

The indicated chemical reactions are not the only ones that because of strongly exothermic performance might be used for chemical dispersion—the decomposition of alkali and alkaline-earth azides being one other example. However, since this subject is at the moment merely speculative in the possible application to space exploration, the examples given should suffice to indicate approaches that will most likely have been realized by the time this book reaches the public.

Part IX

Basic Behavior and Properties of Materials

chapter 31

Introduction

In describing the *phenomena* of pyrotechnics and presenting formulations, numerous materials have been shown to achieve the desired results. The functionality of such materials was explained more or less in detail as the situation warranted. Thus, the specific properties of prime ignition materials, the selection of effective light-producing fuels and oxidizers, and the choice of mixtures to achieve very high temperatures or exact caloric outputs were pointed out. If pyrochemistry were an exact science, one would start with the properties of the materials and arrive by deductive processes at the most suitable formulas. At the present state of the art, such proceedings would not be successful. We are still in a mainly empirical phase of gathering information by trial and error, of drawing conclusions from measurements following qualitative and quantitative variations in ingredients, and building our theories on systematic experimentation.

For these reasons, a description of single materials, their measurable properties, and their behavior on interactions has been appended to the preceding parts. As our knowledge is fragmentary, so are the facts presented, which are in part data and in part subjective, or at least not exactly measured or measurable, observations.

A formula is, however, not based on mere considerations of suitable properties of components. Numerous extraneous factors come into play and are discussed. They include surveillance properties (involving, among others, compatibility and adverse external influences) and also such practical aspects as availability, reproducibility, economy, and the uses of processing aids.

In presenting this information, it is hard to arrange it in a logical sequence. Essentially, I have retained the division used in *Modern Pyrotechnics*. The *general properties* of materials with emphasis on particle size are followed by *phase changes* of materials and products. *Caloric output* follows with more extensive tables of data than formerly offered. *Reactivity*, including the subjects of *ignitibility*, *compatibility*, and negative aspects such as *destructive external influences*, takes another chapter. *Processing* and *processing aids*, and, lastly, *binders* conclude this part. Thumb-nail sketches of specific materials and their individual characteristics are added in Part X.

chapter 32

Nonspecific Properties of Materials

> Specifying the particle size distribution of a subsieve powder by requiring that no more than 5% shall be coarser than the finest testing sieve is akin to agreeing to buy a mongrel dog upon being shown the tip of his tail stuck through a knothole in a board fence.[535]

The pyrotechnist's relation to the pyrochemically active materials in a formula depends on the specific purpose: For the *production* to prescribed specifications of a standardized item, he is mainly concerned with obtaining the materials promptly from commercial sources and in the verification of compliance of the delivered materials with the specifications. In a *developmental effort*, the emphasis is on the experimental study of different grades of known and proven materials in varying proportions. Finally, in *research* as well as in development of new types of devices, he may also consider substances formerly not regarded as useful for pyrotechnics.

The major nonspecific properties of materials and aspects of their usefulness are *purity* and *particle size* as well as *availability* and *economy*. In production of standardized items, the choice may seem to be limited, but in practice the lines are not drawn so tightly as it seems. Specifications are rarely so rigid or so well defined that two competing materials of different price and slightly different quality might not be suitable. In the development phase, the inexperienced formulator frequently succumbs to the temptation to start out by reaching for handy laboratory chemicals. Later on, he may be unable to match his results when using commercial products made to defined physical specifications and pure enough for his purpose, but different from the expensive reaction-grade chemicals. Even more difficulties are expected with metal powders such as magnesium, obtained from a supplier of fine chemicals. In the research phase, the scientist may find that the chemical that gives him the new effect is only available in limited quantities either in the physical state in which he used it or as a chemical individual. This, however, should not deter him from the use of such materials. If the result of research leads to large-scale production and the special effects cannot be obtained by conventional means, then either the higher cost has to be borne or, more likely, the demand for the new materials will stimulate their production and reduce their price.

Purity

Most chemicals, such as metal powders, other elements, and metal oxides, are far from being of highest purity in the form in which they are offered in the trade for pyrotechnic purposes. Such materials are mostly well suited though, exceptionally, some impurities in a commercial product can be detrimental.

Impurities are somewhat more important in the water-soluble salts such as nitrates, chlorates, and perchlorates. Commercial potassium salts are generally of adequately high purity because of the ease with which they can be recrystallized, being little soluble in cold water and much more soluble in hot, and the same applies to barium nitrate. This is fortunate for the production of colored lights where even traces of sodium, a common impurity, would influence adversely the creation of a blue, green, or red effect. Sodium nitrate is less hygroscopic (to a point) the purer it is. Obviously the manufacturer must decide whether his product can tolerate the somewhat more hygroscopic grade or if he has to use the more expensive purer material.

In a recent (summer 1967) production of flares, an unaccountably low light output was observed. After every other reason for this calamitous occurrence had been checked, the most unlikely cause was discovered: The manufacturer of the sodium nitrate, during change-over from processing potassium nitrate to sodium nitrate, had furnished a sodium nitrate that contained several percent of the potassium salt. Thus the subject of purity may come up under unexpected circumstances; it is difficult to safeguard against such events since it is not usual to perform a chemical analysis as part of the quality control of incoming raw materials.

Particle Size

No property of the ingredients of a pyrotechnic formula is more critical than what is loosely called *particle size*, and none is at times more elusive and baffling. Our discussion will concentrate on metal powders, these being literally the focal points of most compositions (the Latin word *focus* means hearth or fireplace). Particle size or, more correctly, the surface area of the metal particles has a great influence on burning rate and ignitibility. The manufacturer is often able to counterbalance a slight change in burning rate, caused by variations in his metal powders, by alterations in the formula, but the major difficulty lies in the fact that his means of discovering the changes in the metal from shipment to shipment or batch to batch are limited. Thus he will not know about it, or at least not exactly, until the end item is tested, and this is most unsatisfactory under production conditions.

If metal powders could be bought as one buys ball-bearings—i.e. as truly spherical entities all of the same diameter and of a true smooth surface area, approaching the geometrical surface as much as is practically possible—the problem of reproducibility would barely exist. (Perhaps such a powder would be unignitible, but that is another problem!)

Spherical particles made by the so-called atomizing process have been developed for the pyrotechnically most important metals—magnesium and aluminum (also their alloys)—and are marketed in average particle sizes from a few microns in diameter to a mixture of particles, all or nearly all of which pass a "30 mesh" standard screen and are retained on "50 mesh." Such powders —even though there is a certain latitude of total surface area, especially for the finer grades—have the best reproducibility. Problems arise not only through poor definition of the particle-size *distribution* but in cases where the globules deviate from sphericity. They may form pear-shaped and other irregular more or less spheroidal particles. Should the powder consist of other types of comminuted, irregular fragments, the definition becomes even more difficult. And while certain types of such materials can still be compared from batch to batch by comparing narrow cuts of screening fractions, this becomes less and less meaningful the more fine "subsieve size" particles are in the powder. It is obvious that, say, 5% by weight of particles of about 40 μ diameter just passing through a 325 mesh screen or the same quantity of much finer particles will cause a significant difference in the total surface area. In the coarser grades, a material that is formed from fragments with a rough, very irregular surface structure from balled-up but not massive spherical globules or from deeply rifted scales would behave differently from an essentially smooth and denser material.

The geometries of the metal powders used in pyrotechnics can be briefly illustrated in connection with manufacturing procedures, starting with those that produce the finest, most active particles.

So-called ultrafine metal powders below 1 μ in diameter can be made by pyrogenic dissociation of the vapors of the carbonyls of iron, nickel, and cobalt.[525] Aluminum of a particle size below the resolving power of the electron microscope has been formed by evaporation and condensation under vacuum in an inert atmosphere.[106] Similarly, evaporated magnesium has been quenched by JP-4 fuel for directly making a slurry fuel.[107] Also, long ball-milling of fine magnesium powder in the presence of a surfactant can lead to particles below 1 μ. Iron[526] or nickel[527] electrolytically deposited on a mercury cathode form very active, often pyrophoric, fine powders that, however, can be stabilized in order to be handled in air.

An unusual way of disintegration is by the conversion of a solid metal into a brittle hydride. Wilhelm and Chiotti[528] used thorium to produce a higher hydride, Th_4H_{15}, and then decomposed this hydride to ThH_2 and finally to pure thorium by heating under vacuum and breaking the metal down to a "30 mesh" material. Massive uranium acts similarly. All metals that can be embrittled by hydrogen, such as zirconium or its alloys (e.g. Zircaloy-2) can be converted into powder via hydriding, comminution, and subsequent degassing.[529]

Reduction processes in which a finely powdered pigment-like oxide is reduced at a temperature below its own melting point and that of the resulting metal furnish the so-called subsieve powders of which titanium, zirconium, amorphous boron, and hydrogen-reduced iron are the most important representatives.

Other powders of very large surface area are the flaked materials, which formerly were made by stamping, but are now produced from coarser grained powder by ball-milling in the presence of an organic solvent, stearic acid, or oleic acid. Except in fireworks items, flakes are now little used in pyrotechnics because they make compositions bulky and hard to consolidate.

Metals of low or moderately high melting points can be converted into particles by forming molten globules that congeal by passing through the air or an inert gas. *Granulating* and *shotting*, the latter from molten metal poured through sieves in a shot tower, produce particles too coarse for pyrotechnics. By blowing compressed air, helium, or argon through molten aluminum, magnesium, or their alloys, the *atomized* powders, ranging from 5—1000 μ particle diameter, are obtained. These may exhibit nearly perfect sphericity, but quite irregular particles that hardly deserve the term "globules," are similarly produced and are called "blown aluminum." A detailed discussion of the powders of aluminum and magnesium will be found in Chapter 41. It is a curious fact that aluminum is commercially available in just about every type of size and shape from superfine to coarsest granules.

In the above, we have not paid any attention to oxidizer salts, metal oxides, and nonmetallic fuels such as sulfur and red phosphorus. If a general statement may be allowed, in all these materials degrees of fineness are important to achieve intimacy of mixing (provided other materials are of comparable size) and hence ease of initiation and smooth reactivity. Thus, a match made with coarsely crystalline chlorate will sputter, a thermate made with a dense type of black iron oxide rather than the rifted coarse flakes or scales may be unignitible, and delay mixtures made with identical amounts of perchlorate of varying fineness may differ little in burning rate but largely in initiability. Zinc oxide, on the other hand, may have a large influence on the burning rate of HC smokes, depending on its mode of manufacture and, presumably, its particle size and shape.[318a,b]

Particle Size Determination

Determination of particle size, size distribution, and surface area increases in difficulty with the degree of accuracy and the validity of the obtained information. In view of the extensive literature on the subject, a few remarks may suffice. The well-known U.S. Standard *sieve* series, mostly used between the screen number 16 with an opening of 1190 μ (1.190 mm) and number 325 with a 44 μ opening, used in series in an automatic shaker such as the Ro-Tap (W. S. Tyler Co.), serves well for powders of spheroidal shapes and containing relatively few fine (—200 or —325) particles. A recent article[530] claims the usefulness of micromesh sieves made by electroforming in extending screening methods into a useful range of 5—50 μ.

Methods based on *sedimentation* rates (Micromerograph) or flotation (Roller Analyzer) aim to offer the actual particle-size distribution. Optical *microscopy* is qualitatively informative but its faults are the extreme smallness of the sample, the tediousness of size tabulation if numerical results are desired, and the inability to discriminate in some cases between discrete particles and agglomerates.

Since, under proper conditions, gases or solutes from liquid solutions are adsorbed on solids in monomolecular layers, methods for surface area determination from the *adsorption* of gases such as nitrogen at low temperature or of stearic acid from a solution in isopropanol are used. They are, however, little suited for routine quality control. Other methods or attempts at correlation are based on chemical reactivity, electrical or heat *conductivity* of a column, *compressibility*, and the *bulk density* of a column of powder produced with a minimum of any kind of consolidation or vibrated to a point where further reduction in volume becomes minimal. The most frequently used indirect method, however, is based on *permeability*. Here, a "fluid," generally air, passes through a bed of particles in the form of a column and the rate of flow has a definite relation to the surface area of the powder. If the particles are spherical, or nearly so, good agreement can be obtained with directly observed size determinations. In other cases, particles of the same type can be compared, even though the calculated average particle size may be a fictitious value. A frequently used and simply operated instrument for such measurements is the Fisher Sub-sieve Sizer.

A goodly number of lucidly written books on fine particles and their measurement have appeared, such as those by Orr and Dallavalle,[531] Rose,[532] Cadle,[533] and Herdan.[534] The booklet by the Metals Disintegrating Company[535] is quite detailed on the practice of the Roller Analyzer, the Fisher Sub-sieve Sizer, and methods of adsorption. The Sharples Micromerograph is evaluated and regarded as more suitable for quality control than for absolute values in a report by Kaye, Middlebrooks and Weingarten.[536] Comyn[537] discusses the inadequacy of particle-size specifications in a technical report and recommends the B.E.T. (Brunauer-Emmit-Teller) nitrogen adsorption method for total surface area determination and the Eagle-Picher turbidimetric method ("Turbimeter") for size distribution. The latter is based on optical permeability during sedimentation. An unusual apparatus is the Coulter Counter.[538] Its measurements are based on the change in electrical conductivity in a liquid by the interposition of particles of different electrical behavior. For calibration, ragweed pollen of 18-19 or mulberry pollen of 11-12 μ size can be used as standards.

The quoted works themselves contain numerous references, many from the special literature of the paint industry and of powder metallurgy. The work by Goetzel[539] offers specific information on the latter, and the treatment of very fine particles mainly in connection with air pollution is found in a newer book by Cadle.[540]

Occasionally, metal powders of seemingly the same particle size purity display variations in pyrochemical behavior that are and disturbing to the user. Such powders may cause and wide varieties of burning rate or, still worse, of most baffling initiability or compactability in identical formulations. Tungsten and iron powder are well-known examples. Since these metal powders are industrially produced for purposes other than pyrotechnics, minor variations in their production process (or more important, a major change in the manufacturing technique) may lead to products of widely varying character as far as pyrotechnic application is concerned, while the other properties for powder metallurgical application are unaffected. Sometimes, one can choose by trial and error from samples of stored production lots and select a suitable lot large enough to satisfy a complete production run—a rather sorry expedient. A similar but lesser problem arose when natural gums and resins were used more frequently and when the *provenience*, i.e. the place of origin, and the specific and not measurable (or at least not attempted to be measured) properties of the substance, were tied to a geographical name, and to the shape, shade of color, and even type of packing of the natural or processed material.

Availability

The foregoing considerations bring up the subject of availability—technical and strategic. The pyrotechnic manufacturer or research scientist relies on a producer of powdered materials of specified particle size, since he is rarely equipped to convert a bulk material into powder. Smaller amounts of brittle substances such as oxidizer salts, are sometimes comminuted by passing coarser crystals or powders through various types of mill. In these, the particles are reduced in size by a flailing impact (hammermill) or by shearing action and removed through perforated plates, the continuous process repeated if necessary until the desired average particle size, say of 5 or 10 μ, is achieved.

Ball-milling will also reduce particle size of brittle substances, while with metal powders or actual pyrotechnic mixtures, wet ball-milling may lead to increased reactivity without change in size. This change may be the result of the breaking up of agglomerates or more likely of stripping the metal particles from inhibiting surface layers, and it must be taken into account when ball-milling is used as a mixing procedure.

Physicochemical methods of precipitation can be used for sample quantities, such as mixing a solution of a water-soluble salt with alcohol. However, in the case of metals that are not offered in the trade in form of powder of suitable grain sizes, it is nearly impossible to achieve with simple means a pyrotechnically useful state of disintegration. If the metal is brittle, lathe-cutting may lead to small enough chips. This technical gap frustrates the research-minded who would like to explore systematically such materials as alkaline earth metals and binary or ternary alloys of a great variety of compositions. Again, with more and more metals becoming available for powder metallurgy—even such former rarities as niobium (columbium) or tantalum powder—it may be by mere chance that the existing material has pyrotechnically valuable properties.

Another aspect is the *quantitative* availability if a "new" material becomes so desirable for extended usage that demand exceeds facilities or domestic deposits of raw materials, especially if the larger natural deposits are in potentially hostile territories. During World War II, under enormously increased demands, such materials as magnesium, aluminum powders, antimony compounds, and perchlorates (the latter because of use and irretrievable dissipation of platinum in the electrolytic manufacture) were *critical materials.* (The term *strategic material* pertains to the requirement in support of the war effort as such, while *critical* implies shortage in any one form).

Economy

This chapter on the general properties of materials will be closed with a consideration of questions of economy. Since fuels may be as low in price as a few cents per pound or higher than $10, while oxidizers and binders are generally in the dime to quarters group, the cost of a pyrotechnic mixture hinges mostly on the choice of the fuels. In civilian pyrotechnics, especially for mass-produced competitive items, economy is of course a decisive factor. Even when most of the materials are generally in the pennies to 25 cents per pound category, as in matches, emergency signals, and many fireworks items, the economic aspect is constantly in the mind of the manufacturer.

At the opposite end of the scale is the use of the really expensive materials costing $10 or more per pound. When the pyrotechnician is faced with an unusual, perhaps really new, requirement, his thoughts must be concentrated on the solution of the technical problem and he must leave the question of economy and all other problems, even that of *manufacturability*, as secondary considerations. We are speaking here strictly of research, but the point must be emphasized since the entirely valid demand for *practical* thinking throughout a developmental phase must not obscure the need for bold approaches along novel lines divorced from considerations of hazard, cost, or limited availability of materials. But even in production practice, we nowadays find expensive rare metals used routinely.

That chemicals costing over $10 per pound are considered at all is due to several facts: Pyrotechnic items and, in particular, the special compositions in which those materials are used, are small, amounting on the average to a few grams per unit. Moreover, costs of chemicals may be only a small fraction of costs for the mechanical parts and for the labor used in making the item. The reliability of the first fire, igniter, or delay train that uses these expensive materials will govern the performance of the much larger, or more costly, main item. Within the complexity of modern warfare, as well as in space exploration, the consequence of a failure or inadequate performance of even one signal, marker, actuator, or power source may be far reaching. Since these devices are integral components of airplanes, rockets, large missiles, and nuclear devices, the materials costs shrink into insignificance, as do the costs of establishing performance limits and statistically valid reliability data. This latter aspect of true economy, i.e. the expenditure of adequate sums to cope with the peculiarities of "one-shot" subassemblies, is not always well understood.

From examples presented in earlier chapters, it should be abundantly clear that every substance (especially every fuel) represents an individual that can only be replaced with difficulty by another with which it has some properties in common. Thus, the cheapest fuels, such as sulfur, carbon, asphalt, sugars, and resins (all in the "pennies per pound" class) behave entirely differently from the metals aluminum and magnesium of roughly ten times the cost; these again are different from boron, titanium, and zirconium in the "$10" group. But even within a group, substitutions are often impossible. To take advantage of the lower equivalent weight of titanium as against ziconium or of the higher heat output of aluminum against magnesium can only be done with elaborate adjustment—if possible at all.

chapter 33

Phase Changes

The physical state—solid, liquid, or gaseous—of the products of reaction, as well as some phase transitions of the reactants, prior to chemical interaction, play an important part in the functionality of the reaction, superseding even the caloric output. Pyrotechnically useful heat must be "located" according to one of three patterns: It can be concentrated at the exact place where it is created and flow mainly into the immediate surroundings; it may stay in a flame pattern above the burning zone but remain essentially stationary; or it will be carried convectively to a designated area by motion of particles or gases. Furthermore, to be properly effective, such heat must be concentrated within suitable temperature ranges that may loosely be divided into "low" for the range from below red heat to moderate red heat; "medium" up to incipient white heat; and "high" up to the highest obtainable limit.

The enormous differences in temperature one can create depend (disregarding energy dissipation) on heat output, rate of reaction, and on the physical properties of the ingredients and the products. Of these factors, the *phases* of the products—solid, liquid, or gaseous—are undoubtedly the most important. We need only think of a certain amount of thermite compared with the same amount of an ammonium nitrate propellant, both of identical heat output per unit weight, to appreciate the extremes of behavior due to phase changes.

In this example, the division seems to be fairly clear cut, since the products from the propellant are unequivocally gaseous even in the standard state or only slightly above it, including water vapor. But in a flare we have to ascertain if or to what extent an oxide or chloride is volatilized, and even in a highly refractory system such as thermite when used in technical quantities, some of the aluminum oxide whose bulk is concentrated in a liquid slag may appear as "Rauch" (smoke).[22]

Dislocation of heat is also caused by the small amounts of gas from adsorbed moisture and minor impuritites, as well as the interstitial air between the particles, especially in unconsolidated masses.

Calculated heat output may be diminished, in practice, and the achievable temperature limited by instability of products at high temperatures. Exceptionally, this may take the form of secondary interaction. In former discussions, examples were cited of secondary reactions yielding additional heat. The classical thermite reaction shows the opposite behavior. Concurrent with the main reaction of aluminum and iron oxide to aluminum oxide, Al_2O_3, and iron, the product Al_2O_3 combines with some unused aluminum to form in endothermic reaction the lower oxide Al_2O. If silicon is also present the gaseous silicon (II) oxide, SiO, is formed. Since Al_2O has the low boiling point of 1450°C, its evaporation as well as that of SiO, if present, causes an additional drain on the caloric concentration in the liquefied products. Eventually, the lower oxides may burn in air to the stable higher oxides, but by then the lost energy is recovered away from the desired place.

The effect of melting and boiling points of original materials rather than of products has been touched upon earlier in one of its facets in describing the aerosolization of the oxides of magnesium and aluminum after prior vaporization of the metals. Aside from the special meaning this has for high-intensity white flares, there is the negative aspect that magnesium oxide, notwithstanding its extremely high melting and boiling points and its stability, can dissipate heat by the dispersion of the oxide as if it were a gas. Similarly, the low sublimation point of the combustion product of phosphorus restricts the latter's usefulness as an incendiary.

The liquid state in the reaction zone causes, as a rule, little concern from the viewpoint of heat loss, since the latent heat of fusion is small. It can, however, cause a form of convective heat loss that may have unfortunate consequences. In a flare that burns with the flame upward, such as a railroad fusee, a molten slag is sometimes normally permitted to drop off. By faulty formulation, an excessive liquefaction may take place so that not only the burned-out slag but the molten material in the burning zone itself may slide off and prematurely end the burning of the flare. Similar conditions may occur with flame-down burning candles under violent motion.

Tables 15, 16, and 17 give figures on melting and boiling points of elements, oxides, and halides—starting materials of formulas or products, as the case may be. Table 18 adds a number of highly refractory compounds that may possibly appear as products of highest temperature reaction but are mainly interesting as structural refractory materials.

Melting and boiling points of substances at high temperatures are quoted among others from Brewer,[542] and from a book by Kubaschewski and Evans,[64] the latter providing many useful tables on heat of formation and heat of transformation, heat capacities, and information on calorimetry.

The difficult problems of measuring the properties of substances at very high temperatures have received attention mainly in connection with the study of modern rockets and the need for highly refractory nozzle materials and other structural parts. A book by Kingery[543] is concerned with these problems, as are two collections of articles, both entitled *High Temperature Technology*.[544,545] The literature on refractory materials is steadily growing as evidenced by a series of handbooks, one of them compiling physical constants,[546]

Table 15
Melting and Boiling Points of Elements

Element	*M.p., °C*	*B.p., °C*
Ag	961	2193
Al	660	2327
As	—	610 s
B	2300	2550
Be	1283	(2400)
C	—	4347
Ca	850	1420
Cu	1083	2582
Fe	1535	2800
Li	180	1326
Mg	650	1120
Mn	1244	2087
Mo	2610	4830
Nb (Cb)	2487	4930
Ni	1455	2800
Pb	327	1750
Pt	1769	4010
S	119	445
Sb	630	1440
Se	217	685
Si	1410	2600
Sn	232	2337
Ta	2977	5430
Te	450	1087
Th	1840	4200
Ti	1812	3260
W	3380	5530
Zn	420	907
Zr	1852	(4750)

s = sublimes

and others reporting on Russian compilations on the properties of high temperature materials.[547,548,549] Referring to Table 15:

Figures at very high temperatures may be only approximate even though they are not rounded off. Data about Mo, Nb (Cb), Ta, and W are from a 1960 Data Sheet;[541]

According to references cited by Gmelin, magnesium *sublimes* in a vacuum below 2 mm Hg at a temperature between 500 and 650°C;

The alloy of 54% magnesium and 46% aluminum, actually the intermetallic compound Mg_4Al_3 melts at 463°C according to Shid-

Table 16
Melting and Boiling Points of Oxides

Oxides	*M.p., °C*	*B.p., °C*	*B.p., °K*
Al_2O_3	2030		3800 d
B_2O_3	450		2520
BaO	1925		3000
BeO	2530	4120	
CaO	2600		3800
CeO_2	2600		
Cr_2O_3	2400		3300 d
Cu_2O	1230		
$Fe_{0 \cdot 95}O$	1371		3400 d
Fe_3O_4	1597		2060 d
Fe_2O_3	1457 d		1730 d
K_2O			1750 d
La_2O_3	2320		4470
Li_2O	1430		
MgO		2770 s	
MnO	1785		3400 d
Mn_3O_4	1560		
MoO_3	795	1280	
Na_2O	920		
Nb_2O_5	1460		
NiO	1960		
P_2O_5	580	605	
PbO	885	(1470)	
Sb_2O_3	656	1425	
SiO_2	1713		2250 d
SrO	2460		3500
Ta_2O_5	1870		
ThO_2	3200		4670
TiO	1760		
TiO_2	1920		
Ti_2O_3	2130		3300
WO_3	1473		
Y_2O_3	2420		4570
ZnO	1975		
ZrO_2	2700	(4300)	

d = decomposes
s = sublimes

lovsky,[3] who also shows the complete phase diagram of the two metals.

Referring to Table 16:

Data given in °C are taken from Kubaschewski and Evans;[64] in °K, from Brewer.[542] The latter also gives boiling points of some oxides in high vacuum that are of the order of 1000° lower than at atmospheric pressure. Mellor, in supplements, cites for K_2O: sublimes at 881°C under 600 mm Hg pressure; for Na_2O: begins to vaporize at 1350°C, boils >1600°C. No data are given for Rb_2O and Cs_2O but the melting points for the higher oxides are near 500°C.

Table 17
Melting and Boiling Points of Halides and Chalcogenides

	M.p., °C	*B.p., °C*
LiCl	614	1382
NaCl	801	1465
KCl	772	1407
RbCl	717	1381
CsCl	645	1300
$MgCl_2$	714	1418
$CaCl_2$	782	(2000)
$SrCl_2$	763	(2000)
$BaCl_2$	960	(1830)
$AlCl_3$		180 s
$FeCl_3$	307	319
$FeCl_2$	677	1026
$ZnCl_2$	275	756
$CuCl_2$	498	d
Cu_2Cl_2	430	1366
$HgCl_2$	277	304
Hg_2Cl_2		383 s
AgI	558	1504
PbI_2	412	872
FeS	1193	
Al_2S_3	1118*	
PbS	1114	
PbSe	1065	
PbTe	905	

d = decomposes
s = sublimes
* Sublimes in vacuo at 1100—1250°C; forms AlS, in inert atmosphere at ordinary pressure, at 2100°C. Al_2S_3 in mixture with 27% Al_2O_3 melts at 997°, according to Biltz and Caspari as quoted by Mellor.[67]

Table 18
Approximate Melting Points of Refractory Compounds

Carbides	°*C*	*Silicides*	°*C*
Ta, Nb, Hf	4000	Ta	2400
Zr	3500	W	2200
Ti	3200	Zr	2100
W	2900	Mo	2000
Al	2800	Ti	1500
B	2500	Mg	1100 d
U	2500		
Si	2200 d, 2600		
Borides	°*C*	*Nitrides*	°*C*
Hf	3300	Hf	3300
Ta	3100	Ta	3000
Zr	3000	Zr	3000
Ti	3000	B	3000 d
Nb	2900 d	Ti	2900 d
W	2900	Al	2200 d
Cr	2000	Si	1900 d

d = decomposes

References (Table 18): among others, *Handbook of High Temperature Materials*,[546] a compilation of literature data showing wide variations. Therefore, the above-given figures are rounded-off values indicating temperature *level* rather than observed temperatures. Formulas are omitted but more often than not various reported component ratios such as WB, WB_2, WB_6, W_2B_5 show the same or closely ranging melting points;

The most diverse data exist about the best-known compound silicon carbide (SiC), perhaps because oxidation (in air), dissociation, sublimation, and phase change obscure the picture. No aluminum silicide has been described.

Mixed compounds exist with properties such as resistance to air oxidation, sometimes specially advantageous.[550]

Anyone concerned with the phase changes of materials at very high temperatures is advised to scrutinize the latest original literature for experimental details and confirmation of data in order to obtain the presumably most reliable figures. Not only are measurements at high temperatures beset with difficulties, but it is sometimes dubious what is actually measured. Thus, in the cases of oxides that appear in several oxidation states, incipient decomposition and subsequent solution of one oxide in another may cloud the issue of melting points.

The cooling effect from phase changes has practical applications that, while not directly pyrotechnical, are closely related enough to warrant a few notes. The *cooling of the gases* from a pyro cartridge so as to make them suitable for instantaneous inflation of flotation bags is one example; *ablation*, i.e. the evaporation of protective insulation material to protect nose cones of missiles and other exposed components has come into frequent use. New methods of prevention of *gun erosion* by having the gun propellant gases pass over certain coolants have had striking effects, greatly surpassing the conventional additives to the propellant powder. One of these methods—some others have not been released for general publication—consists of insertion of a rayon cloth, impregnated with a mixture of 45% titanium (IV) oxide (TiO_2) and 55% wax, in the mouth of a cartridge around the periphery of the propellant charge.

When one molten material dissolves in another without chemical reaction, the melting point is lowered, depending on the added amount. If the "solvent" in turn is soluble in the other material, a point of minimum melting point exists. A mixture having this characteristic is called a *eutectic*. The conditions are more complex if the components, beside being mutually soluble in the liquid state, form solid solutions, if they have crystalline transition points, are only partially soluble in each other, or also form one or more compounds.[551] More than two substances may also form eutectics. Up to now, such behavior, while very important in metallurgy, had little bearing on pyrotechnics, but it begins to draw attention where oxidizer salt mixtures are concerned. The consolidation of fuel-oxidizer mixtures by melting is seldom performed, the main reason probably being the difficulty of controlling the melt temperature below reaction or prereaction temperature and the consequent formidable explosion hazard. However, a study of eutectic mixtures of oxidizers discloses low-melting combinations that can be used reasonably safely under special circumstances. Examples are the consolidation of chlorate mixtures (with small amounts of relatively inert fuels) as oxygen sources or the pelletizing of oxidizers after melting and "shotting," useful for castable propellants.[552] Since melting point and reactivity often seem to be intimately connected, the use of such mixed oxidizers, not necessarily melted together with the fuel, may sometimes be of interest. One of Bowden's patents[370] illustrates this point: The eutectic of the nitrates of potassium, sodium, and lithium is claimed to produuc a black powder burning as much as ten times faster than the standard material. Table 19

Table 19
Melting Points of Nitrates and Their Eutectics

*Weight % of Nitrates**	*°C*
100 Li	255.0
100 Na	308.0
55 Li, 45 Na	198.0
100 K	333.0
34 Li, 66 K	129.0
45 Na, 55 K	225.7
30 Li, 16 Na, 54 K	119.0
100 NH_4	170.0
78.2 NH_4, 21.8 Na	121.1
66½ NH_4, 21 Na; 12½ K	118.5
100 Urea*	133.0
18.8 Li, 81.2 Urea	71.0
50 Li, 50 Urea	98.5
100 Ag	212.0
25 Li, 75 Ag	171.5
20 Li, 61 NH_4, 19 Ag	52.0
8 Na, 55½ NH_4, 36½ Ag	95.0
12 Na, 67½ NH_4, 20½ Ag	94.0
100 Tl (monovalent)	206.0
12½ Li, 87½ Tl	136.5
100 Rb	310.0
18 Li, 82 Rb	154.0
46½ Li, 53½ Rb	179.5
32 Na, 68 Rb	178.5
100 Cs	414.0
32 Li, 68 Cs	174.0
100 Ca	561.0
70 Li, 30 Ca	235.2
30½ Li, 40 Na, 29½ Ca	170.3
23 Li, 62 K, 15 Ca	117.4
11.8 Na, 39.4 K, 48.8 Ca	133.0

* Figures preceding chemical symbol are weight %, while the symbol stands for the metal *nitrate*. "Urea" stands for itself, *not* for the nitrate.

Data are taken from Technical Data Sheet TD-105 (Foote Mineral Company) and Mellor, especially from Supplement II, Part II, p. 1250 ff., where eutectics with alkali dichromates of m.p. 202—216°C, potentially interesting pyrotechnically, are also found.

gives a list of various mixtures of nitrates, together with the melting points of the single salts, taken from several sources such as a Foote Mineral Company report,[553] Mellor,[67] and others.

A special study of interest is the one by Vitoria[554] on mixtures of the chlorates of sodium and potassium. According to these experimental series, the eutectic point lies at 232°C for the mixture with 75% sodium chlorate. Otherwise, information on eutectics of halates or perhalates is sparse. Hogan and Gordon[555] have reported on eutectics of lithium nitrate or sodium nitrate and potassium perchlorate of M.P. 230 and 245°C respectively, and on those with alkaline earth nitrates. Ammonium and lithium perchlorate

($32\frac{1}{2}$: $67\frac{1}{2}$) melt at 182°C; lithium nitrate and perchlorate (36:64) at 172°C. All the percentage figures here and above are weight percent, not mole percent.

The physical integrity of a pyrotechnic formulation may, exceptionally, be threatened by evaporation. This can occur when a rather volatile compound is a necessary component because of unique properties that make it difficult or impossible to replace it with a more stable ingredient. Two such compounds have been described earlier: One is metaldehyde of a sublimation point of about 112°C and found only in novelty items such as repeatedly ignitible matches, hence of little concern in commercial practice; the other is the HC smoke ingredient hexachloroethane. Its vapor pressure, as evidenced by the odor at room temperature, is sufficient to cause concern. In practice, however, with tightly compressed and sealed mixtures in substantial units, the volatility of HC seems to cause no trouble.

Because we have earlier included long-term delays based on the dissolution or softening of celluloid, it should be pointed out that this solid solution of nitrocellulose and camphor alters its physicochemical behavior with time, i.e. with loss of camphor, but a "plateau" of stability seems to be reached after some aging.

Highly hydrated salts have been proposed in mixtures where the water is intended to be an essential reactant as oxidizer.[556] Such salts are obviously not stable, but they may form lower hydrates that are stable even at moderately elevated temperature and might be safely used. Undoubtedly, those hydrates that are the normal products of commerce, such as the vitriols, borax, etc., would be dubious candidates for any pyrotechnic formula.

chapter 34

Caloric Output

Significance of Caloric Output

The requirement for a specified quantity of calories to be delivered and therefore exactly measured is the exception rather than the rule in the applications of pyrotechnic heat production. However, with the mounting interest in the basic behavior of pyrotechnic heat sources and their relation to radiant output, dissemination, and other objectives, the question of how many calories a fuel or a mixture delivers is frequently in the mind of the modern practitioner. Heat output has the advantage that (as with physical constants) it often can be measured exactly. However, as has been emphasized previously, the number of calories as such is only one of the factors that determine usefulness or efficacy of a mixture.

Calorimetry

One determines the heat of combustion by reacting the substance with pure oxygen following electric initiation in a so-called bomb calorimeter and measuring the amount of released heat by the temperature rise of water in which the vessel is immersed. It is a painstaking but not particularly difficult procedure that need not be described here in detail. Normally used for finding the fuel value of coals and other organic substances, the method gives the heat of formation of an oxide formed from an element or the heat of combustion of a compound. In either case, this value, with some possible minor reservations, is also the heat of combustion in air. As a rule, the pyrotechnician has no need to perform such measurements himself, since the data for most substances are well known from scientific investigations.

Chemical compounds furnish on combustion a heat output that is either larger or smaller than it would be if the combustible chemical elements contained therein were burned in elementary state. As a rule the heat output is smaller, and this means, according to the laws of thermodynamics, that the formation of the compound itself from the elements—if this were possible, and it may or may not be in practice—takes place with loss of energy to the surroundings, which is the meaning of heat release. Such a compound is called an *exothermic compound.* In the opposite case, we speak of an *endothermic compound,* one that requires heat input for its formation and will release such heat on decomposition into the elements. By the same token, the heat of combustion of such a compound is then larger than that of its elementary components. Endothermic compounds are usually, but not always, unstable or easily decomposed; their additional heat increment makes them useful in pyrochemical reactions, as discussed for high-temperature flames (Chapter 27). Other endothermic compounds are valuable as primary explosives.

Taking by convention the heat of formation of all elements in the *standard state* (explained below) as zero, the heat of formation of a compound can be calculated from heat of combustion data. Again, the pyrotechnician will use established values from previously quoted sources but must be careful in the use of such figures to observe their positive or negative character: An exothermic compound has, *for the scientist,* a negative heat of formation because the system *loses* energy, and the same applies to an exothermic *reaction.*

Assuming now that we can ascertain by a chemical equation the qualitative and quantitative course of the reaction, we are able to calculate the heat released during the reaction by deducting the sum of the heats of formation of the products from that of the original components of the system. It does not matter whether one uses the "plus" or the "minus" designation for exothermy as long as one is consistent, but since the minus data have now become entrenched in all modern tables, it may be less confusing to use the modern system,* at least for calculations. The scientific data refer to *standard states* of all substances. By definition, or rather by convention, according to Lewis and Randall as quoted in NBS Circular 500, this is the state solid, liquid, or gaseous) in which the substances exist at 25°C and at a pressure of one atmosphere.

To express heat output in reference to the standard state of the products may be proper for essentially gasless reactions where solids are converted into other solids (or, requiring only minimal corrections, into the molten state). However, in those high temperature processes where seemingly nonvolatile reaction products such as

* However, in the equations presented in this book, the author has followed pyrotechnic practice and used the old system with a parenthetical explanation.

sodium oxide, strontium or barium chloride, and even aluminum oxide appear at least partially in vapor form, corrections are necessary that may be of the order of 10% in lowering the expected released thermal energy.

Let us ask why the pyrotechnician should be at all concerned with calorimetric measurements if a few minutes' effort looking up the figures for heats of formation and some simple additions and subtractions seem to yield the desired data. The reason is, first of all, that some mixtures (and more so their products) are so complex that satisfactory equations cannot be written without performing extremely tedious analyses of the products. It may even be impossible to come to proper conclusions about the nature of the products in a solid conglomerate. Moreover, surprising deviations or suspected deviations from expected courses of reactions occur even in fairly simple binary mixtures, as will be shown in examples later in this chapter. We may add that even calorimetry may not provide an adequate answer if the much larger amounts of reactants in practical use raise the level of temperature so much that the reaction takes a different course and secondary interactions change the character of the pyrochemical process from that taking place with the small amounts in a calorimeter.

To return now to practical measurements of heat of reaction, it is obvious that for experimental determination, one need only enclose the reactants in a calorimetric "bomb," initiate the reaction, and measure (as described briefly above) the amount of heat released. One complication is that for a rather inert system, the customary electrically heated wire loop, which requires only a trifling correction, must be enhanced by a first fire mixture, thus reducing precision even if corrections are applied. Some mixtures may be entirely unreactive in the small amounts accommodated in the standard calorimeters. Undoubtedly, everyone concerned with these problems has had some ideas on designing a "55 gallon drum" type calorimeter with or without admission of external air, but the author is not aware of any published experimentation along these lines.

In order to prevent any preferential reaction with oxygen during interaction of solids or liquids in a pyrochemical system, one not only omits pressurization of the vessel with oxygen but removes the air by purging the container repeatedly with argon gas. Even in the small (and therefore generally preferred) peroxide calorimeter (such as the Parr Calorimeter Model 1401), the volume of only 23 ml of air space may give rise to inaccuracies due to competitive oxidation reactions that are small but also unpredictable.

The peroxide calorimeter is so named because routine heat of combustion values are obtained by means of a reaction with solid sodium peroxide in lieu of gaseous oxygen under pressure, which is of course a great convenience and a safety feature. Because of its compactness and small gas volume, the peroxide calorimeter (without peroxide) is now generally used for heat of reaction measurements. HDL scientists who have given much attention to the problem of reliable and reproducible "heat powder" measurements[557,558,559] have added refinements to the above-named Parr Calorimeter, which led to the design marked Model 1411.

The peroxide calorimeters would be little suitable for exact determination of the heat of combustion of pyrochemically interesting fuels using sodium peroxide as the oxygen source because of secondary reactions and the need for too extensive corrections. This calorimeter was ignored in a detailed book on calorimetry,[560] and, to show the gulf that still existed between pyrotechnics and scientific thermochemistry only ten years ago (1956), the chapter on "reactions other than combustion" does not even hint that there could be such a thing as a pyrotechnically interesting reaction that merits inclusion.

Heat of Combustion of Elements and Compounds

The figures for the heat of combustion as such play a subordinate part in pyrotechnics. They are of importance in those formulations that burn in air and contain an excess of a fuel beyond the equivalent amount of oxidizer. The excess fuel is most often magnesium or an organic binder. Also, in many incendiaries, the major fuel such as magnesium or magnesium alloy bodies as well as most of the incendiary filler materials (kerosine, phosphorus) burn in ambient air. Table 20 furnishes a list of elements and Table 21 a list of compounds whose caloric output and other data may be of interest. Figures for the heat output per unit volume are theoretical, realized only if the substance burns as a compact non-porous solid or liquid.

In scanning these tables, one finds two notable facts: one is the superiority of hydrogen, which is reflected in the high heat output of compounds high in hydrogen content, such as certain inorganic hydrides and also of hydrocarbons; the other fact is the good showing that metals of very high density make when considered on a volume rather than on a weight basis. In addition, the oxygen requirements of these elements of high equivalent weight are extremely small.

Table 20
Heat of Combustion of Elements in Oxygen

Material	*Product*	*kcal/g*	*kcal/cm³*
H	H_2O [g]	28.7	(2.0)*
Be	BeO	16.2	29
B	B_2O_3	14.0	33
Li	Li_2O	10.4	5½
C	CO_2	7.8	
C	CO	2.2	
Al	Al_2O_3	7.4	20
Si	SiO_2	7.3	18
Mg	MgO	5.9	10
P (white)	P_2O_5	5.9	11
P (red)	P_2O_5	5.8	13
Ti	TiO_2	4.6	20
Ti	Ti_2O_3	3.8	17
Ca	CaO	3.8	6
Zr	ZrO_2	2.8	18
Na	Na_2O_2	2.6	2½
Nb (Cb)	Nb_2O_5	2.5	21
S	SO_2	2.2	5
Mn	Mn_3O_4	2.0	14½
Ce	CeO_2	1.9	13†
K	KO_2	1.7	1½
Fe	Fe_3O_4	1.6	12½
Ta	Ta_2O_5	1.4	23
Th	ThO_2	1.3	14
Zn	ZnO	1.3	9
W	WO_3	1.1	21
Ni	NiO	1.0	9
As	As_2O_3	1.0	6
Sb	Sb_2O_3	0.7	4½

* Liquid hydrogen of d=0.0709.
† Based on $-\Delta Hf^\circ = 260.18$ kcal/mole CeO_2, according to *J. Am. Chem. Soc.* 75, 5645-7 (1953); older value reported was 233.

Heat of combustion figures, no matter how large, do not imply that the substance can be ignited and will burn in air. Of the fuel metals, aluminum will not burn in air except as finest powder or in flake form, and then only if blown into a flame. Even at 1000°C, the heated powder will otherwise only gradually take up oxygen from the air to about the point corresponding to the formula AlO.[67] The reason is probably a combination of high heat conductivity and the adherence of the refractory oxide. Foil or wire burn, however, in pure oxygen even at diminished pressure as in a flash bulb.

A similarly reticent behavior is shown by amorphous boron that will only burn partially even under compressed oxygen.[561,562] This abnormality is caused by the formation of a glassy boron (III) oxide that envelops the unburnt portion of the element except, according to

Table 21
Heat of Combustion of Various Materials

	kcal/g
Gasoline, Kerosine, Paraffin, Petrolatum	11
Paraffin Oil, Edible Oils, Asphalt, Stearic Acid	9½—10
Anthracite	8½—9*
Bituminous Coal	7—9*
Resinous Wood Extracts	7½—8½
Lignites	7*
Coke	6—8
Charcoal	6½—7
Casein, Animal Glue	6
Wool (air-dry)	5—5½
Peat	4—5½*
Conifer Twigs, Pine Needles	5
Wood (dry)	4½—5
Dry Grass, Sphagnum, Ferns	4
Cellulose, Fine Papers, Cotton, Sugar	4
Wheat Straw (dry)	3½

* Dry, on a mineral matter-free basis. Ash content may vary greatly but is 5—10% for many grades. Because of variations in nature of materials and sometimes in moisture content, only round numbers within about 5% of cited values from various sources are given.

Rizzo,[563] if the oxidation takes place above 900°C, when the oxide volatilizes. The oxidation phenomena of boron are further complicated by the formation of a volatile suboxide, perhaps B_7O, formed above 1100°C. Some form of a suboxide has also been postulated as being part of the technical boron made by reduction of the normal oxide with magnesium. In the *Rare Metals Handbook*,[516] Cooper reports briefly some of these findings and cites original literature. Technical amorphous boron contains about 89% boron, several percent of magnesium, some minor imPurititties, the rest oxygen.

The partial inertness of aluminum and boron on burning in air does not affect the reactions with solid oxidizers, which are prompt and complete.

Aluminum displays another form of unexpected pyrochemical behavior that may affect its heat output on combustion but is more interesting in other aspects. One might think that *alloying* has little influence on the enthalpy of the mixture, but aluminum forms intermetallic compounds in strongly exothermic reactions of which three examples are given, even though only one of them concerns the subject of combustion in air: In alloying magnesium and aluminum, the compound Mg_4Al_3 (54% Mg, 46% Al) is formed. This reaction yields 48.7 kcal/mole or 0.38 kcal/g of components. Thus the heat output from combustion of the alloy in a firebomb will be smaller than that of the two metals in physical mixture though the deficit is not large. In another example, however, the exothermy of the "alloying" itself becomes a heat-producing process in its own right, as described in Chapter 24 for the item called Pyrofuze. It occurs in the reaction between palladium and aluminum described as furnishing 327 cal/g of components. (A ratio of 3 parts of palladium to 1 part of aluminum has been indicated in one data sheet[564]) The reaction proceeds with disintegration of the palladium clad wire, sheet, or granule with white-hot sparks produced not by burning but derived entirely from formation of an intermetallic compound.

Mellor[67] cites the phenomenon of a temperature rise of 30 g of molten gold at 1155°C after addition of only 0.3 g of aluminum. The latter, being cold, caused an initial drop to 1045°, whereupon the temperature rose to 1380°C. A perusal of Circular 500,[63] as well as of the books by Kubaschewski and Evans[64] and Kubaschewski and Caterall,[564a] shows numerous strongly exothermic reactions between metals as expressed by the heat of formation of arsenides, antimonides, bismuthides, aluminides, and others.

Returning now to the subject of combustion in air, a few examples will be given demonstrating the high heat of combustion that is generally achieved with compounds rich in hydrogen. In the case of the boranes, the heat of combustion is of the order of 17 kcal/g. The fact that the boranes are endothermic compounds adds slightly to the heat of combustion. Conversely, the exothermic formation of the silanes (11.3, 36.2, 54.4, and 70.8 kcal/mole for the first four homologous members) deducts a little from their heat of combustion;[565] it is 9—10 kcal/g. Lithium aluminum hydride ($LiAlH_4$), lithium borohydride ($LiBH_4$), and aluminum borohydride [$Al(BH_4)_3$] also display prodigious heats of combustion of 10, 14, and 14 kcal/g respectively. These compounds have no place in present day pyrotechnics, but searchers for high-energy propellants have considered them since the energy output of some exceeds that of the common liquid hydrocarbon fuels.

In the case of the simple metal hydrides, the hydrogen content is small and the exothermic heat of formation of the hydride becomes significant. Thus calcium hydride (CaH_2) will do little better on combustion than calcium metal.[566] In the heavy metal hydrides, the conditions are similar. The exothermic heat of formation of all pyrochemically applicable metal hydrides such as of calcium, zirconium, and many rare earth metals is of the order of 40 kcal/mole, according to Mikheeva*.[79]

The simple hydrides of aluminum, gallium, and indium (AlH_3, GaH_3, InH_3) will be mentioned merely because their names may be entirely unfamiliar, i.e. *alane*, *gallane*, and *indane*, to which must be added the incomparable neologism *beane* (BeH_2) that in any other context is a bit puzzling.

Combustion in Nitrogen, Carbon Dioxide, and Water Vapor

If we extend the term "atmosphere" to include the nitrogen, carbon dioxide, or water vapor in the air, the behavior of the most energetic fuel metals in relation to these oxidizing agents becomes of interest. The atmospheres of the planets Mars and Venus may consist essentially of mixtures of carbon dioxide and nitrogen, which has stimulated a study on the subject.[568] Data therein refer mainly to ignition temperatures, which appear to be higher than in air for carbon dioxide and still higher for nitrogen.

The heat of combustion in nitrogen, carbon dioxide, or water vapor compared with combustion in oxygen and calculated in the same manner, i.e. per gram metal only, is considerably smaller. For magnesium, the figures are 1.5 kcal in nitrogen, 4.0 kcal in carbon dioxide, and 3.5 kcal in water vapor, compared with 5.9 kcal in oxygen. Some implications of these three types of reaction will be found in the following chapter.

Thermitic Reactions

Next to combustion in air, the simplest reactions are those between an element and an oxide, generally called thermitic reactions after the thermite reactions repeatedly mentioned earlier. Table 22 shows a sizable number of such combinations. This table is merely a handy reference for expected heat output on the basis of grams or (theoretical) cubic centimeters of combined reactants and gives no information on the character of the reaction or even if it takes place at all, though the reactions of smallest heat output can be expected to be incomplete or difficult to perform.

An approximate gage for the reactivity of aluminum with various oxides is the difference between the equivalent heats of formation of aluminum oxide (66 kcal) and the oxide to be reduced: If this difference exceeds 40 kcal (as is the case for PbO, Bi_2O_3, CuO, and PbO_2)

* For $TiH_{1.73}$, Δ Hf° is —31.1 kcal/mole according to Sieverts and Gotta,[567] misprinted as 3.1 in the quoted AEC translation.

Table 22
Thermitic Combinations of Elements with Oxides

	PbO_2		CuO		MnO_2		MoO_3		Fe_2O_3	
	kcal/g	*kcal/cm*^{3}	*kcal/g*	*kcal/cm*^{3}	*kcal/g*	*kcal/cm*^{3}	*kcal/g*	*kcal/cm*^{3}	*kcal/g*	*kcal/cm*^{3}
Be	0.88	6.4	1.23	6.3	1.60	6.2	1.51	5.5	1.29	5.3
Al	0.73	5.1	0.98	5.0	1.15	4.6	1.10	4.2	0.95	4.0
Th	0.48	4.9	0.56	4.8	0.53	4.4	0.52	4.1	0.48	3.9
Zr	0.58	4.8	0.74	4.7	0.75	4.2	0.74	3.9	0.64	3.7
Ti	0.59	4.5	0.76	4.4	0.80	3.8	0.78	3.5	0.67	3.3
B	0.53	4.3	0.73	4.1	0.76	3.3	0.73	3.0	0.58	2.7
Mg	0.77	4.1	1.03	4.1	1.20	3.6	1.16	3.4	1.01	3.2
Ta	0.35	3.9	0.41	3.8	0.32	2.9	0.33	2.6	0.27	2.4
Si	0.52	3.7	0.69	3.6	0.70	2.8	0.69	2.6	0.55	2.3
Ca	0.74	3.1	0.96	3.0	1.07	2.6	1.04	2.5	0.93	2.4
Mn	0.34	2.9	0.41	2.8	0.40	2.4	0.31	1.7	0.24	1.5
Li	0.82	2.8	1.13	2.7	1.40	2.3	1.33	2.2	1.14	2.1
W	0.19	2.1	0.21	1.9	0.04	0.4	0.06	0.5	0.02	0.2
Mo	0.19	1.8	0.22	1.6	0.03	0.2				
	WO_3		Fe_3O_4		Pb_3O_4		Cu_2O		Cr_2O_3	
	kcal/g	*kcal/cm*^{3}	*kcal/g*	*kcal/cm*^{3}	*kcal/g*	*kcal/cm*^{3}	*kcal/g*	*kcal/cm*^{3}	*kcal/g*	*kcal/cm*^{3}
Be	0.92	5.0	1.18	4.9	0.56	4.3	0.70	3.7	0.94	3.8
Al	0.69	3.8	0.87	3.7	0.47	3.5	0.58	3.1	0.63	2.6
Th	0.41	3.7	0.46	3.7	0.36	3.5	0.41	3.1	0.34	2.8
Zr	0.51	3.5	0.60	3.4	0.39	3.3	0.47	2.9	0.41	2.3
Ti	0.51	3.1	0.62	3.0	0.39	3.0	0.47	2.7	0.39	1.9
B	0.40	2.4	0.52	2.4	0.32	2.6	0.40	2.3	0.19	0.9
Mg	0.76	3.1	0.94	3.1	0.51	3.0	0.62	2.7	0.72	2.3
Ta	0.22	2.2	0.26	2.1	0.23	2.4	0.28	2.1	0.08	0.7
Si	0.39	2.1	0.50	2.1	0.32	2.4	0.40	2.1	0.20	0.8
Ca	0.72	2.3	0.87	2.3	0.51	2.4	0.61	2.3	0.68	1.7
Mn	0.19	1.4	0.22	1.3	0.21	1.8	0.26	1.7		
Li	0.83	2.1	1.05	2.0	0.53	2.2	0.65	2.1	0.81	1.5
W					0.10	1.0	0.13	1.0		
Mo	0.06	0.5			0.10	0.9	0.13	0.9		

Figures for these and many other data on binary mixtures were taken from tabulations in a report by Raisen, Katz, and Franson,[569] except that the figures were rounded off. Only exothermic reactions are reported. Arrangement according to heat output on a volume basis is approximately in descending order from the left upper corner down to the right.

the reaction is apt to be excessively vigorous, even violent; around 33 kcal (the rounded-off value for the reactions with iron oxides) normal behavior is indicated; much below this value (as for WO_3, Ta_2O_5, TiO_2, or ZrO_2) the reaction is normally not completed. This subject has been more thoroughly discussed in Chapter 30 where the technological aspects of the thermite process are treated.

Reactions with Alkaline Earth Chromates

The reactions of the most active fuel-type metals such as zirconium and of the element boron with the chromates of calcium, strontium, and barium seem to be closely related chemically to the thermite reaction. They are essentially gasless and can be regarded as reactions between a metal and metal oxides. However, they do not lead to separable metals.* As mentioned in Chapter 26, these mixtures are very important for the delivery of an exact number of calories and are therefore useful for the delicate heat transfer to molten-electrolyte cells and for other purposes. The suitability of these combinations stems from a number of advantageous properties such as very low gas evolution; formation of a firmly sintered residue; a combination of great ease of ignition and more or less rapid (but mostly not violent) reactivity; and, best of all, delivery of an exact, predeterminable amount of heat in unbroken series in the most useful 300—500 cal/g range (and also in higher ranges up to 850 cal) from binary mixtures through mere change in the ratio of components.

Mixtures of this type were originally developed as first fires and delay compositions. They contained 5-10% boron, and barium chromate. As actual heat sources, the mixtures of zirconium and barium chromate were developed and were known under designations such as Z-2 or Z-3. These are often used as "heat paper" on a substratum of silica fibers.

The stoichiometric relations of the fuel and the oxidizer are obscured by two facts: Both boron and zirconium are impure and (especially in the case of boron) the actual composition of the fuel is unknown; uncertainty exists also as to the products of the reactions and the enthalpies of possible or probable secondary products such as borates and zirconates. The primary and textbook reactions might be written as follows:

* This is not true under all conditions. Compare the previously reported reactions of all alkali chromates with an excess of zirconium that produce alkali metal vapors (Chapter 30).

$$2B + 2BaCrO_4 \longrightarrow B_2O_3 + 2BaO + Cr_2O_3 \quad (1)$$

$$2B + BaCrO_4 \longrightarrow B_2O_3 + BaO + Cr \quad (1a)$$

$$3Zr + 4BaCrO_4 \longrightarrow 3ZrO_2 + 4BaO + 2Cr_2O_3 \quad (2)$$

$$3Zr + 2BaCrO_4 \longrightarrow 3ZrO_2 + 2BaO + 2Cr \quad (2a)$$

With allowances for the actual metal content of zirconium (a small correction is also necessary for the hafnium content of about 2¹/₂%) and postulation of barium zirconate formation (see Table 27), reasonable accord between calculated and measured caloric output is established.[569a] However, the situation is more complex with boron mixtures where one encounters increase of heat output with increase of the percentage of boron in the mixtures much beyond the amounts of Equation (la). Thus, even with the reasonable assumption of secondary barium borate formation, the stoichiometry and heat output of the mixtures with more than about 10% of technical boron (theoretical 8%) is obscure. Chromium boride formation may be a factor.

In a series of experiments and measurements performed in the writer's laboratory, it was shown that the range of heat outputs can be vastly extended by using strontium and (even better) calcium chromate in lieu of the barium chromate (Table 23). Because of the lower equivalent weight of the alkaline earth "ballast" in the calcium chromate, the peak calories per gram for series of mixtures with boron

Table 23
Heat Values of Boron or Zirconium with Barium Chromate or Calcium Chromate

	Remainder:			*Remainder:*	
% B	*$BaCrO_4$, cal/g*	*$CaCrO_4$, cal/g*	*% Zr*	*$BaCrO_4$, cal/g*	*$CaCrO_4$, cal/g*
4	360	—	15	300	—
5	410	630	20	387	375
8	470	685	25	437	475
10	510	700	30	470	570
12		770	35	490	610
15	567	785	40	505*	630
17	573*		45	485	650
18		830	50	455	670*
20	537	860*	55		655
25	515	820	60		605
30	507	800	65		540
35		775			

* Peak values. Because caloric output of production quantities in many of these mixtures can be controlled within ± 5 cal/g the values are given in small calories.

or zirconium lie at 0.86 and 0.67 kcal compared with 0.57 and 0.50 for barium chromate. It may be that earlier students of the problem were deterred by the formula $CaCrO_4 . 2H_2O$ given for the calcium chromate in the handbooks. It turned out, however, that the commercial pigment is the anhydrous compound. Traces of water and perhaps other volatile impurities are easily removed by heating the calcium chromate to bright red heat for several hours. (A trifling superficial reduction, causing a greenish discoloration, may occur.) It should also be noted that on subsequent slurrying with water, the calcium chromate remains anhydrous—an interesting parallel to the behavior of calcium sulfate.

The high caloric mixtures with calcium chromate are excellent fire transfer compositions and are used in squibs and as "gasless" heat powders in general.

Reactions of the Chalcogens

According to the dictionary definition, the chalcogens ("ore-formers") are oxygen, sulfur, selenium, and tellurium, but in chemical usage oxygen seems to be generally omitted from the group. Several metals and sulfur combine quite vigorously under glow. The best known example is that of fine iron powder and powdered sulfur. This conversion of the mixture under bright glow into something that is not magnetic and from which sulfur is not extractable by carbon disulfide has long been the basic object lesson for demonstrating the difference between a mechanical mixture and a chemical compound. Less well known is the fact that flake aluminum and sulfur behave quite similarly, in fact, under a heat evolution that on a weight basis is nearly as large as for thermite.

Lead and selenium or tellurium, of practical importance in blasting cap initiation,[570,571] show such low heat output that one wonders that they actually are pyrotechnically useful. However, the calculated figures are somewhat distorted because of the high equivalent weight of lead. The enthalpies per gram mole are as large as for iron and sulfur, i.e. of the order of —22 kcal. Table 24 shows calculated values from heat of formation data for interesting metal/chalcogen combinations.

While the chalcogens play the part of oxidizer in these reactions, they are also useful as fuels. Here, instead of being reduced to the minus two valence state, they are oxidized to the plus four or plus six state.

Table 24
Exothermy of Metal/Chalcogen Reactions

	S		Se		Te		
Metal	*kcal /g*	*kcal /cm³*	*kcal /g*	*kcal /cm³*	*kcal /g*	*kcal /cm³*	*Products*
Mg	1.5	2.8	0.60	2.1	0.30	1.4	MgS, MgSe, MgTe
Al	0.8	1.8					Al_2S_3
Zn	0.5	2.0	0.20	1.3	0.20	1.0	ZnS, ZnSe, ZnTe
Mn	0.6	2.1	0.20	1.2			MnS, MnSe
Mo	0.4	1.3					MoS_2
Fe	0.3	1.0					FeS
Co	0.2	0.9	0.10	0.4			CoS, CoSe
Ni	0.2	0.8	0.10	0.4			NiS, NiSe
Cu	0.1	0.5					CuS
Pb	0.1	0.7	0.06	0.5	0.05	0.4	PbS, PbSe, PbTe

Note: Data reported by different authors vary, some considerably. A high value for the molar heat of formation of Al_2S_3 is cited by Kubaschewski and Evans[64] from a Russian source, which also should cast doubt on the corresponding values for Al_2Se_3 and Al_2Te_3.

In such mixtures of practical importance as selenium or tellurium with barium peroxide, secondary neutralizing reactions increase the heat output due to selenate and tellurate formation. Johnson[572] has postulated the following equations:

$$Se + 3BaO_2 \longrightarrow BaSeO_4 + 2BaO + 95.3 \text{ kcal (exothermic)}$$

$$2Se + 2BaO_2 \longrightarrow BaSeO_4 + BaSe + 121 \text{ kcal (exothermic)}$$

Per gram of reactants, the heat outputs are 0.16 and 0.24 kcal respectively.

High Calorific Reactions with "Salt-Type" Oxidizers

Nitrates, chlorates, and perchlorates are high in oxygen content and more or less low in the amount of "ballast" they carry in the form of metal oxide and nitrogen, or as chloride. The energy requirement for their decomposition is small for nitrates and zero (actually, very slightly negative) for perchlorates. With chlorates, the decomposition proceeds under energy release. For these reasons, the mixtures of these salts with certain metals furnish the highest heat outputs of all combinations—a fact utilized in flare formulas, ignition pellets, and numerous other applications.

Tables 25 and 26 give a selection of such mixtures, though without indication of actual usefulness. The practitioner will notice that the mixtures of the more common materials he deals with yield from 2.0—2.5 kcal/g or 5.0—6.5 kcal/cm³ (theoretical) for the most energetic combinations. The scientist will take note of the considerably higher

Table 25
Elements and Nitrates

Nitrate Product	N_2 plus	Li Li_2O		Na Na_2O		Na Na		K K_2O		K K		Ba BaO	
		kcal/g	*kcal/cm*^{3}	*kcal/g*	*kcal/cm*^{3}	*kcal/g*	*kcal/cm*^{3}	*kcal/g*	*kcal/cm*^{3}	*kcal/g*	*kcalcm*^{3}	*kcal/g*	*kcal/cm*^{3}
Fuel	*Product*												
Be	BeO	3.7	7.8	2.9	6.1								
Al	Al_2O_3	2.5	6.3	2.1	5.1			1.8	4.0	1.8	4.2	1.6	4.9
Mg	MgO	2.4	4.9	2.1	4.1	2.0	4.0	1.8	3.4	1.8	3.4	1.6	4.1
Zr	ZrO_2	1.5	6.0	1.4	4.7	1.3	4.7	1.2	3.8	1.1	3.9	1.1	4.7
Ti	Ti_2O_3	1.8	5.6	1.5	4.4			1.2	3.5			1.1	4.4
B	B_2O_3	2.4	5.7	1.9	4.2*			1.5	3.2			1.3	4.2
Si	SiO_2	3.3	4.9	1.7	3.7			1.3	2.9			1.2	3.7

* 5.4 for borate formation

Table 26
Elements with Chlorates or Perchlorates

	$Mg(ClO_4)_2$		$LiClO_4$		$NaClO_3$		$KClO_4$		$KClO_3$	
Fuel	*kcal/g*	*kcal/cm³*	*kcal/g*	*kcal/cm³*	*kcal/g*	*kcal/cm³*	*kcal/g*	*kcal/cm³*	*kcal/g*	*kcal/cm³*
Be	4.0	9.5	4.0	9.1	3.4	7.9	3.3	7.9	3.0	6.6
Al	2.9	7.7	2.9	7.4	2.6	6.6	2.5	6.5	2.3	5.6
B	2.9	7.4	2.9	7.0	2.5	6.1	2.4	6.0	2.2	5.0
Zr	1.8	7.3	1.8	7.0	1.6	6.2	1.6	6.2	1.5	5.4
Ti	2.1	7.0	2.1	6.7	1.9	5.9	1.8	5.9	1.7	5.1
Mg	2.8	5.9	2.8	5.7	2.5	5.2	2.4	5.2	2.3	4.7
Mn	1.2	5.1	1.2	4.8	1.1	4.4	1.0	4.3	1.0	3.8
W	0.8	4.9	0.7	4.6	0.7	4.1	0.7	4.0	0.7	3.4
Mo	1.0	4.4	1.0	4.1	1.0	3.7	0.9	3.6	0.8	3.1
Fe	1.0	4.3	1.0	4.0	1.0	3.7	0.9	3.6	0.9	3.1
Ca	2.3	4.2	2.2	4.1	2.1	3.9	2.0	3.8	1.9	3.6
Li	3.6	4.1	3.5	3.8	3.0	3.6	2.9	3.6	2.7	3.3
C	1.4	3.6	1.4	3.3	1.2	3.0	1.2	2.9	1.1	2.5
S							0.7	1.7	0.7	1.5

Products are the normally stable oxides and chlorides—for carbon, CO_2; for sulfur, SO_2. Figures for sodium perchlorate would be about halfway between those for lithium and potassium perchlorate.

output for mixtures that include the metal beryllium or the oxidizers not normally employed, magnesium or lithium perchlorate. Secondary reactions may occur, more likely in confined systems than in flares. What the actual reaction products are and hence the true stoichiometry and enthalpy changes at the moment when the useful pyrophysical phenomena take place may be dubious.

Iodates, though formally analogous, are not comparable to chlorates in thermochemical behavior since the conversion of an iodate into the iodide requires considerable heat input. The expected caloric effect of a mixture with a metal would be more comparable to the reaction of a nitrate. The behavior of the periodates is rather obscure and no figures about their heat of formation in solid state appear in Circular 500. Some metaperiodates ($Me^{I}IO_4$) have been reported as explosively decomposing on heating.[60,67] Stable, solid iodine pentoxide (I_2O_5) is an exothermic compound (heat of formation —43 kcal/mole).[60] Its pyrochemical possibilities seem not to have attracted any attention.

The reactions of the nitrates, chlorates, and perchlorates with sulfur, carbon, and numerous organic substances, while of practical value, are difficult to put into meaningful equations, and caloric output will vary with reaction conditions because the products of the reactions are not always the same. Notwithstanding wide variations in composition, the heat output for these types of mixtures stays generally between 0.5 and 1.0 kcal/g. For sucrose and potassium chlorate, the basic heat source of many colored smokes, the figures are 0.7 and 1.0 kcal/g, depending on whether the reaction goes to CO or CO_2 formation.* Ammonium nitrate in propellant-type combinations with charcoal, starch, or nitroguanidine yields about 0.9 kcal/g and black powder between 0.6 and 0.8 kcal/g.

Miscellaneous Reactions

Reactions of zinc or aluminum with organic halogen compounds as used in certain white smoke formulations also fall into the intermediate (0.5—1.0 kcal/g) range. In order to present meaningful figures, the considerable heat of vaporization of the aluminum (III) chloride ($AlCl_3$) and the still higher one of zinc chloride ($ZnCl_2$) must be considered. The reaction of aluminum with hexachloroethane (C_2Cl_6) furnishes 0.98 kcal/g in standard states but only 0.88

* On p. 208 of *Modern Pyrotechnics*, the figures given were not only wrong and in reverse order, but an erroneous conclusion was tacked onto the false data!

for Al_2Cl_6 [g]. The corresponding figures for zinc and carbon tetrachloride are 0.58 and 0.45 kcal/g.

In conclusion, Table 27 is added in order to given an indication of the heat increment that may enhance a reaction due to neutralization of an alkaline and an acidic oxide if they are reaction products. In contradistinction to other tables, the molar heat of reaction between the respective oxides is presented. Per gram of reactants, the figures vary between 0.10 and 0.80 kcal/g.

The foregoing covers the major points of interest in calorific output data. Not touched upon were reactions that are either unusual and indeterminate such as with sulfates; obsolete as with permanganates; mainly of theoretical interest as with fluorine compounds. Some specialized figures such as for the decomposition reactions of chlorates can be found in the appropriate chapters.

Table 27
Neutralizing Reactions

	Acidic Oxide					
Alkaline Oxide	SiO_2	B_2O_3	MoO_3	WO_3	P_2O_5	ZrO_2
Na_2O	58.0	105.0*	88.0	95.0	118.0	
CaO	20.1	29.4	39.6††	49.2††	170.5†	
BaO	20.7				238.0†	23.0**
PbO	1.0		33.1			
Al_2O_3	38.0					
MgO	8.7	12.9				

Figures apply to the reactions between the oxides as indicated and are calculated from Circular 500, except as noted. The reactions are all exothermic.

* Also $Na_2O + 2B_2O_3 \longrightarrow Na_2B_4O_7$ + 74 kcal (exothermic).

† For reaction leading to compound $Ca_3(PO_4)_2$ and $Ba_3(PO_4)_2$, respectively.

†† Data from NASA report.[573] They differ considerably from applicable figures in Circular 500.

** Calculated from heat of formation of barium zirconate as cited by DOFL.[569a]

chapter 35

Reactivity—Wanted and Unwanted

Scope of Chapter

Continuing the characterization of materials by measurable attributes, the ignitibility of single substances exposed to air or other gases will be described, as well as the behavior of fuel-oxidizer mixtures under various conditions leading to initiation—heat, static electricity, impact, and friction. Destructive influences, especially of moisture and other external factors comprising the broad field of *surveillance* and surveillance testing, will be described.

Mechanism of Reaction of Solids

While the combustion of gaseous fuels can boast a sizable literature, the ignition of solid fuels and the phenomena of solid/solid reaction are probably the least explored. In certain model reactions, the finer points of interaction have been more or less well established and the subject emerges sporadically in the general literature.[574-579] An increased amount of empirical data is a hopeful sign but the subject, admittedly difficult to approach theoretically, has to the author's knowledge not led to a unified and generally applicable theory.

Remy's Treatise[60] offers a chapter on reactions of solid substances and a subchapter on reactions between solids only. It is pointed out that chemical interaction between two solid powders in intimate mixture may occur as a true solid-phase reaction, not dependent on lower-melting impurities or gaseous decomposition products—a viewpoint taken from investigations by Tammann (1927) and Hedvall (1929).

A noteworthy statement is that the inception of a reaction coincides almost always with the temperature at which one or the other of the reactants undergoes recrystallization. This may occur by partial fusion on the surface (sintering) that can take place (as proved by Hedvall) below actual fusion temperature accompanied by diffusion processes, sintering, and recrystallization with consequent shrinkage. These observations appear to derive from ceramic mixtures but may be applicable to pyrochemical combinations.

In greater detail, though with barely a reference to exothermic processes, the scientific effort in the physicochemical behavior of solids has been compiled in the book by Hauffe.[71] In Garner's book,[70] intramolecular exothermic decomposition and other solid reactions are treated by several authors, but again exothermic solid/solid reactions get short shift.

Measurement of Stability and Reacitivity

Considerable progress has been made in devising new methods or in refining older ones for the determination of various aspects of reactivity of materials and mixtures. For continuous monitoring of exothermic and endothermic changes that occur during heating of an element, compound, or mixture, the method called Differential Thermal Analysis (DTA) is widely used—by no means exclusively for pyrochemically interesting substances. In fact it was designed to study the behavior of clays at different temperatures by LeChatelier in 1887.[580,581] Briefly, the method consists of comparing the temperature rise caused by a carefully controlled steady heat input with that of an equally treated sample of a "thermally inert" substance, i.e. one known not to undergo chemical, structural, or phase changes in the range of observed temperatures. An easily accessible, short description of the procedure and a cross-sectional schematic drawing of the apparatus is found in an article by Campbell and Weingarten[582] incidental to a study on black powder. Thermogravimetric analysis (TGA) is the measurement of weight changes during such heating.[583] A refinement of DTA has been called Derivative Differential Thermal Analysis (DDTA).[584,585] A comprehensive presentation of the different thermal methods of analysis, though not specifically slanted toward pyrochemical reactions, is found in a book by Wendlandt.[585a]

DTA will detect events caused by rise in temperature, such as loss of moisture, change in crystal structure, fusion, evaporation, boiling, and thermal decomposition. In the case of fuel-oxidizer mixtures, a gradual steepening of the heating curve over that of the compared inert substance indicates certain pre-ignition reactions between fuel and oxidizer that are not yet self-sustaining. As a terminal result, the spike in the temperature curve will show the ignition or initiation temperature. Such figures, especially on initiation, may be different from those obtained by other methods where the heating occurs abruptly by exposing the material suddenly to elevated temperature or by keeping it at any one level of temperature for some time. From the latter method, one may learn that the material or mixture is indefinitely stable at a certain temperature—an important factor in environmental surveillance; that a mixture decays and becomes unsuitable for its purported use but does not "go off"; and finally, that it ignites or reacts instantaneously. In the latter case, the instantaneous ignition point would be extrapolated from small ignition delays at several temperatures observed slightly below the theoretical point. This type of test is usually performed by immersion of a thin-walled copper tube containing 100 mg of material into a carefully controlled constant temperature bath of Wood's metal using a different sample for each 5°F step and by measuring time to ignition down to a 2-sec interval.[586] Recently, Sinclair[587] has reported a technique that permits the use of extremely small samples.

The mere fact of the existence of pyrophoric metal powders and of the relatively low ignition temperatures of dusts (while the same substance in coarsely powdered form has a much higher ignition temperature and in bulk form may be unignitible) shows the dependence of the ignition temperature on particle size. It carries over into the problem of initiation of mixtures. The ease of initiation is generally dependent on the particle size or, to put it more precisely, on the surface area of the fuel and (though in most cases to a much smaller degree) of the oxidizer. The word "ease" is of course a subjective term. By effecting the transfer of so many calories at certain temperature levels and over a certain time interval, one initiates release of energy on the surface of the fuel-oxidizer mixture that may or may not be sufficient to become a self-sustained reaction. It may be better to leave it at that and let better minds develop the mathematics of these processes. An auspicious beginning by Johnson[393] has been discussed in Chapter 23.

The sensitivity of many finely powdered fuel-type materials to self-ignition greatly concerns the pyrotechnician from the viewpoint of safety. Many fuel powders, even those not classed as of "subsieve size," contain enough very fine particles in, say, the 1—5 μ diameter range to present a spontaneous ignition hazard as outlined in Chapter 9. They may be ignited by electrical discharge from the human body or by electricity created through motion, especially in contact with paper, plastic sheets, or when the powder is poured from one container into another. In order to establish standards of hazardousness or to test the sensitivity of materials and of mixtures, the deliberate input of measured static electric charges in stationary dusts or in dust clouds has been performed with a variety of apparatus. An extensive literature on the subject is available in Bureau of Mines reports and publications by Picatinny Arsenal, Naval Ordnance Laboratory, and others. For our purpose, it suffices to point out the severe and ever-present static hazard for zirconium in the finest commercially available form and the potential hazardousness of the finer fractions of powders such as aluminum, magnesium, their alloys, and titanium. All these metals retain their susceptibility to spontaneous ignition under nitrogen or carbon dioxide.[135,139]

From certain preliminary data for a series of tests on the electrostatic sensitivity of mixtures,[588] it can be concluded that most commonly encountered fuel-oxidizer mixtures, disregarding first fires, require for successful initiation a minimal electrical input of the order of 0.1—15 joules. As a guide, the Bureau of Mines quotes a figure of 0.015 joules as the possible (maximal) discharge from a human body.[589]

The processor is equally concerned with the sensitivity of mixtures to impact and friction. Devices that permit a weight to drop from measured heights on a small sample (mostly 20 mg) have been designed among others by the Bureau of Mines and by Picatinny Arsenal.[590] Both differ in the arrangement of the sample, also in the units of measurement since BM data are in centimeters, PA data in inches. The values, if no other indication is given, represent the minimum height at which spark or flash, smoke or sound is observed, at least once in ten trials, a fresh sample being used for every trial. Other figures such as the all-negative maximum height or the 50% firing level may be indicated. Tests of this type use often the "staircase" method (Bruceton test) whereby the drop-weight height is lowered if the preceding test was negative or raised if it was positive. Under Picatinny Arsenal practice, a starting height of 12 in. is usual and a variation of 2 in. down or 4 in. up follows in order to find the proper range. This scheme leads quickly to the critical range of heights and reduces the number of samples required. When as many as 250 samples are available, a "rundown" procedure of testing 25 to 50 samples at any one of at least five heights within the critical range is an alternate procedure. It must be obvious and it is borne out by practice that numerous factors, some difficult to control, such as the even spreading of the sample on the anvil, will influence the result. The practitioner can often satisfy his curiosity as to classification of a mixture in regard to impact sensitivity by testing a sample on a steel block and tapping it with a small peen hammer. He will soon gain sufficient skill to grade at least the more sensitive mixtures. His main thought, however, should be that even the mixture that yields no positive results on roughest treatment may still be hazardous in other respects.

Basically identical drop-test methods have been devised for determining the sensitivity of stab or percussion primers.[45] Stevenson discussed the method used at Frankford Arsenal[591] with special modifications for an investigation of variations in a tracer igniter mixture and their relation to impact sensitivity.

Elaborate and bulky test equipment is required for a standardized friction test using either a fiber or a steel shoe.[592] The results are more subjective than with the impact machine and are expressed as explosions, crackles, sparks, or no reaction.[593] Friction tests in an improvised style such as in a little mortar have probably little value except for a comparison of highly sensitive mixtures.

The susceptibility of a solid to the igniting effect of an actual flash or flame would be one of the most desirable tests if it could be quantitatively measured. Probably everyone in the field has pondered over this problem or even devised some kind of test setup, but the results seem generally to have been discouraging. A related method, which has also been tried in one form or another for solid propellant ignition, is exposure of a sample to a weighed quantity of black powder[594] at a certain distance.

Russian researchers, quoted by Shidlovsky (and undoubtedly others elsewhere), have attempted to achieve quantitative measurements from minimum and maximum distances between composition and the spit of a flame from the end of a length of safety fuse, but reproducibility seems to be poor.

Ignition, Initiation, and Decomposition

Table 28 gives ignition temperatures in air collected from several sources. Because of the great variety of particle sizes as well as differences in methods of determination of ignition point (not always disclosed with quoted figures), these data must be taken with reservations.

The reactivity of magnesium with nitrogen, carbon dioxide, and water has been mentioned in Chapter 32 and it warrants some additional remarks regarding the specific conditions under which this reactivity is evidenced or nonexistent. If magnesium is burned with insufficient air, such as by accidental ignition in a closed ball mill, the strong odor of ammonia after exposure of the product to ambient air is evidence of magnesium nitride (Mg_3N_2) formation, which hydrolyses easily in contact with moist air:

$$Mg_3N_2 + 6H_2O \longrightarrow 3Mg(OH)_2 + 2NH_3$$

However, this reaction of magnesium and nitrogen takes place only above 1000°C, considerably above the melting point of the metal. Therefore, magnesium, used in the fabrication of certain automobile crankcases can be melted under nitrogen pressure.[597] It might be added that liquid magnesium in a melt kettle is quite easily handled without a protective gas by merely covering it with some molten flux. If unprotected, ignition starts slowly at one point and is easily suppressed. When pouring the metal into a mold, a little sulfur, which burns off without leaving a residue, provides the protection. These observations date from the author's efforts with miniature incendiaries ("magnesium matchheads") in World War II when casting magnesium slugs (and sad-looking specimens they were!) was the only way to beat the long delivery times for the extruded rod.

Burning magnesium on which a stream of dry carbon dioxide (as from a fire extinguisher) is directed will continue to burn under heavy soot formation, attesting to the reaction

$$2Mg + CO_2 \longrightarrow 2MgO + C$$

Similarly, steam supports the combustion of ignited magnesium:

$$Mg + H_2O \longrightarrow MgO + H_2$$

More dramatic is the reaction of liquid water with burning magnesium where the physical violence caused by the disturbance of the magnesium metal by steam and the reaction leading to hydrogen formation is spectacular. However, if a relatively small magnesium body is quenched in a large excess of water, the cooling effect stops the reaction safely. The reactivity of magnesium and water at room temperature may play a part in sensitizing it for spontaneous ignitibility.

Shidlovsky[598] has studied the explosive behavior of metal/water mixtures and regards magnesium in the presence of water as very shock-sensitive. Aluminum powder and water or magnesium and methanol are less sensitive but their explosive potential is enhanced

Table 28
Ignition Temperature*

	°*C*	*Remarks*
Lithium	200	in bulk
Sodium	125	in bulk
Beryllium	>2000†	<45 μ
Magnesium	475—550	fine powder
	600	coarse powder
	540	ribbon
	618	cast pieces
Calcium	(300)	in oxygen
Boron	(400—900)	gradual oxidation
Aluminum	585	flake
	>700	atomized, foil, etc.
	660	in bulk in oxygen
Carbon	300—400	charcoal
	555	pure carbon black
	313—535	various blacks
	>700	graphite
	700—800	diamond in oxygen
Silicon	950	
Phosphorus, white	45	
Phosphorus, red	260	technical
	360	technical, stabilized
	430	purest
Arsenic	>400	in vapor form
Antimony	700—800	
Sulfur	260—300	fine powder, rhombic, "flour"
	>190	"flowers"
Titanium		
	250	finest powder
	400	powder
	700—800	massive
Zirconium	250—300	finest powder
Zirconium Hydride	270	
Cerium	160	

by the addition of PETN.

The *protection* of fine zirconium or titanium powders by moistening with water requires the presence of at least 25% of water. If less water is present, the metal might not only be ignitible but the combustion may be extremely violent because of steam formation and reaction of metal and water. Actual use of water in the form of hydrated salts has been claimed for pyrotechnic purposes.[556]

The scarcity of systematic data on the initiation temperature of

Table 28 (continued)

	°C	*Remarks*
Thorium	280	fine powder
Iron	290	H-reduced
Manganese	450—492	fine powder
Zinc	500—650	oxidizes>150°
Uranium	150—170	
Plastics		
Ethylcellulose	296	
Polyethylene	349	
Urethane	416	
Polyamide (Nylon)	424	
Phenolic Paper Laminate	429	
Polyvinyl Chloride	454	
Polystyrene	496	
Polyester (+glass fiber)	510	
Miscellaneous Substances		
Organic Dusts	145—200	
Paper	233	
Paraffin Wax, Gasoline, Kerosine, Fuel Oil #1-4, Wood Shavings, Cotton Batting	245—260	
Glycerine	393	
Various Starches, Sugar	380—422	
Fuel Oil #6	407	
Cylinder Oil	417	
Peat	475	
Various Coals	480—660	
Asphalt	485	
Safety Matches	185—200	
SAW Matches	120—160	

* In air, except when noted.

† Figures on beryllium run the gamut from this extremely high temperature as given in a NASA report[596] to pyrophoricity for superfine material.[568] A BM report[139] gives 540°C for an impure, 1 μ material in a dust layer.

Other figures quoted from numerous sources such as company pamphlets, NBS data reported in handbook tables, also Costa et al.,[595] and Fisher.[450] Figures on plastics from *Modern Plastics*.[138]

binary fuel-oxidizer mixtures restricts general statements. If we scrutinize Table 29 from Shidlovsky's book, it is apparent that with the relatively most reactive oxidizer salt—potassium chlorate—either the ignition temperature or the melting point of the fuel may govern the point of initiation. Conversely, when the oxidizer has a higher decomposition temperature or a higher melting point or both, as is the case with the perchlorate and nitrate in relation to the melting or ignition point of the fuel, those higher temperatures are responsible

Table 29
Initiation Temperatures of Binary Mixtures

	M.P.	*Sulfur* 113	*Lactose* 202	*Char-coal*	*Magnesium* 650		*Aluminum* 660
Potassium Chlorate	360	220	195	335	540	670*	785
Potassium Perchlorate	588	560	315	460	460		765
Potassium Nitrate	337	440	390	445	565	670*	>890
Sodium Nitrate	310					610*	
Barium Nitrate	596					660*	
All figures in °C.							

Data from Shidlovsky (3rd ed.) Table 9.1, Stoichiometric relations. Cup method except for figures with *. The latter are extrapolated from controlled heating in an oven according to U.S.S.R. Government Specification GOST (Gosudarstvenny Obshchesoyuzny Standart) 2040-43.

for higher initiation temperature. Similar relations apply to reactions between more or less active fuels, say, zirconium versus aluminum and a very inert oxidizer such as iron oxide. A few applicable data are shown in Table 30.

It may happen that the difference in reactivity between fuel and oxidizer is large enough to prevent interaction. Thus, in a mixture of sulfur and sodium nitrate the sulfur may burn off without furnishing enough heat to bring the nitrate to the reaction temperature. How-

Table 30
Initiation Temperatures of Various Mixtures

Ignition Mixtures	*Formula No.*	*°C*
Zr/Fe_2O_3/SiO_2 (A1A)	181	300
B/KNO_3	184*	370
Mg/Teflon†		493
Si/PbO_2/CuO	172	540
Mg/BaO_2, 12/88%††		570
Mg/PbO_2, 20/80%		600
Mg/$PbCrO_4$, 20/80%		620
Delay Powders		
Se/BaO_2	193	>265
Cr/$BaCrO_4$/$KClO_4$	191	340
W/$BaCrO_4$/$KClO_4$	190	430
Zr-Ni Alloys/$BaCrO_4$/$KClO_4$	188	495
Mn/$BaCrO_4$/$PbCrO_4$ (D-16)	189	382—522
B/$BaCrO_4$, 10/90%	187	685
B/$BaCrO_4$, 5/95%	186	700

* From Data Sheets of Flare-Northern Division, Atlantic Research Corp., same figure for 18/82% mixture without binder.

† From Data Sheet of Flare-Northern Division, Atlantic Research Corp.

†† Ignition point is depressed to 320, 360, or 395°C, respectively, after addition of 2% of sulfur, rosin, or asphalt.

ever, in compacted mixtures and with a strong enough initiating heat-source this does not occur.

Because of the intimate relation of melting point and decomposition temperature of oxidizers with reaction temperature, a number of such figures have been compiled in Table 31. It should be no surprise that even creditable figures of melting points of these types of compounds show some variations and quoted decomposition temperatures vary even more. When decomposition sets in at a temperature below the phase change, then what is called the melting point will always be uncertain.

Table 31
Melting and Decomposition Temperatures of Various Oxidizers

Substance	*M.p.,* °*C*	*Decomp. Temp.,* °*C*
$LiClO_3$	127.6	
$NaClO_3$	256	350
$KClO_3$	364	565—620
$KClO_3$ catalyzed	∽350	335—415
$KClO_3/NaClO_3$, 25/75%	232	—
$KClO_3/NaClO_3$, 85/15%	—	305
$AgClO_3$	231	270
$BaClO_3$	∽400	>300
$LiBrO_3$	248 d	
$NaBrO_3$	381	
$KBrO_3$	370	
$RbBrO_3$	430	
$CsBrO_3$	420	
KIO_3	560	
$AgIO_3$	>200	
$LiClO_4$	236	440
$NaClO_4$	473	490—527
$KClO_4$	588	510—619
$RbClO_4$	606	635
$CsClO_4$	575	630
$BaClO_4$	469	504
KIO_4		256—288
Pb_3IO_5		170
$LiNO_3$	252	365
$NaNO_3$	304	(520)
KNO_3	332	628—800
$CsNO_3$	417	
$AgNO_3$	214	305
$Ba(NO_3)_2$	588	605

Table 31 (continued)

Substance	*M.p., °C*	*Decomp. Temp., °C*
Sr $(NO_3)_2$	645	
Tl^INO_3	206.5	
Na_2CrO_4	800	(stable 1000)
K_2CrO_4	984	
$K_2Cr_2O_7$	395	>500
CrO_3	180—190	300—550
$(NH_4)_2$ CrO_4		>204
$CaCrO_4$		>1000
$PbCrO_4$	844	(starts 600)
PbO_2		375—460
Pb_2O_3		460—520
Pb_3O_4		585—678
PbO	890	
SrO_2		460
BaO_2		795
$LiMnO_4$		190
$NaMnO_4$		170
$KMnO_4$		240
$RbMnO_4$		259
$CsMnO_4$		320

Destructive Reactions

Negative aspects of ignitibility and reactivity fall under the headings of *hygroscopicity* and *self-destructive interaction* (which are mostly intimately connected), extremes of *temperature*, and *mechanical stresses*.

All solid substances of large surface area, such as the ingredients of pyrotechnic formulas, adsorb moisture from the air, the amount depending on the relative humidity. Such moisture is of the order of some hundredths or tenths of one percent in weight and is generally of little concern, since it does not adversely affect ignition or burning behavior. By heating the materials or their mixture (if this is practical) prior to loading, exposing the material as little as possible to normal ambient conditions, and by loading under specially (and expensively) created low humidity conditions, the presence of no more than minimal amounts of moisture in the final product can be assured. Exceptions are charcoal with about 5% water and uncalcined infusorial earth with 6%.[599] When cardboard tubes, paper and cardboard disks, felt pads, and the like are used, one should realize that fibrous material, such as chipboard, will absorb 6-8% water in equilibrium with common ambient humidity conditions of the air, say 50 to 60% RH, and still more in really damp surroundings. A preheating of such component parts can be beneficial, in particular, when the items are subsequently protected from moisture by canning, sealing in plastic or plastic/foil laminate bags or at least partially sealing them through waxing or judicious packaging. Aspects of the problem of sealing have been discussed in some detail in the chapter on matches.

The pyrotechnician is frequently forced to use substances that can be handled well under low or moderately high humidity conditions but that attract several percent of water above a certain point of ambient humidity. The most frequently encountered are sodium nitrate and strontium nitrate. If influx of moisture occurs in storage. the crystals will agglomerate, first loosely, which necessitates only a mild grinding or forceful screening action, but eventually an irreversible lump formation occurs. In salts that do not attract more than minimal moisture, such as potassium chlorate or perchlorate and barium nitrate, a less pronounced but in the end just as damaging accretion occurs, perhaps mainly as the result of time and pressure under the material's own weight. Of the fuels, the organic materials such as carbohydrates, hydrophilic binders, and charcoal are affected by ambient humidity.

The term *hygroscopicity* is rather unsatisfactory since it leaves one guessing if the material will attract moisture to a damaging degree under average humidity conditions or merely at extremely high humidity, whether it will *deliquesce* and thus irreversibly lose its physical integrity as discrete particles, or remain physically intact. But in any case it signifies that at some point the material will attract enough moisture to make its proper pyrochemical functioning uncertain, whether such moisture is picked up prior to final processing or has a chance to infiltrate in storage. Substances of high solubility in water are generally suspected of such behavior and, as a rule of thumb, as tabulated by Shidlovsky,[3] the relative humidity over a saturated solution of the salt can be called an indicator of hygroscopicity. Shidlovsky states that when this relative humidity at 20°C is 92.5% or above, the salt can be safely used. There may be trouble if the relative humidity over the saturated solution drops to 80% or lower.

Measurements of water content after exposure until equilibrium is reached at various constant humidities yield objectively the practical information of the safe exposure time during processing or up to a certain point of ambient relative humidity for extended periods. A few such studies are found in the literature, some reported in AMCP 706-187.[32] The considerably lesser moisture attraction of the purer grade in the cases of sodium or strontium nitrate points toward the impurities as a contributing factor of moisture attraction. How reliable such figures from laboratory tests are, is difficult to evaluate. A formidable obstacle is the maintaining of an unequivocally defined level of humidity, especially in the critical 90-100% RH region where a minute temperature drop or unevenness of temperature in the test vessel may lead to condensation. U.S.P. grade sodium nitrate seems to attract only traces of moisture at 45% RH whereas at 70% RH, 11% moisture pick-up has been reported. The difference in behavior between grades of different purity disappears in the region of severe exposure. Considering the importance of the subject to the pyrotechnic manufacturer, the scarcity of well-defined experimental data is surprising.

If metal and oxidizer salts in mixture are exposed to high humidity, the normal deterioration of a metal, such as magnesium, is intensified by the action of the moist salt on the metal surface. Oxidation/reduction will take place, as a rule, though perchlorates resist reduction even by some strong reductants, a notable exception being iron powder. The powerful reducing agent is ferrous hydroxide, $Fe(OH)_2$, an intermediate in the "rusting" process. Aluminum, being protected by a passivating oxide skin, will withstand water-wet processing in a neutral medium but will deteriorate from contact with either acidic or alkaline substances. A nitrate in the presence of an active metal may undergo reduction to ammonia if moisture is present, even though otherwise the mixture—e.g. aluminum and barium nitrate—may be stable.

In order to counteract a possible surface deterioration of metal powders, protective coatings have long been used. No such protection is absolute, it merely allays and delays the attack. Coating with molten paraffin has been recommended for steel fillings by Weingart.[7] The irregular shaped, lathe-cut magnesium "powders" of World War II were treated with linseed oil or mixtures of linseed and castor oil. (If the manufacturer of such a powder was also a fabricator of signals, he first sweated blood to keep his powder free from all traces of machine oil, which made it unacceptable, then proceeded to add 3% of oil to protect it!) Zirconium/nickel alloys, which do not share the excellent passivity of the elementary zirconium by itself, are "dichromated" by boiling in a slightly acidified sodium dichromate solution,[416] Dichromating of manganese or cobalt powder may be followed by a stearic acid treatment according to Comyn,[600] though this appears undesirable in some applications.[601] "Inactivation" of powders of aluminum, magnesium, or zinc has been patented,[602] and consists of pickling followed by stearic acid adsorption. All these operations are tedious. Magnesium in bulk, as in incendiary casings, can also be somewhat protected from corrosion by a variety of dichromate treatments or better by anodizing,[603] whereby, in one method, dichromate together with fluoride and phosphate come into play. Such treatments also form the basis for better adhesion of paint coatings.

It may be added that the pyrotechnic manufacturer should not rely on vague notions of corrosion resistance when faced with problems such as attack by sea water, burial in the ground, etc. of completed items. Sea water is particularly detrimental even to stainless steels because of the combined optimal effect of 3-4% sodium chloride and of dissolved oxygen.[604] Even bacteria, acting as depolarizers and activators,[605] contribute to corrosion of steels. On the other hand, certain aluminum and magnesium alloys are resistant in varying degrees to sea water.[606]

Protection of solid metal and even of metal powder, especially of iron, can be obtained by certain nitrites of organic bases classed as vapor phase inhibitors (VPI), such as dicyclohexylammonium nitrite (Shell Oil Co.), which is effective in the form of coated paper (MIL-P-3420B) for wrapping of sensitive metal parts.

In the case of an acid-forming nonmetallic fuel such as sulfur the consequences of moisture infiltration can be the most serious, especially if the oxidizer is a chlorate. By the combined effect of air and water, strong acids of sulfur are formed that in turn liberate the free chloric acid from the chlorate. The latter is an active oxidizer that converts more sulfur into sulfuric acid. Thus a "vicious circle" (or cycle) is established, leading to possible complete breakdown of the composition and under unfavorable conditions of heat retention and accumulation causing spontaneous ignition.[159,160,169,172] However, by the addition of neutralizing oxides or carbonates, the mechanism of self-destruction can be effectively aborted. The materials used are zinc oxide, calcium carbonate (as ground limestone or, more active, as precipitated chalk, called whiting), magnesium carbonate, aluminum oxide (alumina), and, in the case of certain smoke mixtures, sodium bicarbonate. Since red phosphorus oxidizes also with acid formation, it should never be used except in the presence of a few percent of such a neutralizer.

Incompatibilities may be caused by the presence of base metals and of salts of a relatively nobler metal such as copper or mercury. This incompatibility may extend to the metal body with which the mixture is in intimate contact. The situation does not arise frequently, but one might face it with flare formulas for blue flame.

A special and also unusual case of destruction through contact with metal, in this case with copper or copper bronze, is that of red phosphorus, as mentioned in Chapter 12. The oxidation in the presence of air and moisture is also strongly promoted by impurities such as iron or copper within the material (not necessarily as metals). Stabilized red phosphorus, JAN-P-670A, is not only low in iron and copper but particles of less than 10 μ diameter are mostly removed and 2.5% of alumina is added as a stabilizer. From experiences in the match industry, the author would confidently say that the special manipulations of red phosphorus leading to the stabilized variety are quite unnecessary since several percent of effective neutralizers such as zinc oxide prevent any acidification of the system in storage.

Recently, McLain and Lewis[607] have discovered a remarkable spontaneous explosive reactivity of chlorate and sulfur in the present of copper ions. They voice a warning that copper or bronze screens, etc. may under some circumstances lead to a "doping" of chlorate that increases tendencies to breakdown even at room temperature.

Occasionally, other baffling and obscure destructive reactions take place. They are especially harmful when they emanate from the main body of the system and exert their influence on a smaller but vital part, such as a first fire or a priming system, thus destroying the chain of fire transfer at the very beginning. Magnesium mixed with black powder has been found to cause formation of hydrogen sulfide, which destroys primer material containing lead compounds.[48] Residual solvent or unreacted organic monomers of plastic binders have been suspected of such obscure influence when an unexplainable failure has occurred.

Survival and Surveillance of Finished Items

When all the critical properties of an item have seemingly been adequately explored and proper functioning has been established, the question arises how well the object of the development or production will fare after it has been shipped to its final destination. It may undergo long intermediate periods of storage, possible exposure to extremes of environmental conditions, and handling and mishandling in transit. Exceptionally, surveillance tests are carried out in ambient storage for years, as on propellant powder for visible decay or for corrosion resistance on packaged hardware.[608] This, however, is rarely possible or practical.

A number of physical exposure tests involving the item proper or in packaged or packed-for-shipping condition, such as repeated drops from a prescribed height in several attitudes and upon a defined object, e.g. a steel plate, are fairly straightforward. If the drop height is very large (40 ft), the purpose of the test is normally the safety of the item, not the concern for the survival and subsequent usability. In fact, these aspects of testing should always be clearly stated: Must the item function properly after completion of the test or is it the aim to demonstrate that some extreme of abuse causes no damage to the surroundings and danger to personnel? Jolt and jumble tests (separate and so named) simulate continual rough handling expected to occur in transit.

Transportation vibration of powder mixtures may cause some unmixing—an improbable occurence but one that must be tucked away somewhere in the trouble—shooter's mind, especially when powders of highly diverse specific gravities are involved. Even the single materials in storage, such as metal powders in slack-filled drums, may show a certain degree of separation into finer and coarser particles—a fact that can sometimes be visually observed if a powder is poured in the form of a conical mound, larger grains appearing at the base and periphery of the pile.

Tests of exposure to high humidity and temperature, and resistance to salt spray or to attack by fungi must be performed by trading time for exaggeratedly severe conditions. These are not necessarily fully valid survival tests, but they attempt to give a fair picture of the behavior of material.

Applicable U.S. Government specifications are:*

MIL-STD-300	Jolt	MIL-STD-314	Waterproofness
MIL-STD-301	Jumble	MIL-E-5272C	Environmental Testing, Aeronautical & Associated Equipment

* The quoted MIL-STD surveillance tests of the 300 series now appear combined under MIL-STD-331 of 10 January 1966, entitled "Fuze and Fuze Components, Environmental and Performance Tests For."

MIL-STD-302	40-ft Drop		(Temperature and Humidity Cycling—Temperature Shock and Exposure and Performance at High & Low Temperature)
MIL-STD-358	5-ft Drop	MIL-STD-306	Salt Spray
MIL-STD-304	Temperature & Humidity Cycling (popularly called "JAN Cycle")	MIL-STD-356	Salt Spray
		MIL-A-8625A	Salt Spray
MIL-STD-305	Vacuum Steam	MIL-STD-810A	Environmental Test Methods for Aerospace & Ground Equipment
MIL-STD-303	Transportation Vibration	MIL-F-8261A	Fungus Resistance

Disregarding overtesting where severity is increased while time is shortened, a valid question is how realistic within the limits of normal terrestrial conditions is the frequent range of test conditions from —65°F to +160°F. The low end of the scale is approached or even exceeded on airplane transportation at high altitude and in arctic areas. As to the high end, an illuminating study available to the public is by Porter,[609] conducted at Yuma, Arizona. Under the most adverse conditions, a maximum temperature of 152°F was encountered under and near the roof of railroad freight cars. While munitions will wherever possible be packed and protected so that ambient maximum temperatures are not reached or built up due to slowness of heat transfer and limits of exposure time under the diurnal cycle, it must also be considered that under emergency conditions in the field the opposite may take place—maximum exposure when munitions are openly exposed to the weather because there is neither time nor the means available for protected storage. A temperature of 215°F (101°C) has been measured on an aircraft wing in the direct sun in Arizona and elsewhere 177°F (81°C) on an aircraft where the internal temperature had reached 167°F (75°C).[610]

In addition, entirely new problems have arisen with clandestine "cold war" and "limited war" activities, such as the underground and partisan actions of World War II and their more recent counterparts. Burial of material in the ground or under fresh or salt water poses severe surveillance conditions for the outer envelope, while long-delayed activation under tropical rainy season weather exposes the more or less unprotected item itself to infiltration of moisture that jeopardizes its eventual proper functioning.

A new factor pertaining both to stability and reactivity is the behavior of pyrochemical components under the influence of certain types of radiation. One aspect is the exposure to neutrons in an extraterrestrial environment or perhaps in the proximity of nuclear devices. An example of a material sensitive to such exposure is the element boron. Ordinary boron consists of 80% of isotope B_{11} and 20% of B_{10}. Of these, B_{10} captures neutrons at a rate about 10,000 times higher than that of B_{11}. Boron-containing compositions are therefore regarded with suspicion where extreme conditions of exposure to certain types of radiation are possible. A way to avoid the difficulty would be the use of a B_{11}-enriched (up to 98%) boron. Unfortunately, this raises the price to $25 per gram, which might be a bit steep for use in a pyrotechnic application.

Oxidizers such as ammonium perchlorate, chlorates, and periodates have been exposed to UV radiation, and in many cases chemical activity and decomposition rates could be increased. The preirradiation of ammonium perchlorate has been studied by Freeman and Anderson,[611] the decomposition of potassium periodate by Phillips and Taylor,[612] and NavOrd 7147[508] quotes several authors in connection with work on chlorates. A growing literature in this special field is to be expected and present interest is indicated in several articles in the book *Reactivity of Solids*.[613]

Note to Chapter 35

Aspects of *safety* in the pyrotechnic laboratory and in production have been touched upon previously in connection with specific subjects, and in Chapter 2 a number of pamphlets treating this matter have been cited. It might not be amiss at this point, i.e. under the chapter heading Reactions—Wanted and Unwanted, to add a few general points on safety in the pyrotechnic laboratory with some side glances on production.

In explosives technology, the hazards encountered with primary explosives or with high explosives are rather well defined. There we deal most frequently with chemical compounds whose properties, and especially whose response to friction, impact, static sensitivity, or plain ignition, can be ascertained by fairly straightforward measurements. In pyrotechnics, however, the nonhomogeneous mixtures in their infinite varieties of weight ratios, surface properties of particles, sensitizing or phlegmatizing additives, and the great dependence in behavior on the amount of material itself create much uncertainty as to actual performance under conditions conducive to accidental initiation. Moreover, the constantly increasing number of "new" substances, i.e. not previously employed pyrotechnically, such as rare metals of small particle size, makes it impossible for the experimenter to predict the behavior of a novel combination. One can of course draw some conclusions from the speed of reaction and other behavior on willful initiation and from comparing the functioning of a small amount—say, of a few grams—with that of a pound of unconsolidated materials. And there are the standardized sensitivity tests—for what they are worth—described earlier in this chapter. Unfortunately, all such tests are mere warning signs if the results prove to be unfavorable from the viewpoint of safety, but a negative result does not exclude hazardous behavior under slightly different conditions.

This uncertainty should lead one to the conclusion that every pyrotechnic operation must be regarded as possibly dangerous in some manner even after safety precautions aiming at prevention of accidental initiation have been taken. To express it somewhat differently, the emphasis lies on the ever-present and undeniable possibility of accidental initiation, rather than on the means for diminution of the chances of a mishap. This shifts the attitude from one of complacency ("everything has been done to prevent an accident") to a more realistic viewpoint of reflecting in this manner: "Now what happens if the mixture goes off notwithstanding all measures taken?"

Under this aspect, and since complete remote control of weighing, mixing or pressing is almost out of the question, the protection of personnel is the first point of departure. In the laboratory where relatively small amounts are handled, this involves first and foremost the protection of the eyes; next, exposed skin surfaces—face, hand, and forearms; then the prevention of ignition of clothing.

In these measures, compromises are necessary in order to make the worker accept the encumbrance of eye glasses, gloves, safety shields, etc., and also in order not to hinder the operator in fulfilling his task adequately. The idea that an operator, be he scientist or workman, would protect himself out of self-interest is unfortunately entirely false. Why this is so would take a whole chapter to attempt to analyze and explain. Because of such unwillingness to obey mere exhortations, absolutely no choice must be given to the individual whether or not he will wear his safety glasses if he is in a work area. There are no difficulties with the permanent wearer of prescription-type hardened lenses and fire-resistant frames. But the person who does not wear glasses away from work must be forced to wear them in the work area *all the time* and must not be permitted to use his own judgment as to his situation. True, he may merely handle some inert material at a certain moment, but he may be near a co-worker engaged in a hazardous job.

Side-shields on eye glasses or complete eye-shields fall in the category of additional desirable protection, but one that is often foregone in order to make the somewhat lesser protection by ordinary glasses acceptable and hence enforceable.

A similar compromise can be achieved with *gloves.* Suede leather gloves (which are washable and must be kept clean*) permit mobility of the fingers while affording protection against searing flashes. Rubber or plastic gloves must never be worn, since they melt from heat and stick to the damaged skin, thus being worse than no protection at all.

Of outer clothing, anything that is reasonably flame-resistant and does not lend itself to build-up of static electricity is permissible. Wool clothing, though it may be suspected of creating static, is an excellent protective covering.

The bare face is best protected if the operator stays behind a heavy glass or plastic shield with the hands and arms working around it. Face shields protect only in certain attitudes, though some are designed to protect the chin and throat area. They may be needed where stationary shields are impractical.

Safety showers are most desirable, safety blankets of dubious value. A bucket of sand or water or (in the case of handling red phosphorus) merely a large wet rag, may be a good thing to have handy. In order to stop the flaming and spontaneous reignition of phosphorus, especially after it has become attached to the skin, an aqueous solution of copper vitriol (cupric sulfate) should be readily available. Such a solution causes precipitation of metallic copper on the surface of (white) phosphorus, thus excluding air sufficiently to prevent reignition.

The subject of fire fighting will not be treated further. Certain

* They are easily washed in lukewarm water while on one's hands. When clean, a rinse after the final soaping is *omitted.* This keeps the glove flexible after drying.

special problems, such as those concerning sodium or magnesium fires, have been treated earlier.

One additional point pertaining to the safety in the laboratory must be mentioned since it is so often disregarded. This is the presence of flammable solvents on shelves above the work area or on the laboratory table. A small flash or explosion can be converted into a raging fire by the proximity of a bottle containing a few ounces of acetone, etc. that is there because it plays a part in the preparation of the pyrotechnic mixture. The same applies to other reactive materials that so often clutter the laboratory area.

Aside from the preparation and handling of overly large production batches under conditions of inadequate shielding or absence of remote control, the potentially must hazardous pyrotechnic operation seems to be the *pilot production* that is too small or too intimately connected with the laboratory group to be put in the production area and is performed in the crowded and insufficiently policed area of the laboratory rather than in a specially designed and suited experimental pilot-production station.

It is a valuable point to remember that what is in front of the operator on the assembly line is probably less of a hazard than what is on one side (loose powder) or on the other (accumulated, open, half-finished units). On the really large production line, the flow of processed items may obviate this condition, but not always. And while we have the picture of the production line before us, it is the duty of the supervisor and the safety committee members to see that every worker is located so that he can instantaneously back away from his work area and leave the room or building without interference from objects or persons.

It would go too far to delve deeper into the aspects of safe production engineering. Mixing operations are mostly remotely started and stopped, but it is often necessary to enter the mixing bay for intermittent manipulation or to empty a mixer. The use of aluminized asbestos suits and special safety helmets for this operation should be the rule rather than the exception. The old-fashioned method of mixing by forcing the materials through a framed screen with a gloved hand has the virtue of gentle action and no investment costs, but it is slow and unsuitable for those mixtures that tend to violent or explosive burning in powdered form when present in the quantities handled in such an operation.

Pressing behind a barricade, by use of a rotary index table of which only a small part protrudes into the loading and unloading area, is a good and safe arrangement. Much depends of course on the size of the item.

Accidents in fireworks or other pyrotechnic processing plants are only a minute fraction of industrial or private accidents. They capture the headlines because of their spectacular nature. Seldom are they adequately reported or investigated so that one could judge if neglect of ordered safety measures, an unfortunate concatenation of circumstances, ignorance about the hazardousness of the material, or plain absence of proper procedures was the cause of the mishap.

chapter 36

Binders and Other Accessory Materials

Compositions that are not used in loose powder form are consolidated in one of three ways: They can be mixed with a sizable amount of a strong adhesive to form firm beads, bulbs, or elongated rods supported by wires, wood, or paper splints; they can be compressed or loosely aggregated with the aid of adhesives to serve as overall burning stars; or they may be pressed into a protective envelope with or without a binder.

In the first category belong electric ignition beads, matches, sparklers, and ignition or expulsion mixtures slurried and painted on surgical gauze pads, and ignition-sensitive end-seals of safety-fuse trains consisting of a paste of black meal powder and NC lacquer. Here the binder often becomes an important intrinsic component of the mixture. It is the only substance that holds the mass together and may also be a major fuel and flame-former. Its rheological properties in solution can be the decisive factor in shaping the mass and may even determine its reactivity, as explained in the chapter on matches.

Though many mixtures compact quite well under compression, a certain amount of adhesive material is desirable in overall burning stars, be it that the rate of consolidation is slight, as is the case in many fireworks items, or that the pellet is subjected to strong adverse influences such as transportation vibration.

In partially enclosed items, even though most of the mixture may be protected from physical disintegration by a strong cardboard or metal cylinder, binders are regularly added for increased strength and also for flame formation. The exception is the item whose proper function demands minimum gas development.

Among water-soluble or water-dispersible binders are dextrins, gum arabic, casein, and animal hide glue. The latter has received great attention from the match manufacturers as well as from other industries.[614,615] Hide glue is marketed in a great variety of viscosities and "jelly strengths." These properties determine the quality and price. Anyone using a high-grade animal-hide glue should make himself thoroughly familiar with the peculiarities of its solutions, e.g. the formation of a reversible gel (at about 85°F), the skin formation of the liquefied colloid, its foaming properties, etc.

Fortunately, the less specialized formulations do not employ such a difficult material; in fact, only rarely does one find a water-soluble colloid. A frequently found binder material is lacquer-grade nitrocellulose (MIL-N-244A, Grade D, sometimes also Grade A), dissolved in acetone (O-A-51) or other low-boiling solvents such as ethyl acetate. Celluloid (MIL-B-10854) solutions in acetone are sometimes preferred. Usual amounts are 1.8-4.0% on a dry basis.

The binder most often specified in large military flare candles belongs to the class of thermosetting alkyd-styrene resins (OS 8819, Class 3) more commonly called unsaturated polyesters (with added styrene). It is best known by the tradename of Laminac 4116 (MIL-R-7575A) and is made by the American Cyanamid Company. Since the liquid resin is relatively low in viscosity, it has the virtue of being able to be added to the solid powders without any volatile solvent. For activation, 1-2% of methyl ethyl ketone peroxide (tradename Lupersol DDM) are mixed with the resin. This "catalyst," actually a curing agent that reacts chemically, converts the resin first into the gel state, then in strongly exothermic reaction into a hard solid. The reaction is accelerated by the promoting agent cobalt naphthenate. It is inhibited by air, so that curing in thin layers may leave a tacky mass, but the type 4110 is claimed to be less influenced by air.

Laminac is usually admixed as 4-9% of the formula weight. Notwithstanding its relatively low viscosity, the incorporation, usually together with the metal powder in a preliminary mixing phase, can be somewhat difficult. The addition of a wetting agent, specifically of Pluronic F-68 (the tradename for a polyoxy alkylene polyol), has been recommended with the claim that it reduces friction (presumably by facilitating spreading out of the liquid resins) and thus shortens mixing time.[273a] It is also possible to add more styrene as a diluent and this additional copolymerizing plastic monomer will become part of the solid resin. The practice has been to add some acetone or trichloroethylene whenever it appears desirable to facilitate distribution of the resin and complete wetting of the metal powder. One may have qualms about the addition of such solvents, but both evaporate quickly during mixing and it seems that their hazard is small compared with the chance that unmoistened metal powder and subsequently added dry oxidizer salt are subjected to the strong pressures and frictional forces of the mix muller. Presence of residual solvents definitely retards setting, which may be desirable to a point, but after some aging following mixing, no solvents should remain. Afterwards, one expects that the setting and consequent hardening will proceed at room temperature under the self-accelerating influence of the exothermically progressing polymerization.

The specific gravity of Laminac 4116 has been given as 1.12 before and 1.17-1.20 after cure with a shrinkage of 6.5%. No figures on the actual heat of polymerization of these resins seem to have been published anywhere. The given figure (from the manufacturer's data sheets) of 350°F for "peak exotherm," i.e. for the resin mixed only with curing agents, is of no help except for comparison with other cold-setting resins and may relate to curing time as well as to heat of reaction. One may take a cue from the heat of polymerization of various "plastics" monomers such as methylmethacrylate, vinylidene chloride, styrene, acrylonitrile, and vinylacetate, which have been given as 13.0, 14.4, 16.1, 17.3, and 21.3 kcal per mole of monomer,[616] from which values between 130 and 325 cal/g can be calculated.

Epoxy resins have been used successfully as strong binders in some special mixtures. A suitable type is Epon 828 (Shell Oil Co.) diluted with the active additive AGE (allyl glycidyl ether) that reduces the viscosity to less than one-tenth and about 8% DETA (diethylene-triamine) as a cold-curing "catalyst" (actually a cross-linking agent). This mixture can be used very much like a polyester resin. For less rigid compositions, a polyfunctional mercaptan such as LP-2 and LP-3 (Thiokol Chemical Corp.) can be used and such a resin can be combined with epoxy resin. A number of these types of resins are claimed as pyrotechnic binders in a patent by Hart, Eppig, and Powers.[617]

Wherever "plastic binders" appear on the scene, the words "cast-able" and "extrudable" are not far away. However, the amount of liquid resins in pyrotechnic mixtures seldom reaches the minimum weight at which the compositions become pourable or reasonably well extrudable, i.e. 20-25%. By dilution with solvents, the pourable state can be attained, but the problem of solvent removal and the fact that a low-density sponge or a collapsing mass will then remain would restrict this technique to special cicumstances.

The plastic binders that are used in monomeric or low polymeric form and are cured by additives must be distinguished from fully polymerized synthetic resins, or from natural solid resins, which require the presence of a solvent in order to display their full binding power. The exceptions are such materials that are plastic in the rheological rather than the vernacular sense, i.e. that flow easily under pressure. Gilsonite, a natural, pure asphalt, can be stored in finely powdered state and thus is easily incorporated in a mixture, but will also display good binding power on pelletizing. It was popular in World War II formulas, but it tends to leave a gummy residue in molds and rams after some period of continual production.

Natural gums, such as red gum (gum accroides), shellac, or rosin, are found in some formulas, but in general the tendency has been to get away from natural products, which, however, seem to persist in fireworks formulas. Rosin is not only unpleasant to comminute because it gums up the grinding equipment, but the particles lump together in storage. Limed rosin or calcium resinate, however, remains stable as do various other rosin derivatives and a cheap, related natural resin called Vinsol.

Of the synthetic resins that are found in pyrotechnic formulas, we find vinyl derivatives such as polyvinyl acetate and chloride, the latter as a chlorine donor rather than as a binder; rubbery resins such as polyacrylic resin Hycar 1052 × 3 in MEK solvent have been successfully incorporated in pellets.

A survey of the applications of plastic binders to smoke formulations, the pros and cons of their addition, and a literature search has been performed by a contractor to CRDL.[618]

Shidlovsky[3] describes and gives formulas with *Iditol*, a lacquer-type condensate of phenol and formaldehyde and with Bakelite A of the same general class as to composition but requiring as much as 180°C for the formation of the stable B- or C-type high polymer. Federoff[48] has reported that Bakelite A was also found in German World War II tracer compositions. Undoubtedly, a large number of synthetic polymeric materials can be incorporated in pyrotechnic mixtures, but in actual use the number seems to be strangely limited, perhaps because strictly comparative experiments pointing toward special advantages are rarely performed. However, modern requirements for high temperature stability have brought silicone resins and especially fluorinated polymers into the pyrochemical orbit, though mostly for use in propellants or explosives.[405,476,477]

All resins perform several tasks in compressed columns or pellets.

They are first a compaction aid, since as liquid or easily deformed plastic substances they reduce internal friction and friction against the enveloping mold or casing and tend to fill interstices between particles. For this reason, oils, waxes, paraffin, and dry lubricants such as graphite and, exceptionally, molybdenum disulfide, exert a beneficial influence and are in older formulas often the only "binder." Even moistening with water or a volatile organic solvent alone can lead to considerable improvement of compactability. Were it not for the difficulty of complete removal, such a vehicle would perhaps be more often used in low-gassing, highly compacted columns since no effective inorganic, nongassing binder has yet been found. Sodium silicate has been mentioned in an Italian patent,[456] but the formula given appears to be an impossible one. Alkali silicates mixed with borax or gelatinous ferric hydroxide, $Fe(OH)_3$, are shown in a British patent.[457] Colloidal silica in aqueous dispersion has weak binding power. Perhaps the silicate and phosphate cements point the way toward better inorganic binders. They are described in an encyclopedia article on "Dental Materials."[24]

In the typical flare formula, the oxidizer does not furnish enough oxygen to satisfy even the demands of the metal fuel; therefore the organic binder acts mainly as a ballast and diluent in regard to calorific output in the self-contained state. In the air, the evaporated binder or its pyrogenic decomposition products will burn. There are indications that at least a part of the chemically contained oxygen in a binder will react in the burning zone with excess of a strong reductant such as magnesium. Since the CH_2 group requires a minimum of 2.3 parts of oxygen per unit weight, hence roughly five to six parts of the most oxygen-rich salts, it is obvious that one does not find oxygen balanced formulas that include the binder. (Nitrocellulose, nearly oxygen-balanced by itself, would be an exception.)

A number of accessory materials and processing acids other than the foregoing have been mentioned in various places earlier; some more must be added here to complete the picture. *Manufacturing aids* that remain with the finished materials are stearic or oleic acid in flaked aluminum or stainless steel flakes. Surface protection of metal powders has been cited in the preceding chapter. It may be performed by the manufacturer or by the user.

Burning-rate modifiers have also been mentioned. They may be true catalysts or influence the initiation or the progress of reaction in some indirect manner. An instructive as well as baffling demonstration of such an influence, which can be used as a "parlor trick," is setting a cube of sugar on the edge of an ashtray and attempting to light it with a match; no flame appears, the sugar merely melts. The demonstrator then proceeds to show that the same or any other cube will be easily ignited and burn with a flame. The trick consists in applying surreptitiously a speck of cigarette ash to one corner of the cube, and the thus created "hot spot" acts somewhat like a wick.

In propellants, increase in burning rate has been demonstrated by embedment of pieces of metal wires. They act physically by conducting the heat from the flame front deeper and faster into the propellant grain. In a somewhat similar vein, it has been found that radiative heat transfer in more or less translucent propellants can create hot spots in dark specks of impurities and thus cause erratic burning. This is of course undesirable. Uniform darkening of the grain by carbon black is the answer to this deficiency.

Substances that keep powdered materials in free-flowing condition —*anticaking aids*—are used daily by everyone who is not on a salt-free diet. Such additives are found in table salt for this purpose. The salts of pyrotechnics, especially potassium chlorate and confectioner's sugar, are treated with 3% of magnesium carbonate and the effectiveness of such additions in preventing agglomeration can be startling. Sulfur tends to form loose aggregates, which can appear in a mixture as visible specks and little balls. These disperse easily on addition of some infusorial earth. Even more efficient is pyrogenic silica such as Cab-O-Sil of 15-20 mμ particle size, of which as little as 0.04% is effective.[444]

The anticaking properties of acid magenta and other dyestuffs have been described by Whetstone,[619] who recommends an 0.10% coverage of ammonium nitrate crystals during the crystallization process from solution as the preferred method. Butchart[620] claims the addition of certain water-soluble anthraquinone derivatives to solutions or to wet crystals to promote free-flowing "low setting" tendencies in potassium nitrate crystals. Similar improvements are claimed by Marti[621] for the addition of 2-3% CuO or Cu_2O to ammonium nitrate.

A processing aid may be psychologically rather than physically acting if it serves to identify a mixture by means of an addition of color. Toluidine Red Toner as a dry pigment, about 0.5% of the total, has been used in tracer compositions for this purpose. A soluble dye admixed to a protective lacquer coating might have the added advantage of indicating thickness and evenness of coating by the depth of coloration.

Mix-ups on processing lines may occasionally create a real problem and the formulator might be advised to tint mixtures in a series of increments that look alike but are functionally different. Red iron oxide pigment should be considered for this purpose. And, speaking of identification, the most disastrous mix-up is the one of potassium perchlorate and chlorate in the pyrotechnic laboratory, which at one time caused a serious accident to an associate in the writer's laboratory. By the same token, the acidic and relatively unstable form of sulfur, the "flowers of sulfur," should be banned from the shelf entirely. Its only place in pyrotechnics would be in items of the field-expedient type.

Preservatives are sometimes added to hydrophilic binders such as glue and dextrins. Sodium ortho phenyl phenoxide (tradename: Dowicide A) is used in matches. Copper 8-quinolinate ("Quindex") is an effective fungicide and mildewicide, 1% of the commercial 10% dispersion being added to dipping was according to MIL-W-10885. This wax is a high-melting microcrystalline coating material for cardboard and paper. There are, of course, numerous other effective fungicidal or fungistatic agents and other preservatives, many described in a book on deterioration of materials.[610] While everyone knows that paper, textiles, and cordage are easily destroyed by microorganisms, it should be realized that rubber and "plastics," seemingly not nourishing substrata for bacteria, etc., are attacked under tropical and other unfavorable conditions.

Part X

Specific Materials

chapter 37

General Remarks

The following notes on the materials most frequently employed as active ingredients of pyrotechnic formulas are written as supplements to the dispersed remarks on specific properties, not as comprehensive descriptions. Extensive evaluation of each single material from the pyrochemical viewpoint has never been attempted and the lack of such information is probably often keenly felt, An example is the confusing multiplicity of specifications and lack of even unambiguous terminology for aluminum powder. Being in the position of one who asks questions, the writer leaves it to better-informed persons to provide the answers. AMCP 706-187[32] is a beginning, but it leaves out many desirable specific facts. And inorganic chemical textbooks are nowadays rarely factually detailed, emphasizing "understanding" while supplying a bare minimum of those "disjointed, incoherent facts" that, alas, are the life-blood of developmental research and of technology in general.

chapter 38

Sulfur, Selenium, Tellurium

Sulfur is one of the old-time fuels of fireworks—abundantly available, cheap, safe to handle, nontoxic, of good stability, easily ignitible. It is not useful for achieving very high temperature or high caloric output, and dissipates heat, since the products of combustion are gaseous.

When raw sulfur is distilled for purification, it liquefies on cooling and becomes a crystalline (rhombic) solid, completely soluble in carbon disulfide. After grinding, such *flour* of sulfur is the material of MIL-S-00487A (1965) and the only type used correctly in pyrotechnics proper. On the other hand, if the sulfur vapor is chilled to solidify directly from the gaseous state, a fine powder is obtained consisting of various molecular aggregates, part of which are insoluble in carbon disulfide. It is called *flowers* of sulfur, is much easier oxidized in air than the other kind, and always contains acid (up to 0.25%). Because of its greater activity, it is used as a fungicide and therapeutic agent and may, because of convenient access from the drug store, often find its way in the laboratory. Only in field-expedient devices should such sulfur ever be considered pyrotechnically.

Sulfur ignites at about 260°C (data as low as 190°C pertain to sulfur flowers) and burns to sulfur dioxide (SO_2), but some trioxide (SO_3) may appear, especially in the presence of iron oxides or other catalysts. Among minor products, sulfide, thiocyanate, thiosulfate, and sulfate may occur in alkaline residues.

Sulfur and its congeners *selenium* and *tellurium*—the three forming the group called the *chalcogens*—can either be fuels or serve as oxidizers, in the latter case mainly in reactions with lead or iron. Selenium and tellurium, while used only in some specialties (delays and igniters for blasting caps), are highly toxic when ingested. Tellurium, in particular, causes a noisome odor in exhaled air and perspiration as long as the element is in the human body.

The major uses of sulfur are for black powder, railroad fusees and similar low-powered flares, thermate, matches, and colored smokes ("Chemical Corps" type).

chapter 39

Carbon and Carbonaceous Fuels

Carbon shares several significant properties with sulfur: It is easily ignited, safe, fairly stable, abundant, cheap, and its products of combustion are gaseous. It can, however, hardly be called versatile. In fact, with the decline of black powder, it has been mostly relegated to some fireworks effects to produce a prolonged glow of reddish sparks.

The so-called amorphous carbons—charcoals, or more rarely, the "blacks"—are the only ones considered, graphite being much too difficult to ignite, though it is combustible above 700°C. Graphite is sometimes used as a mold release agent and its effect on the reactions is either indifferent or adverse.

Charcoals are made by the pyrolytic decomposition of wood and are often not even remotely close to being pure carbon if manufactured at relatively low temperatures. Alder charcoal, used in some black powders, contains only about 70% pure carbon. The best calcined grades retain hydrogen and oxygen equivalent to about 5% water, and some nitrogen. Therefore, no composition containing charcoal can be safely regarded as stable over long periods. Porosity reduces the apparent specific gravity to below 1 (it floats on water), while its true specific gravity is about 1.8. The specification for wood charcoal, JAN-C-178A, permits a moisture content of 5%; black powder, consequently, may contain 0.70%.

The author has not found *Carbon Black*, MIL-C-11403A, or *Lamp Black*, TT-L-706, in ordnance formulas, but a "*Carbon Black*, dry for use in explosives" appears under MIL-C-306 and has been specified for and used in a *Nuclear Air-Burst Simulator*.

The ignition points of charcoal and other carbons vary widely and are influenced by the presence of catalysts such as lead salts. Nebel and Cramer,[622] in a study concerning the behavior of carbon in gasoline engines, reported an ignition temperature of 555°C for pure carbon, but the addition of lead compounds depressed the ignition temperature, the lowest figure being 266°C. The ignition temperature of charcoal varies with the kind of wood and the duration and maximum temperature of pyrogenic decomposition—300 to 400°C is given in textbooks.[60,61]

Charcoal is used mainly in black powder, the other carbons for fireworks effects and as a darkening agent in match striking strips and certain propellants.

Of carbonaceous fuels, only those that are not also binders will be mentioned here. Common sugar (sucrose, saccharose), of specification JJJ-S-791F, mostly used in powdered form, and the somewhat more expensive lactose monohydrate, specification MIL-L-13751, are examples of this small group. Sucrose is an important cool-burning fuel in colored smokes (Navy type) and for dissemination of toxic agents. Hardwood flour is probably the cheapest fuel in pyrotechnics and is used in commercial fusees. Rosin and rosin derivatives belong here and are occasionally found in formulas.

Because pyrotechnic formulas favor retention rather than dissipation of heat, carbonaceous fuels are not much used and often only unavoidably if the material functions as an adhesive.

chapter 40

Boron and Silicon

Boron and silicon are elements and nonmetals like sulfur, carbon, and phosphorus. In pyrotechnic behavior, they differ widely from these three, in part because of nonvolatility of their oxides. They act like metal fuels to a certain extent and are often incorrectly called "metals," though they form acidic oxides that appear as borates or silicates in secondary exothermic reactions in the "ash."

Boron surpasses in heat output every element except hydrogen and beryllium. By itself it burns only partially, even under pure oxygen at higher pressure, forming a glassy, low-melting oxide that envelops the unburnt residue.[561] Reactivity in mixtures, however, is complete and borate formation may contribute significantly to heat output.

Boron, also called amorphous boron, is a very fine (about 1 μ average particle size) somewhat difficult to disperse dark brown powder, for which purchase description PA-PD-451 and specification MIL-B-51092 have been issued. The commercial product may contain only 84-96% boron, the main impurity being oxygen, said to be in the form of a suboxide of boron, (B_4O).[516]

Boron is the most expensive of all regular pyrotechnic ingredients and is only used in small, accessory items such as first fires, rocket igniters, or delay compositions. Its costliness is somewhat lessened because of a very low equivalent weight, so that it is contained in suitable compositions at as little as 3% up to a maximum of 25%. The great value of such compositions is good stability, ease of initiation, persistence of burning even at greatly diminished atmospheric pressure, and excellent fire transfer properties.

The combinations of boron with oxidizers other than lead compounds seem to be less subject to accidental ignition than comparable mixtures of sensitive metals and oxidizers. Boron is used extensively in first fires and delay compositions and may be useful as a specialty heat source. Because of the small amount of boron required in combination with heavy metal oxides, certain mixtures can be used as nonconductive plugs between two electrical leads. On initiation of the mixture by an ignition source, the plug becomes electrically conductive, so that the device operates like the earlier described "explosive" switches but without moving components.

Silicon, MIL-S-230A, is a much longer used ingredient of pyrotechnic mixtures, mainly in ignition compositions and delay trains and in variants of the thermite process. It is abundantly available, stable, inexpensive, and very safe to handle. The crystalline form, finely ground to about 5-19 μ average particle size, is the commercial product. An amorphous type of about 2 μ exists but is not recommended.[23] Though widely used in mass-produced ordnance items such as hand grenade fuzes, it seems never to have been examined in detail for pyrochemical properties such as the influence of impurities contained in technical products.

Silicon is a dark grey powder, difficult to ignite by itself, though its heat of combustion is high. Its normal reaction product is silicon dioxide (SiO_2) that in secondary reactions may form silicates, but the additional heat from these reactions is very small. Under certain conditions at high temperature, a gaseous SiO that forms a solid dark powder on sudden cooling may appear.

With metals, silicon combines in exothermic reactions to silicides such as Mg_2Si. It can also combine exothermically with nitrogen at 1000°C.

chapter 41

Magnesium and Aluminum

Magnesium and aluminum are the most important pyrotechnic fuels in quantity of consumption as well as in effectiveness for creating light and heat. Magnesium, since its introduction into fireworks about 1865, has been in the forefront of the development of radiation effects—white and colored, visible, ultraviolet, and infrared, while aluminum, used pyrochemically since about 1895, has been primarily a heat source in the thermite process but has also taken an increasing part in light production such as in flash charges.

Both metals are now readily available in a variety of powdered forms, also alloyed with each other and forming the compound Mg_4Al_3, and are moderately costly, magnesium more so than aluminum. While safe in storage or if handled by themselves (with the possible exception of the very finest grades as dust ignition hazards), nearly all their pyrotechnic mixtures are enormously hazardous. The danger increases with lowered particle size, reactivity of the oxidizer, and amount of mixture. Little exact knowledge exists about some environmental conditions such as the purported sensitization by water. In recent decades, accidents have occurred with magnesium/sodium nitrate/binder mixtures during the mixing process and with photoflash mixtures with aluminum during loading. An exploding charge of 50 lb or even less will not only demolish a mixing apparatus and blow off the roof and doors of an armored bay, but can crack and dislocate slightly a 12-in. wall of reinforced concrete. Under some conditions, flare mixtures in the finished, consolidated form can explode following normal initiation.

The terminology of the powders and the profusion of old and new specifications and of commercial grades, especially for aluminum is confusing. Here is a list of military specifications:

JAN-M-382A (1959): Only specification given for magnesium in AMCP 706-187.[32] Three types including stamped and atomized material are described. The various grades are designated by numbers such as the nominal mesh size 30/50 by #18, but an average particle size of 350 ± 50 μ is also given.

MIL-P-14067A (1963)—with minor amendments in 1964—contains Type I, atomized *magnesium*; Type II, *aluminum* "spheroidal or granular"; Type III, Mg/Al (65/35) *alloy*, spheroidal. Granulations are five close cuts from 20/50 to 200/325 for magnesium; one only, 200/325, for aluminum; and two for the alloy.

JAN-A-667 (cancelled 1964), quoted in the AMC pamphlet, with three types for aluminum: flakes; "granular or spheroidal manufactured by the atomizing process or grinding"; and "granular or spheroidal by the atomizing process."

MIL-A-512A (1961) shows two grades of aluminum, a very fine one and a coarse one of "irregular granules."

JAN-A-289, superseded by the preceding one.

MIL-A-23950A (1966) shows three grades of spherical aluminum powders, all very fine: 4.5-9.0 μ, 12.0-18.0 μ, and 25.0-30.0 μ.

JAN-M-454 (1952) contains a 50/50 and a 65/35 magnesium/aluminum alloy of different granulations.

A few notes from literature and older correspondences with manufacturers tying in commercial grades with specifications may be helpful: Metals Disintegrating Company's atomized aluminum of designations MD 101 and 201 are in the range of MIL-P-14067A "200/325" material and also apparently of MIL-A-512A, Grade I. The latter specification applies also to Reynolds Metals' "120," while their grade LS-985 corresponds to the same specification, Grade II, "granular."

Of fine commercial aluminum powders, Reynolds # 40 has an average particle size of 7 μ; # 1-511 of 25 μ; Metals Disintegrating Company's # 209 of 18 μ; # 201 of 33 μ; and Alcoa's # 123A of 10 μ.

Going back to descriptive definitions, the confusion continues: Granular metal would seem to be a very coarse material and Edwards and Wray[624] speak of "granulated" for the coarsest grade (up to ½ in. in diameter). *Grained* aluminum, quoting the same source and other information, consists of irregular-shaped grains 1/64 to 1/4 in. in width or length derived from molten aluminum stirred slightly below the temperature of consolidation. It is suitable for commercial thermite. However, the specifications for thermite and thermate define grained aluminum as atomized, 200 mesh material (in thermate only) and "granulated" (in both thermite and thermate) as a —12/+140 mesh coarse powder.

Outside of military specifications seems to be "pyro" aluminum, a dark powder,[7] called by Izzo[9] *alluminio nero* (black) or *oscuro* (dark), of which the grade *piro* (pyro) is "very fine." An American commercial grade consists of extremely fine flakes suitable for and used in flash crackers. Such powder is made from burned paper-backed foil scrap and contains a little carbon that promotes ignitibility.

As a lower-priced substitute for atomized magnesium, a material is offered in the trade that is made by the ball-milling under an inert gas of chips cut from ingots.[624a] The flattened particles are compacted during the milling to form more or less globular shapes that compare in screen "cuts" with corresponding atomized material, but are of course of lower apparent density.

A British sample of a "blown" aluminum consists of irregular more or less spindle-shaped, fairly small particles obviously made by a process similar to "atomizing."

Commercial aluminum flakes, little used in ordnance items because of difficulty in compacting, come in various particle sizes, all of large surface area and formerly made by stamping, but now by wet ball-milling. Either stearic acid or a lubricant similar to oleic acid is added, which determines the ability to film-forming ("leafing") for use in paints or the absence of this property. Flakes not only retain 1-4% of the organic material but their oxygen content can go as high as 15%.

German terms for aluminum and magnesium powders, necessary for the understanding of World War II formulas in CIOS and BIOS reports, are *Griess* (coarse cereal, farina) for very coarse powders; *Flitter* for a coarse, flaky material; *Schliff* for flakes for metal inks; and *Pyroschliff* for finest flakes for fireworks.

Recently, undoubtedly in order to reconcile the demands for high reactivity and burning rate without the difficulties encountered in working flake material into mixtures, new types of aluminum powders called *acicular* or *micro-acicular* have been brought out by the Valley Metallurgical Processing Company. The material resembles wire segments.

To reiterate briefly the uses of these metals in pyrotechnics: Magnesium is the main ingredient of high intensity military flares and a minor ingredient in colored lights. Aluminum is the fuel in most flash compositions and in thermite and thermate. The alloys find limited use in some flash mixtures and those with low aluminum content as bodies of fire-bombs.

chapter 42

Titanium and Zirconium

Titanium and zirconium have gained much favor in pyrotechnic accessory devices of highest reliability, such as first fires, igniters, and heat sources. Both metals combine very high stability with low ignition and initiation temperatures. They are high in calorific output, form white-hot sparks, and display excellent fire-transfer properties. Finely powdered zirconium is exceedingly sensitive to accidental ignition by static electricity.[135,625,626] Both elements are very expensive.

The most useful commercially available titanium powders are made by calcium hydride reduction, and are furnished in a variety of particle sizes, generally coarser than zirconium. The latter is made by calcium metal reduction and is of an average particle size of only 2-5 μ, measured by the permeation method, though coarser grades also exist. The differences of manufacture and grain sizes do not permit a fair comparison of the two as to intrinsic hazardousness, but at present in this country, zirconium is frequently the cause of serious accidents, while to the author's knowledge titanium rarely is. Conversely, the subsieve-size zirconium affords advantages in performance that cannot be duplicated by the other metal.

If the average particle size is at least 10 μ, the hazard in handling zirconium in the dry state is regarded as small. The fine powders are shipped, and wherever possible, processed under water. If, however, the amount of water present is small (less than 20%),* any accidental ignition has been said to have more serious consequences than results from burning in a dry state because of the violently explosive action caused by steam formation and reaction with the water as the oxidizer.

Specifications for the two metals are of limited usefulness for the fabricator. MIL-T-13405B for technical powdered titanium super-

* The stored zirconium metal slurry must be homogenized before removal from its container since separation of water and metal powder may cause a severe hazard.

sedes an old Chemical Corps specification (196-131-422) and refers to one grade for which burning rate limits are given. The burning rate is determined on unconsolidated powder in narrow paper tubes. Such simple comparative tests (sometimes performed on a split metal tube forming a shallow trough perhaps one foot long) can indicate a consistency of the material from container to container or lot to lot.

Zirconium is still designated, as far as the finest commercial grade is concerned, as JAN-Z-00399B with several exceptions. Grade 120A of the Foote Mineral Company and a comparative grade made by Ventron Corporation, Metal Hydrides Division are the ones to choose from. Both conform also to particle-size distributions and surface-area recommendations that came out of DOFL (now HDL) investigations.[569a]

Coarser (but still very fine) zirconium is found in two purchase descriptions: FA-PD-MI-2364, Grade A or B, and the older PD-FED-1655, Grade B. There is also a PA-PD-464.

The ignition temperature of these metal powders is low. It has been given for zirconium as 180-200°C in air. Zirconium always contains traces of hydrogen and its ignition temperature is lower the less hydrogen is present.

Both metals form hydrides of the approximate formula MeH_2, but actually hydrogen is bound to the metal in an unbroken series of proportions as interstitial hydrides. The hydrogen can be removed by heat and vacuum. The hydrides of titanium and zirconium are much safer to handle than the metals themselves and are shipped in dry state. Unfortunately, their greater inertness shows itself also in a lower ignition sensitivity and a slower burning rate, which limits their usefulness. Conversely, the removal of the last traces of hydrogen increases the burning rate and sensitivity to ignition. It has been stated that the purest, very fine zirconium powders are unmanageable because of ignition points as low as 85°C.[627]

Titanium forms several stable oxides—TiO_2, Ti_2O_3, Ti_3O_5, TiO. While the product of combustion of the metal is TiO_2, some calculations of reactions have been based on Ti_2O_3 as the product, undoubtedly because of dissociation of TiO_2 into lower oxides at high temperatures. TiO is a commercially available "easily vaporized, grey-golden" solid used for optical purposes. Zirconium forms only ZrO_2 on oxidative reaction but a lower oxide, ZrO, apparently exists in solid and in gaseous state.

Zirconium is always associated in nature with the element hafnium from which it is difficult to separate. Should hafnium become more available, it promises to have interesting pyrochemical properties. Because of its high density (11.4), nearly twice that of zirconium, its moderate heat of combustion of 1.5 kcal/g becomes on a volume basis nearly as large as that of zirconium.

Titanium and zirconium powders, other than the very fine powders described above, are on the market for powder metallurgical purposes and will undoubtedly play a part in future pyrochemical reactions other than for primers or first fires. Zirconium foil is also available and is replacing aluminum foil in oxygen-filled flash bulbs.[628,629]

Hafnium powder as well as foil of 0.0022-in. thickness has recently been offered for sale but at high cost, the foil being priced at $4.00 per square inch (*Newsletter* of Carborundum Metals Climax, Inc. of June 1966).

chapter 43

Halates and Perhalates

These two groups include chlorates, bromates, iodates, perchlorates, and periodates. Of these, only potassium chlorate and potassium perchlorate are of pyrotechnic importance because they are the most stable nonhygroscopic salts in this class. Barium chlorate is found in fireworks but has been abandoned for military purposes; ammonium perchlorate is rarely found but is important in propellants.

The similarity of the words chlorate and perchlorate is unfortunate and a new nomenclature would be desirable. Misquotations occur even in formulas in the literature and accidents in the laboratory can happen due to mislabeling or because of misreading of correct labels.

Greater hazardousness or activity is often erroneously imputed to the compounds that are higher in oxygen content, namely the perchlorates. Even chemists are sometimes guilty of this dangerous misconception. Actually, all chlorates are much more of a potential hazard than perchlorates because of the considerably greater friction and impact sensitivity of mixtures of chlorates with fuels, especially with nonmetallic solids such as phosphorus or sulfur, but also with sulfides and metal powders.

In forming anions from chlorine and oxygen atoms, the stability of the anion increases with the number of oxygen atoms, from hypochlorite to chlorite, to chlorate, and to perchlorate. This difference is still pronounced at the last step between chlorate and perchlorate. The lower decomposition temperature and lower melting point of the chlorate, its exothermic decomposition, the fact that this reaction is easily catalyzed, and the instability of free chloric acid are all factors that contribute to the greater reactivity and sensitivity of chlorate mixtures.

It must not be concluded, however, that the lower friction and impact sensitivity of perchlorate mixtures makes them "safe" in a general sense. Should a condition of accidental ignition arise at all, such as in mixing and loading operations, then the perchlorate/fuel mixtures are as potent and destructive as chlorate mixtures. The ability of such mixtures in the form of dry powders or even in consolidated state to react instantaneously and be destructively explosive under some conditions is often underrated. Because the volume of the gases and gasified solids from the reactions is much smaller than in "real" explosives, the destructive effect of pyrotechnic explosions is essentially a local one.

From the safety standpoint, neither chlorate nor perchlorate present a hazard by themselves, except that due to danger of spillage they should be located away from fuel-type materials and if stored in very large quantites must never be exposed to any potential major conflagration.

While the production of oxygen from chlorate is an instructive and useful laboratory experiment, chlorate should not be included in so-called chemistry sets for children.[630]

Potassium chlorate is the subject of MIL-P-150B (1962) that includes three grades and seven classes of fineness. Class 6 or Class 7 (the finest) approximate the granulation used for matches. Grade A and B differ in bromate content, which is regarded as possibly exerting a degrading influence. Grade C contains 3% magnesium carbonate (MIL-M-11361B, Class A, Grade B). This prevents agglomeration and promotes free-flowing and thus better dry-mixing behavior. *Potassium perchlorate* is covered by MIL-P-217A in two grades and five classes, the finest specified as 20 ± 5 μ particle size, the same as in the older PA-PD-254, Grade A (for delays). In some delay formulas, this particle size is not small enough and custom grinding must be resorted to.

Chlorates and perchlorates agglomerate in storage, as it seems mainly under the influence of pressure. Up to a point, mild regrinding or screening will restore the original particle size.

Ammonium perchlorate carries the specification MIL-A-192B. Because of the importance of ammonium perchlorate, propellant producers have set up their own specifications.[508]

Chlorates find use in primary explosives and igniter mixtures for colored lights in commercial fireworks, in colored smokes, and for oxygen production.

Potassium perchlorate is an important component of various light producers, especially colored flares and of photoflash powder, and is a burning-rate controlling and accelerating accessory in low gassing ignition and delay mixtures. Potassium and ammonium perchlorate are the major ingredients in composite rocket fuels in combination with organic polymers.

The stability of the chlorates declines with the increase of valence of the cation. Barium chlorate, normally forming a monohydrate that loses its water above 120°C, is somewhat less stable than potassium chlorate. Sodium chlorate would be a desirable substitute for the potassium salt were it not for its hygroscopicity. It has a lower melting and decomposition temperature than potassium chlorate, and mixtures of both oxidizers melt at even lower temperatures.

The heavy metal chlorates are all very soluble and hydrated salts except for monovalent mercury and thallium salts. The organometallic base mercarbid of K. A. Hofmann, which is explosive by itself, forms a chlorate that explodes violently when shaken, even in aqueous suspension.

Bromates, iodates, and periodates have hardly attracted the attention of the pyrochemist, though they could be of considerable theoretical interest. Bromates and iodates would be expected to be more stable than chlorates, and the few data that are available for the iodates point in this direction.

Of the perchlorates other than the potassium and ammonium salt, mainly lithium perchlorate has found interest as an oxidizer for composite propellants and for making elementary oxygen because of its very high oxygen content. It is, however, extremely hygroscopic in the anhydrous state and will deliquesce at normal ambient humidities. Even at an ambient relative humidity of only 25%, it has been said to take up 5% water.

Sodium perchlorate, which is also hygroscopic but less than the lithium salt (though the information is scant), is rarely mentioned and even its specific gravity in anhydrous state, though reported in 1931 as 2.499, came to light only when quoted in Mellor's Supplement I to Volume II (issued as "Supplement II, Part I"). It might deserve more attention and study than given previously.

Hygroscopicity need not be an unsurmountable hindrance to its use in large composite propellant grains, where accidental influx of humidity from the atmosphere can be readily controlled, but it imposes heavy burdens on economy of handling and processing the material up to the point where it is enclosed in the rocket motor. Similar considerations may apply to future pyrotechnic items when presence of larger amounts of plastic binder, in which the hygroscopic salt is embedded, is permitted and hermetic sealing of units prior to initiation is performed.

The greater stability of perchloric acid and its alkali salts, compared with chloric acid and chlorates, applies also to its alkaline-earth and heavy-metal salts, but all are either highly hydrated or extremely hygroscopic when prepared in anhydrous state. This does not apply to some cationic complex salts that have been proposed for colored light production. Others deserve attention because of their explosive properties.

An intriguing property of potassium perchlorate is its ability to form isomorphous crystal mixtures with potassium permanganate in an apparently continuous series, reported from 3-96 mole-% potassium permanganate and cited by Mellor.[67] A similar behavior is reported with rubidium permanganate and with potassium periodate (KIO_4). To the author's knowledge, this possibility of influencing the reactivity of the perchlorate by "built-in" permanganate has never been pyrochemically exploited.

chapter 44

Nitrates

The nitrates are more economical and more abundantly available in war time and safer to handle but more difficult to initiate under comparable conditions than chlorates or perchlorates. They are practically equal to chlorates in usable oxygen but also release nitrogen. This is generally not objectionable in flare candles, star compositions, etc., but makes them undesirable for fully confined applications other than for pressurization.

Potassium nitrate, formerly the only practical nitrate because of its low solubility and nonhygroscopicity is now frequently replaced by the sodium salt where the latter's spectral characteristics are not objectionable or are even desirable, as in high-intensity flares.

Barium nitrate would be the best of the three from the standpoint of physical stability and also of heat concentration because barium oxide has considerably better refractory properties than either potassium or sodium oxide. However, the high equivalent weight of barium limits its usefulness. Barium and strontium nitrates are indispensable for green and red effects and can also be used combined for white light. In combination with potassium perchlorate and aluminum, barium nitrate is part of standard photoflash powder. Ammonium and guanidine nitrate are sometimes used because of their gas-forming properties, and lead and silver nitrate can be found in special applications.

Sodium nitrate carries the specification number MIL-S-322B, covering two classes, "technical" and "double refined," and two granulations. The purer material, ground to below 50 μ average particle size (30 ± 15 μ), is prescribed for flare formulas of the "Rita" type and the material predried prior to mixing.

Potassium nitrate is described in MIL-P-156B for one purity and three granulations. Aside from use in black powder and some starter mixtures, its main pyrotechnic application is as the oxidizer in boron/potassium nitrate pellets. In the latter, a very fine granulation, 10 ± 5 μ

is used.

Strontium nitrate is found under MIL-S-20322A and barium nitrate in MIL-B-162C. Of little concern to the pyrotechnician is ammonium nitrate, MIL-A-175A.

Heavy metal nitrates are all highly hydrated but the anhydrous salts can be synthesized by special methods[631] or, exceptionally, obtained by careful dehydration. Perhaps they belong to the "pyrotechnics of the future." Thorium and cerium nitrate are actually found in a patented flare formula[632] and in some old flash powders for civilian use, purportedly because of lower smoke formation of the residual oxide and perhaps with the thought of selective spectral emission from the rare earths.

Mixtures of metal powders and nitrates are deceptively inert when tested in small quantities for properties of initiability. Sometimes it is quite difficult to ignite them with an ordinary match in an "ash tray" experiment. As has been mentioned repeatedly, such misleading demonstrable lack of activity is no guarantee for safe behavior under different circumstances.

chapter 45

List of Government Specifications

While numerous government specifications have been cited in previous chapters and are also found in the formulary, it may be helpful to the practical operator in the United States to have a handy reference of the more frequently applicable specifications, brought up-to-date (Fall 1966) as much as possible. Joint Army/Navy (JAN) specifications are being abolished in favor of the MIL-designation, but some older ones still exist. Details signifying which branch of the Defense Department is custodian of a certain specification have been omitted. Military specifications are generally obtained through a contracting officer or through the Naval Supply Depot, Philadelphia, Pennsylvania.

Specifications are subject to minor revisions expressed by addition of letters to the numbers. While the author has tried to secure the most recent information, the given designation may have been superseded. Also, the reader may find the specifications given with the formulas are the original specification or an earlier revision applied at the time the formula was reported.

Acid, Stearic (technical)	MIL-A-271B (1962)
Adhesive, Dextrin	MIL-A-13374B (1963)
Aluminum, Powdered, Flaked, Grained, and Atomized	MIL-A-512A (1961)
Aluminum, Superfine	JAN-A-667 (cancelled 1964)
Aluminum, Powder, Spherical	MIL-A-23950A (1966)
Ammonium Nitrate, Tech.	MIL-A-175A (1963)
Ammonium Perchlorate, Tech.	MIL-A-192B (1965)
Antimony Sulfide	MIL-A-159C (1964)
Anthracene	JAN-A-202A (1958)
Asphaltum (Gilsonite)	JAN-A-356 (1946)
Auramine Hydrochloride (for colored smoke mixtures)	MIL-A-3664 (1952)
Barium Chromate	MIL-B-550A (1958)

Barium Nitrate, Tech.	MIL-B-162C (1964)
Barium Oxalate	JAN-B-660 (1948)
Barium Peroxide	MIL-B-153A (1964)
Binder—Cellulose Nitrate/ Camphor (for pyro mixtures)	MIL-B-10854 (1951)
Boron, Amorphous, Powder	MIL-B-51092 (1962)
Calcium Carbonate, Precipitated (pigment)	MIL-C-15198A (1952)
Calcium Oxalate (monohydrate)	JAN-C-628 (1948)
Calcium Phosphide	MIL-C-3539 (1951)
Calcium Resinate (Limed Rosin)	MIL-C-20470A (1964)
Calcium Silicide, Tech.	MIL-C-324A (1958)
Carbon Black, Dry (for use in explosives)	MIL-C-306 (1965)
Carbon Black, Tech.	MIL-C-11403A (1958)
Cellulose-Nitrate-Plastic (Celluloid or pyroxylin type, for use in ammunition)	MIL-C-15567 (1950)
Cement, Pettman	JAN-C-99 (1945)
Charcoal, Wood, Powdered	JAN-C-178A (1949)
Cupric Oxide	MIL-C-13600A (1962)
Cuprous Oxide	MIL-C-15169 (1964)
Diatomaceous Earth	MIL-D-20550A (1962)
Dyes	
1-(2-Methoxyphenylazo) -2-Naphthol Sudan Red G	MIL-D-3179 (1950)
1, 4 Diphenyl Toluidino Anthraquinone	MIL-D-3277A (1962)
2-(4-Dimethylamino Phenylazo) Naphthalene	MIL-D-3613 (1954)
1-Amino Anthraquinone Fast red A1	MIL-D-3698 (1952)
1, 4 Dimethylamino Anthraquinone, Fast blue B	MIL-D-21354 (1959)
4-Dimethylamino Azobenzene (for colored smoke mixtures)	PA-PD-367 (1953)
1-Methylamino Anthraquinone	MIL-D-3284A (1963)

Indanthrene Dye, Golden Yellow GKAC	MIL-D-50029B (1963)
Fuse, Blasting, Time	JAN-F-360 (1946)
Fuse, M700, Loading, Assembly, Packaging	MIL-F-45144A (1964)
Fuse, Firecracker, 30 sec	MIL-F-18405 (1955)
Fuse, Safety "Bickford Type"	MIL-F-20412 (1951)
Fuse, Time, Plastic-Coated	MIL-F-18215 (1955)
Graphite, Dry	MIL-G-155A (1962)
Glass, Ground (for ordnance use)	JAN-G-479 (1947)
Gum Arabic	JJJ-G-821 (1954)
Hexachloroethane, Tech.	MIL-H-235A (1956)
Hexachlorobenzene	JAN-H-257 (1945)
Iron Oxide, Black	MIL-I-275A (1962)
Iron Oxide (Red), Ferric	JAN-I-706A (1962)
Lactose, Tech.	MIL-L-13751 (1954)
Lead Chromate	JAN-L-488 (1947)
Lead Dioxide, Tech.	MIL-L-376B (1965)
Lead Monoxide (Litharge, dry)	MIL-L-1147A (1952)
Linseed Oil (Boiled)	TT-L-190C (1964)
Linseed Oil (Raw)	TT-L-215A (1960)
Magnesium Carbonate	MIL-M-11361B (1962)
Magnesium Powder	JAN-M-382A (1959)
Magnesium/Aluminum Powder	JAN-M-454 (1952)
Manganese, Powdered (for use in ammunition)	JAN-M-476A (1948)
Manganese Delay Composition (D-16 Powder)	MIL-M-21383 (1958)
Manganese Dioxide	MIL-M-3281 (1950)
Matches	MIL-M-1513 (cancelled 1963)
Matches, Safety and Matches, Non-Safety, SAW	EE-M-101H (1964)
Striker for Firestarter M1 (Match Head Mixture V)	MIL-STD-MS-585
Nitrocellulose (12.6% N)	MIL-N-244A (1965)
Iron Oxide, Black (Pigment, synthetic, dry)	TT-P-390 (1963)

Iron Oxide, Red (Indian Red and Bright Red) (Pigment, dry, paint)	TT-P-375 (1957)
Parlon, Chlorinated Rubber	FA-PD-FED-1657
Plaster of Paris (Gypsum, Calcined)	SS-P-402 (1945)
Phosphorus, Red-Yellow	MIL-P-211A (1957)
Phosphorus, Stabilized	JAN-P-670A (1966)
Photoflash Powder	MIL-P-466A (1951) (Cancelled Jan. 1964)*
Pigment, Toluidine (Red Toner, Dry)	TT-P-445 (1961)
Pigment, Lampblack	TT-P-350A (1961)
Polychlorotrifluoroethylene (Kel-F Oils & Waxes, plastic molding material)	L-P-385A (1964)
Polytetrafluoroethylene [Teflon, Fluon (British)]	PA-PD-614
Polyvinylchloride (PVC)	MIL-P-20307 (1951)
Potassium Bicarbonate	MIL-P-3173 (1950)
Potassium Chlorate	MIL-P-150B (1962)
Potassium Nitrate	MIL-P-156B (1956)
Potassium Perchlorate	MIL-P-217A (1962)
Powder, A1A (Gasless Ignition)	MIL-P-22264, Amend. 2 (1962)
Powder, Black	JAN-P-223B, Amend. 1 (1962)
Powder, Metal, Atomized [Mg, Al, Mg/Al]	MIL-P-14067A, Amend. 1 (1964)
Quickmatch	MIL-Q-378B (1963)
Red Lead (Dry and paste-in-oil)	TT-R-191D (1966)
Resin, Polyester (Low Pressure Lamination)	MIL-R-7575B (1961)
Resin, Thermosetting (Alkyd-Styrene)	OS-8819
Shellac, Orange Gum	MIL-S-20526 (1951)
Silicon, Powdered	MIL-S-230A (1959)
Sodium Bicarbonate	O-S-576C (1961)

* Still being issued (1967) on request for Photoflash Powder Specification.

Sodium Nitrate	MIL-S-322B (1962)
Sodium Oxalate, Tech.	JAN-S-210 (1945)
Strontium Nitrate (Nonhygroscopic)	MIL-S-20322A (1964)
Strontium Oxalate	MIL-S-12210A (1956)
Strontium Peroxide	JAN-S-612A (1964)
Sugar, Beet or Cane (refined and brown)	JJJ-S-791F (1965)
Sulfur, Ground	MIL-S-00487A (1965)
Tetranitrocarbazole (TNC)	MIL-T-13723A (1959)
Thermate, TH3	MIL-STD-572 (1950)
Thermit (burning charge Mk I)	MIL-T-18508 (1955)
Thermite, plain, incendiary	MIL-T-407A (1952)
Thickener, M1	MIL-T-589A (1953)
Thickener, M2	MIL-T-0013025B (1964)
Thickener, M4	MIL-T-50009A (1960)
Titanium, Tech.	MIL-T-13405B (1964)
Tungsten, Powder	MIL-T-13827 (1954)
Tungsten Delay Powder	MIL-T-23132 (1961)
Wax, Impregnating, Waterproofing	MIL-W-10885B (1962)
Wax, Paraffin, Tech.	VV-W-95 (1953)
Zinc Carbonate, Basic	MIL-Z-12061 (1952)
Zinc, Dust	JAN-Z-365 (1946)
Zinc Oxide (for chemical warfare, pyro, or stabilizer)	MIL-Z-291C (1961)
Zinc Oxide, Tech.	MIL-Z-15486A (1952)
Zirconium, Powdered	MIL-Z-00399B (1964)
Zirconium (coarse in FA-878 mix)	TT-P-445 (1961)
Zirconium (coarse)	FA-PD-MI-2364 (1964)
Zirconium	PD-FED-1655
Zirconum Hydride	MIL-Z-21353 (1959)
Zirconium/Nickel	MIL-Z-11410A (1952)

Packaging Materials, Volatile Corrosion Inhibitor Treated, Opaque (VCI)	MIL-P-3420B, Amend. 1 (1964)
Desiccant in Bag (for static dehumidification)	MIL-D-3464C (1963)

Processes for Corrosion Protection of Magnesium Alloy	MIL-M-3171B (1965)
Candlepower, Pyrotechnics, Method of Measuring and Recording	MIL-C-18762 (1955)
Environmental Test Methods for Aerospace and Group Equipment	MIL-STD-810A (1964)
Environmental Testing, Aeronautical and Associated Equipment	MIL-E-5272C (1960)
Fungus Resistance Tests	MIL-F-8261A (1955)
Jolt Test for Use in Development of Fuzes	MIL-STD-300 (1951)*
Jumble Test for Use in Development of Fuzes	MIL-STD-301 (1951)*
40-ft Drop Test for Use in Development of Fuzes	MIL-STD-302 (1951)*
Transportation Vibration Test for Use in Development of Fuzes	MIL-STD-303 (1951)*
Temperature and Humidity Cycling Test for Use in Development of Fuzes	MIL-STD-304 (1951)*
Vacuum-Steam-Pressure Test for Use in Development of Fuzes	MIL-STD-305 (1952)*
Salt Spray (Fog) Test for Use in Development of Fuzes	MIL-STD-306 (1952)*
Waterproofness Test for Use in Development of Fuzes	MIL-STD-314 (1954)*
Salt Spray (Fog) Test for Use in Production of Fuzes	MIL-STD-356 (1953)*
5-ft Drop Test for Use in Production of Fuzes	MIL-STD-358 (1963)*

* Now incorporated in MIL-STD-331 (1966), entitled "Fuze and Fuze Components, Environmental and Performance Tests For."

Salt Spray Tests	MIL-A-8625A
Pyrotechnics, Sampling, Inspection, and Testing	MIL-STD-1234

Part XI

A Formulary of Pyrotechnics

chapter 46

Introduction

"Counsel is no command. That is, I advise you so; but you may do as you please."

James Kelly (1721)[633]

Formulas or recipes belong in the category of "good advice." Such advice, whether freely given or hard to find, may be of dubious value, but even if it is fully valid, it does not guarantee success—rather it is a stepping stone on the way to it.

Commercial formulas are rarely disclosed, at least not the up-to-date ones, and if they are, there is much more to them than meets the eye. The given formulas on matches, for instance, coming out of the practice of the author's company, are examples of valid but complex formulas—complex not only in the physical sense, but because of the difficulty involved in making them workable to full satisfaction. Military formulas with a reasonable amount of detail appear mainly on production drawings available to manufacturers of flares, etc. They suffer, as has been shown in the previous discussion on particle size, from some vagueness of definition of the materials and also from the fact that many imponderables stand between the formula and the acceptable performance of the completed item. Notes may have been added on some phases of processing, compaction pressure, etc., but altogether the beginner and even the experienced pyrotechnician who encounters the specific formula is in the unenviable position of the young bride who bakes her first cake.

The author has tried to get together an even larger number and variety of formulas than in the previous book and has added, where he could and thought it to be of value, the military specification of materials.

What the professional worker in the field desires above all are sets or families of formulas in which the nature of one material, its quantity, particle size, or other variables are changed one at a time and in small steps. Of course, strictly speaking, one cannot change the quantity of any one material without altering several factors of behavior. It is also possible to obtain significant information by so-called "factorial" experimentation where several parameters of conditions are changed at the same time. The resulting changes in performance, such as light output and burning rate, are not only of inestimable value for the practitioner but also afford greater insight into the theoretical aspects of pyrochemical and pyrophysical phenomena. Unfortunately, the number of systematic studies of this kind that reach the public is sparse, though their number is increasing.

Very few formulas in the pyrotechnic field are strictly privately owned. Where commercial contractors develop new items for the Department of Defense and with government funds, only the non-military applications create proprietary rights for the patent owners. Conversely, patents obtained by government employees can be used royalty-free for military purposes. The reader who wishes to find out about restrictions in the use of any of the given formulas will have no difficulties in ascertaining the status from the patents quoted in the references.

chapter 47

Formulas

FORMULAS 1, 2, AND 3
Armstrong's Mixture

	1	2	3
	%	%	%
Potassium Chlorate	81	68	67
Phosphorus	8	12	27
Sulfur	5 1/2	9	3
Precipitated Chalk	5 1/2	11	3

None of these formulas gives the amount of gum arabic or dextrin used in the "gum water" with which the extremely hazardous ingredients are moistened *before* mixing. Formulas 1 and 2 are from German World War II manufacture; 3 from Davis.[8]

FORMULAS 4, 5, AND 6
Stab Primer Compositions

	4	5	6
	%	%	%
Mercury Fulminate	—	—	32
Lead Azide	5	20	—
Basic Lead Styphnate	—	40	—
Tetracene	—	5	—
Potassium Chlorate	53	—	45
Barium Nitrate	—	20	—
Antimony Sulfide	17	15	23
Lead Thiocyanate	25	—	—

Formula 4 is PA-100, used in stab primers as well as detonators; #5 is NOL-130, of higher sensitivity than PA-100 and also higher heat resistance; it may also be formulated with normal instead of basic lead styphnate. #6 is Pom Pom #74, much inferior to both in humidity resistance and now obsolete.

FORMULAS 7, 8, 9, 10, AND 11

Percussion Primer Compositions

	7 %	8 %	9 %	10 %	11 %
Lead Styphnate	—	—	—	—	35.0
Basic Lead Styphnate	—	53	—	60	—
Tetracene	—	5	—	5	3.1
Potassium Chlorate	37.05	—	53	—	—
Barium Nitrate	8.68	22	—	25	31.0
Antimony Sulfide	—	10	17	10	10.3
Lead Thiocyanate	38.13	—	25	—	—
Powdered Glass	10.45	—	—	—	—
Powdered Aluminum	—	10	—	—	—
Powdered Zirconium	—	—	—	—	10.3
Lead Dioxide	—	—	—	—	10.3
TNT	5.69	—	5	—	—

Military Designations: #7—M-39; #8—PA-101; #9—FA-70; #10—NOL-60; #11—FA-959.

FORMULAS 12, 13, 14, AND 15

Electric Primer Compositions

	12 %	13 %	14 %	15 %
Potassium Chlorate	8.5	55	25	60
Lead Mononitro Resorcinate	76.5	—	—	—
Nitrocellulose (1/2 sec—dry base)	15.0	—	—	—
Lead Thiocyanate	—	45	—	—
Diazodinitrophenol (DDNP)	—	—	75	20
Charcoal	—	—	—	15
Nitrostarch	—	—	—	5

Formula 12 for electric matches; #13 widely used in M-59 electric igniter; #14, which is milled with a 2.4% solution of nitrostarch in butylacetate, is a modern squib mixture (Mk 1 Mod 0); #15 is for combined prime ignition and fire transfer.

FORMULAS 16, 17, AND 18

Conductive Primer Mixtures

	16 %	17 %	18 %
Zirconium, fine	71/2	6—9	15
Zirconium, coarse	321/2	30—35	—
Zirconium Hydride	—	—	30
Lead Dioxide	25	18—22	20

Barium Nitrate	35	15—25	15
PETN	—	15—23	20

Formulas 16 and 17 according to U.S. Pat. 2,970,047;[198] #18 after U.S. Pat. 3,090,310.[199] Ciccone (#16 and #17) defines the fine zirconium as of <5 μ particle size; the coarse grade as 100%—100 mesh; 50% —200 mesh; max. 35% —325 mesh and average particle size to exceed 10 μ.

FORMULAS 19, 20, 21, AND 22

Special Purpose Safety Matches

	19 %	**20** %	**21** %	**22** %
Dextrin	—	20	2	*
Gum Arabic	11 1/2	—	—	—
Potassium Chlorate	60	50	88	45
Wood Charcoal	6	—	10	3
Wood Flour	—	—	—	8
Antimony Sulfide	—	30	—	—
Red Iron Oxide (Fe_2O_3)	—	—	—	5
Powdered Glass	22 1/2	—	—	39

FORMULA 23

Commercial Safety Match, U.S.A.

	%
Animal (Hide) Glue	9—11
Starch	2—3
Sulfur	3—5
Potassium Chlorate	45—55
Neutralizer (ZnO, $CaCO_3$)	3
Diatomaceous Earth	5—6
Other Siliceous Fillers (powdered glass, "fine" silica)	15—32
Burning Rate Catalyst ($K_2Cr_2O_7$ or PbS_2O_3)	to suit
Water-Soluble Dye	to suit

The best striking qualities are obtained by use of very high grade glue (such as Peter Cooper Grade IIa extra with foaming properties), leading to a match paste of d=1.20-1.35 at a water content of 30-32%.

* 6 3/4 parts (dry base) of NC as 25% solution added to 100 parts.

FORMULA 24

European Safety Match

	%
Animal Glue	11
Sulfur	5
Potassium Chlorate	51
Zinc Oxide	7
Black Iron Oxide (Fe_3O_4)	6
Manganese Dioxide	4
Powdered Glass	15
Potassium Dichromate	1

This formula is taken from Shidlovsky.[3]

FORMULA 25 AND 26

SAW ("strike-anywhere") Match Tip and Base Compositions

	25 %	**26** %
Animal Glue	11	12
Extender (Starch)	4	5
Paraffin	—	2
Potassium Chlorate	32	37
Phosphorus Sesquisulfide (P_4S_3)	10	3
Sulfur	—	6
Rosin	4	6
Dammar Gum	—	3
Infusorial Earth	—	3
Powdered Glass and Other Filler	33	21 1/2
Potassium Dichromate	—	1/2
Zinc Oxide	6	1

Formula 25 represents the tip formula, which ignites on any hard surface. Formula 26 is its base, loaded with combustibles for strong billowing flame but of low friction sensitivity.

FORMULAS 27, 28, AND 29

Safety Match Striker

	27 %	**28** %	**29** %
Animal Glue	—	16*	9.3
Dextrin	20	—	7.0
Red Phosphorus	50	50	37.2
Antimony Sulfide	—	—	33.5
Iron Oxide (Fe_3O_4)	—	—	7.0
Manganese Dioxide	—	—	3.4

Calcium Carbonate	—	5	2.0
Powdered Glass	—	25	0.6
Sand (clean sharp edges)	30	—	—
Carbon Black	—	4	—

* Casein and animal glue plus hardener of U.S. Pat. 2,722,484.[227] Formula 29 is taken from Shidlovsky.[3]

FORMULA 30

Waterproof Coating of SAW Matches

	%
NC 1/2 sec RS in ethyl acetate/butyl acetate (plus ethanol) as 25% solution	70.0
Cellulose acetate-butyrate in acetone as 19% solution	13.7
Dibutyl Phthalate	16.3
Dye Soluble in Organic Solvents	optional
Toluene and Isopropanol (added in 2:1 ratio to achieve ca. 1000±200 cP viscosity in 20% solution)	

The grade of cellulose acetate-butyrate used in the years 1944-45 was designated as 161-2.

FORMULA 31

Older High-Intensity Flare

	%
Magnesium	36.80
Aluminum, flake	5.62
Barium Nitrate	41.16
Sodium Oxalate	12.15
Paraffin	2.16
Linseed Oil	1.00
Castor Oil	1.00
Cobalt Naphthenate	0.11

Used in *Mk* 5 and *Mk* 6 *Aircraft Parachute Flares.* Light output in terms of candlepower per square inch of burning area or candlepower seconds per gram of composition is substantially lower than the output achieved by Formulas 32 through 35.

FORMULAS 32—35

High-Intensity White Flares

	32	**33**	**34**	**35**
	%	%	%	%
Magnesium "30/50" JAN-M-382A,* Type III, Gran. 18	58±2	45 1/2	53	55

Sodium Nitrate MIL-S-322, Grade B	37 1/2±2	45 1/2	39	36
Laminac	4 1/2±1/2	9	8	9

* Now MIL-P-14067A

Formula 32 is the illuminant in the *Mk* 24 *Flare*, #35 in the *M*301*A*2 *Illuminating Shell*, 81 mm. Formula 33 was used during the Korean War in the *M*314 *Howitzer Illuminating Shell*, 105 mm; #34 was the formula in a similar, experimental projectile. Laminac #4110, which is less air-inhibited than #4116, otherwise quite similar, was employed in the latter.

FORMULA 36

Underwater Flare

	%
Magnesium	16
Aluminum	12
Barium Sulfate	40
Barium Nitrate	32

As binder, 8 parts of linseed oil and 1 part of manganese dioxide were added. This experimental formula is given here mainly to demonstrate the capability of sulfates to substitute for the conventional oxidizers.

FORMULAS 37 AND 38

White Light Compositions with Sulfate as Oxidizer

	37 %	**38** %
Magnesium/Aluminum Alloy	41	—
Magnesium "Griess"	—	40
Sodium Nitrate	11	13
Plaster of Paris	32	40
Water	1	7
Calcium Carbonate	15	—

German W. W. II formulas. Plaster of Paris sets with the added water and is a substitute for nitrate.

FORMULA 39

Mild, White Light Source

	%
Red Gum	21
Dextrin	6
Potassium Perchlorate	73

This mixture is the illuminating "paste" used in the *M*118 *Booby Trap Simulator*.

FORMULA 40

White Star

	%
Magnesium	25
Aluminum	14
Barium Nitrate	42
Strontium Nitrate	11
Asphaltum	5
Linseed Oil	3

FORMULA 41

Combined Smoke and Light for Drift and Float Signals

	%
Magnesium	8
Red Phosphorus	51
Magnese Dioxide	35
Zinc Oxide	3
Linseed Oil	3

Pyrotechnically, an unusual combination, effecting mainly the evaporation of the phosphorus that afterwards ignites at an orifice and burns in air.

FORMULA 42

Standard Photoflash Powder

	%
Aluminum, Atomized, JAN-A-289 Type C, Class C (with exceptions), $20 \pm 5\ \mu$	40
Potassium Perchlorate PA-PD-254,24 μ	30
Barium Nitrate PA-PD-253,147 μ	30

This mixture, Type III, Class A, contains a sizable excess of fuel. Its calculated heat of reaction including complete burning of excess aluminum in air is 2.8 kcal/g. Depending on the amount of vibration (tapping) its actual density varies from 1.7-1.8 or from 60-65% compaction.

FORMULA 43

Gunflash Simulator

	%
Magnesium	45
Potassium Perchlorate	35
Barium Nitrate	15
Barium Oxalate	3
Calcium Oxalate	1
Graphite	1

Used in a Bakelite tube in the *M110 Gunflash Simulator*. If the magnesium powder contains particles larger than —50 mesh, visible sparks and streamer effects are observed.

FORMULA 44

Projectile Ground-Burst Simulator

	%
Magnesium	34
Aluminum	26
Potassium Perchlorate	40

This mixture, Photoflash Powder, Type I, Class A, was used at the start of W. W. II for *M46 Photoflash Bombs* but is now relegated to small items such as the *M115 Projectile Ground-Burst Simulator* in which the flash and sound follow a pyrotechnic whistle.

FORMULA 45

Flash for Booby Trap Simulator

	%
Magnesium, Grade A, Type I	17
Antimony Sulfide, Grade I or II, Class C	33
Potassium Perchlorate	50

FORMULA 46

Tank Gun Simulator

	Parts
Navy SPD* Propellant Powder	33 1/3
Magnesium/Aluminum Alloy (50:50)	33 1/3
Potassium Perchlorate	33 1/3

The cartridge contains, in addition to 90 g of this mixture, some fine black powder mixed with calcium oxalate and barium oxalate.

* This designation is said to stand for *S*olid *P*ropellant, *D*iphenylamine.

FORMULAS 47—50
High Altitude Flash Charges

	47	**48**	**49**	**50**
	%	%	%	%
Aluminum, fine	40	31	20	—
Calcium Metal	—	—	30	80
Sodium Perchlorate	—	—	—	20
Potassium Perchlorate	60	49	50	—
Calcium Fluoride	—	20	—	—

In these formulas, the aluminum must be very fine (less than 17 μ), otherwise light output is small. Formula 49 is superior at high altitude but #48 would be preferable from the viewpoint of availability and stability. Formula 50 is vastly superior to all of them but it would be difficult to preserve the easily oxidized metal and the highly hygroscopic salt of this experimental mixture. A more stable formula with calcium uses 10% potassium perchlorate and 10% sodium nitrate.[283a]

FORMULAS 51—59
High Intensity (Colored) Flare, Star and Tracer Mixtures (For Star Signal)

	Red			Green			Yellow		
	Fl	St	Tr	Fl	St	Tr	Fl	St	Tr
	51	**52**	**53**	**54**	**55**	**56**	**57**	**58**	**59**
	%	%	%	%	%	%	%	%	%
Magnesium	29	23	46	26	15	48	26	19	49
Gilsonite	2	8	3	2	—	3	2	9	5
Oil	2	—	—	2	2	—	2	—	—
Hexachlorobenzene	4	6	4	7	15	6	5	7	—
Copper Powder	—	—	—	—	2	2	—	—	—
Cupric Oxide	—	—	—	2	—	—	—	—	—
Barium Nitrate	—	—	—	45	66	16	29	—	—
Strontium Nitrate	34	41	18	—	—	—	—	—	—
Potassium Perchlorate	29	22	29	16	—	25	23	50	31
Sodium Oxalate	—	—	—	—	—	—	13	15	15

W. W. II formulas of high energy, brilliant, relatively slow burning mixtures for end-burning ground flares (Fl), star signals (St), and tracers for double star signals (Tr.) The magnesium was lathe-cut "powder"—prior to availability of atomized powder.

Mixtures containing so much magnesium display too much white light for real depth of color. Also, everything else being comparable, hexachlorobenzene is not as good a color intensifier as PVC in more modern formulas.

FORMULAS 60, 61, AND 62

Red Signal Flare

	60	**61**	**62**
	%	%	%
Magnesium	21	17 1/2	30
Strontium Nitrate	45	45	42
Potassium Perchlorate	15	25	9
Hexachlorobenzene	12	—	—
Polyvinylchloride	—	5	12
Gilsonite	7	7 1/2	—
Laminac	—	—	7

Formulas 60 and 61 are alternate formulas for the *Mk 13 Mod 0 Day and Night Signal* and might be called the prototype formulas for red burning ordnance items. Formula 62 is a fairly new formula for a hand-held rocket-type signal,[634] which differs from most colored flame formulas by including a polyester resin (Laminac) as the binder.

FORMULA 63

Red Burning Railroad Flare

("Fusee")

	%
Strontium Nitrate	74
Potassium Perchlorate	6
Sulfur	10
Grease or Wax	2—6
Sawdust, Hardwood Shavings	8—4

Variations of this type of commercial flare are rather limited because of cost and the need for a very low burning rate. However, different, inaccessible private formulations exist. The formula presented is the result of an analysis of a good commercial mixture.

FORMULAS 64—69

Artillery Tracers

	64	**65**	**66**	**67**	**68**	**69**
	Red	Red	Green	Yellow	Yellow	White
	%	%	%	%	%	%
Magnesium	28	28	41	33	43	34
Barium Nitrate	—	—	28	—	41	60
Strontium Nitrate	40	55	—	40	—	—
Potassium Perchlorate	20	—	—	—	—	—
Barium Oxalate	—	—	16	—	—	—
Strontium Oxalate	8	—	—	—	—	—
Sodium Oxalate	—	—	—	17	12	—

Sulfur	—	—	—	—	2	—
Polyvinylchloride	—	17	—	—	—	—
Binder and Fuel	—	—	15*	10†	2††	6†
Calcium Resinate	4	—	—	—	—	—

Formula 64 is the R-257 NATO deep red tracer, notable for its inclusion of perchlorate, #65 is R-284, Cal. .30, also deep red.[635] The others are foreign, as reported by Fedoroff.[48]

Formulas 70 and 71
Slow Burning Red Flare

	49 %	50 %
Magnesium	3—5	8
Charcoal	—	1
Strontium Oxalate	—	8
Strontium Nitrate	68—70	52
Hexachlorobenzene	—	18
Polyvinylchloride	23	—
Linseed Oil	5	—
Stearic Acid	—	13

These formulas differ from #60 and #61 by the absence of perchlorate and the considerably lower magnesium content, which favors slow burning. They were reported to the writer about twelve years ago as experimental NOL formulas.

Formula 72
Drill Mine Signal—Red Flame

	parts
Magnesium	8
Calcium Silicide	2
Strontium Oxalate	10
Strontium Nitrate	38
Ammonium Perchlorate	15
Polyvinylchloride	17
Stearic Acid	6

Used in *Mk 43* and *44 Mods 0 Drill Mine Signals*, a rather unorthodox, long-flame formula with ammonium perchlorate. This formula (and any red one in general) is easily adaptable for green flame.

* Unspecified
† Phenolic resin
†† Unaccounted

FORMULA 73

Drill Mine Signal—Green Flame

	%
Magnesium	20
Barium Nitrate	50
Potassum Perchlorate	10
Polyvinylchloride	16
Asphaltum	4

For *Drill Mine Signal MK 39 Mod 0.*

FORMULA 74

Green Flare

	Parts
Magnesium	35
Barium Nitrate	221/2
Potassium Perchlorate	221/2
Polyvinylchloride	13
Laminac	5

This formula corresponds to #62 and is from the same source. Notwithstanding its higher magnesium content, it has been reported as yielding only about 40% of the light output of the red flame of same B.T.

FORMULAS 75 AND 76

Yellow Light

	75 Parts	**76** %
Aluminum	31/2	3—20
Magnesium	—	0—11
Potassium Nitrate	151/2	—
Strontium Nitrate	151/2	—
Barium Nitrate	—	63—67
Sodium Oxalate	64	8—17
Sulfur	—	4—5
Castor Oil	2	2—3
Rosin	5	—

These formulas were taken from OP 2793[50] where they are marked as pertaining to numerous diverse items. Formula 75 is rather unusual and only occurs in one item.

FORMULAS 77 AND 78
Blue Lights

	77	**78**
	%	%
Potassium Perchlorate	39.8	40
Ammonium Perchlorate	—	30
Barium Nitrate	19.5	—
Basic Copper Carbonate	—	15
Paris Green	32.6	—
Red Gum	—	15
Stearic Acid	8.2	—

Formula 77 is one of three alternates for the *Mk 1 Mod 1 Distress Signal* (*Blue Light*), the other two being fireworks formulas with potassium chlorate. Formula 78 was given to the author by the Rev. Ronald Lancaster (cf. #86). Formula 77 has been characterized[636] as whitish, smoky, unevenly burnng with copious ash, and toxic.

FORMULA 79
"Ashless Blue Flare"

	%
Ammonium Perchlorate	74.2
Copper Dust	11.1
Stearic Acid	11.1
Paraffin	3.7

This formula has been described in detail in an NAD Crane report[636] and has been characterized as much superior in color to #77 though lower in cp.

FORMULA 80
Blue Fusee

	Parts
Ammonium Perchlorate	10
Potassium Perchlorate	20
Barium Nitrate	20
Cuprous Chloride (Cu_2Cl_2)	17
Mercurous Chloride (Hg_2Cl_2)	10
Powdered Sugar	15
Castor Oil	5

This expensive, toxic, high ash-forming, unorthodox formula developed by the author for a fusee-type signal flare has nothing to recommend itself except the excellence of its color. It also burns very slowly (about 16 sec/in.) and with a long torch-like flame.

FORMULAS 81 TO 86
Illuminating Fires
(For Burning in the Open)

	Red		Green		Blue	
	81	**82**	**83**	**84**	**85**	**86**
	%	%	%	%	%	%
Potassium Perchlorate	66	—	—	—	—	26
Potassium Chlorate	—	13	26	14	33	—
Ammonium Perchlorate	—	—	—	—	—	46
Barium Nitrate	—	—	54	72	—	—
Strontium Nitrate	—	66	—	—	—	—
Strontium Carbonate	20	—	—	—	—	—
Barium Chlorate	—	—	—	—	55	—
Red Gum (Acaroid)	14	—	20	—	12	15
Shellac BS* 30/200 mesh	—	17	—	14	—	—
Charcoal BS 150 mesh	—	4	—	—	—	—
Copper Carbonate	—	—	—	—	—	10
Stearine	—	—	—	—	—	3

In firework formulas 81-124, unless stated otherwise, it is assumed that the chemicals are ground to pass a British Standard 120 mesh sieve (124 μ). Shellac 30/200 has a wide range of particle size and so gives a longer burning time. Red Gum or Acaroid Resin has a similar wide range but usually passes 80 mesh (190 μ).

FORMULAS 87—90
Bengal Illuminations

	Red		Green	
	87	**88**	**89**	**90**
	%	%	%	%
Strontium Nitrate	65	63½	—	—
Barium Nitrate	—	—	68½	70
Potassium Chlorate	20	—	—	16
Potassium Perchlorate	—	16	15	—
Shellac BS 30/200	15	—	—	—
Red Gum	—	9½	15	13
Sawdust	—	9½	—	—
Lampblack	—	1½	—	—
Antimony Metal Powder	—	—	1½	—
Paraffin Oil	—	—	—	1
Approximate burning time in sec/in.	45	30	20	40

* The difference between U.S. Standard sieve numbers and British Standard is negligible for all practical purposes. Several books on particle size[534,637] have comparative tables and detailed figures on the two series and on others.

Formulas 91—95
Lances

	Red 91 %	Green 92 %	Blue 93 %	Amber 94 %	White 95 %
Potassium Perchlorate	70	30	64	73	—
Barium Chlorate	—	60	—	—	—
Red Gum	12	—	—	—	—
Shellac	—	—	—	15	—
Gum Copal	—	10	4	—	—
Paris Green	—	—	32	—	—
Strontium Carbonate	18	—	—	—	—
Sodium Oxalate	—	—	—	10	—
Charcoal BS 150 mesh	—	—	—	2	—
Potassium Nitrate	—	—	—	—	65
Sulfur	—	—	—	—	20
Antimony Metal Powder	—	—	—	—	10
Meal Gunpowder	—	—	—	—	5

Formulas 96 and 97
Torches

	96 %	97 %
Potassium Perchlorate	52	—
Aluminum BS 150 Mesh "Bright"	24	—
Aluminum Flitter BS 30/80 mesh	20	—
Dextrin	4	—
Barium Nitrate	—	76
Sulfur	—	4
Petrolatum	—	2
Aluminum, Fine Dark Pyro	—	18

Beside these two formulas, the above-described lances, with some modifications, are useful for torches. A red-burning torch formulation is the following, #98.

Formula 98
Red Torch

	%
Strontium Nitrate	45
Potassium Perchlorate	25
Shellac BS 30/200	14
PVC	6
Linseed Oil	1
Magnesium, Grade O	9

FORMULAS 99—103

Stars

	Blue **99** %	Green **100** %	Red **101** %	Yellow **102** %	White **103** %
Potassium Chlorate	68	30	—	—	—
Potassium Perchlorate	—	—	70	70	—
Barium Chlorate	—	55	—	—	—
Red Gum	—	10	9	6	—
Shellac	—	—	—	6	—
Colophony Resin	6	—	—	—	—
Charcoal BS 150 mesh	—	5	2	—	3
Dextrin	4	(4)	4	4	3
Strontium Carbonate	—	—	15	—	—
Sodium Oxalate	—	—	—	14	—
Paris Green	22	—	—	—	—
Potassium Nitrate	—	—	—	—	51
Sulfur	—	—	—	—	18
Meal Gunpowder	—	—	—	—	15
Antimony Metal Powder	—	—	—	—	10

When the amount of binder is in parentheses, it is in addition to the rest of the formula, which adds up to 100%, as in Formulas 100, 104, and 105.

FORMULAS 104—108

Stars

	Silver **104** %	Gold **105** %	Electric Streamer **106** %	White Glitter **107** %	Yellow Glitter **108** %
Potassium Chlorate	60	—	—	—	—
Aluminum "Bright"	20	—	—	8	7
Aluminum, Dark Pyro	20	—	18	—	—
Mealed Gunpowder	—	58	—	66	70
Charcoal BS 150 Mesh	—	21	—	—	—
Charcoal BS 40/100 Mesh	—	—	20	—	—
Lampblack	—	14	—	—	—
Dextrin	(6)	(7)	7	4	5
Antimony Sulfide, Black	—	7	5	14	8
Strontium Oxalate	—	—	—	8	—
Sodium Oxalate	—	—	—	—	10
Potassium Nitrate	—	—	40	—	—
Sulfur	—	—	10	—	—

FORMULA 109

Magnesium Star

	%
Barium Nitrate	60
or	
Strontium Nitrate	
PVC	20
Magnesium Powder	20

These proportions vary by about 5%. Much depends on the particle size of the nitrate and the magnesium. Press dry.

FORMULA 110

Roman Candle Fuse

	%
Potassium Nitrate	50
Mealed Gunpowder	22
Charcoal 40/100 Mesh BS	11
Charcoal 30/60 Mesh BS	11
Sulfur	6

FORMULA 111

Waterfall

	%
Potassium Chlorate	72
Aluminum "Bright"	28

The composition is damped with a 10% solution of shellac in alcohol before charging.

If potassium perchlorate is used, the amount of aluminum needs to be increased to almost equal amounts of aluminum and potassium perchlorate and the aluminum changed to approximately half "bright" powder and half mixed flitters.

FORMULA 112

Rocket

	%
Potassium Nitrate	62
Charcoal BS 150 Mesh	32
Sulfur	6

The spindle, for example, in a 3/4-in. bore tube would be about 41/4 in. long and with a diameter of 3/8 in. at the base and 1/8 in. at the top. Smaller rockets frequently contain a small amount of gunpowder and larger rockets invariably substitute some of the fine charcoal for a more coarse grade.

FORMULAS 113—115

Sparklers

	113 %	114 %	115 %
Potassium Perchlorate	60	—	—
Barium Nitrate	—	50	17
Potassium Chlorate	—	—	46
Aluminum, Dark Pyro	30	8	—
Dextrin	10	10	6
Steel Filings	—	30	—
Charcoal Fine	—	1/2	—
Neutralizer	—	1 1/2	—
Strontium Carbonate	—	—	12
Cryolite	—	—	8
Shellac	—	—	11

Formula 115 derives from an older U.S. patent 1,936,221 (1933). The outside of the composition is coated with magnesium/aluminum alloy grit.

FORMULAS 116—118

Wheel Turnings

	116 %	117 %	118 %
Meal Gunpowder	88	80	70
Charcoal BS 150 Mesh	12	—	—
Aluminum "Bright"	—	20	—
Fine Iron Turnings	—	—	30

The above compositions are for small drivers. Larger sizes contain charcoal also.

FORMULAS 119—121

Flash Report

	119 %	120 %	121 %
Potassium Perchlorate	66	70	—
Barium Nitrate	—	—	68
Aluminum, Dark Pyro	34	30	23
Sulfur	—	—	9

FORMULAS 122—124
Gerbs

	122	**123**	**124**
	%	%	%
Meal Gunpowder	80	42	72
Charcoal BS 150 Mesh	20	—	—
Potassium Nitrate	—	21	7
Sulfur	—	11	—
Iron Turnings or Titanium	—	18	—
Charcoal BS 40/100 Mesh	—	8	7
Aluminum, Dark Pyro	—	—	7
Aluminum BS 80/120 Mesh	—	—	7

FORMULA 125
White Smoke, Type E

	%
Carbon Tetrachloride	45.9
Aluminum, Grained	5.6
Zinc Oxide	48.5

This improved Berger's mixture for the *AN-M8 Grenade* is now obsolete.

FORMULAS 126 AND 127
White Smoke, Type A

	126	**126′**	**127**	**127′**
	%	%	%	%
Zinc Dust	36	38½	36	38½
Hexachloroethane (HC)	43	46½	44	46½
Ammonium Chloride	6	3	10	6
Ammonium Perchlorate	15	—	10	—
Potassium Perchlorate	—	12	—	9

These older mixtures for the *AN-M8 Grenade* consist of a faster burning upper increment (126 or 126′) and a main charge of the slower burning mixture (127 or 127′), each giving a choice of the use of oxidizer.

These and the following HC formulas were specified to be consolidated at 2500-3000 $lb/in.^2$

FORMULA 128
White Smoke, Type B

	%
Calcium Silicide	9.0
Hexachloroethane (HC)	45.5
Zinc Oxide	45.5

FORMULA 129

White (Gray) HC Smoke, Type C

	%
Aluminum, MIL-A-512A, Type II Grade C, Class 4	9.0
Hexachloroethane (HC) MIL-H-235	44.5
Zinc Oxide, MIL-Z-291C, Grade A, Class I	46.5

Aluminum may be lowered to 5.5% to increase B.T. Zinc carbonate and even more effectively zinc borate lengthen burning time when added in small amounts (0.5-1.5%). The given ratio of zinc oxide to HC of 1.04:1.00 may be changed to a 1:1 ratio.

FORMULAS 130 AND 131

Colored HC Smoke

	130	**130′**	**131**
	%	%	%
Magnesium	19	19	16
Hexachloroethane (HC)	62	55	48
Anthracene	19	26	—
Iron Oxide (Fe_2O_3)	—	—	36
Color	Black		Yellow to Orange

Formulas 130 and 130′ show two slight variations of a U.S. formula, while 131 is a German formula reported by Fedoroff.[48]

FORMULAS 132, 133, AND 134

Inorganic Colored Smokes

	132	**133**	**134**
	%	%	%
Potassium Dichromate	66	35	—
Bismuth Tetroxide	20	—	—
Lead Dioxide	—	50	35
Cupric Oxide	—	—	50
Magnesium Powder	14	15	15
Color	Yellow	Orange	Brown

FORMULA 135
"Pink" Inorganic Smoke

	%
Calcium Silicide	37
Potassium Chromate	9
Potassium Iodate	54

While this mixture has been described as producing pink smoke, other mixtures containing an iodate and a fuel (such as magnesium) form purple smoke because of the prevalence of free iodine.

FORMULAS 136 AND 137
Standardized Colored Smokes

	136 %	**137** %
Potassium Chlorate JAN-P-150, Grade II, Class C or E	20—35	22—30
Sugar, JJJ-S-791 Type 1, powdered	23—35	—
Sulfur	—	8 1/2—12
Dye	30—54	38—47
Sodium Bicarbonate	0—15	18—31 1/2

These are composites of MIL-STD formulas of the following numbers:

Red—504, 518, 527, 528, 529; Yellow—505, 506, 519, 520, 521, 532, 533, 542; Green—516 & 517; Violet—507, 522, 523.

Another red formula specified $18 \pm 4\%$ of lactose and $5 \pm 2\%$ of magnesium carbonate. MIL-STD-503 green specifies 5% of potassium bicarbonate. 4% kerosine is permitted in all formulas but the old MIL-STD-505 (1949) contains the cryptic enjoinder "No oil shall be used with this smoke mixture to prevent dusting."

FORMULA 138
Plastic Bonded Colored Smoke

	Parts
Potassium Chlorate	23
Sugar	18
Green Dye, MIL-D-3277	51
Potassium Bicarbonate	8
combined with	
Polyvinyl Acetate (Plasticized)	2.2
Dichloromethane (solvent for PVA)	50

The combined slurry is "cured" by evaporation of the solvent in situ.

FORMULA 139

Plastic Bonded Colored Smoke

	%
Potassium Chlorate	28 1/4
Sugar	19 1/2
Red Dye, MIL-D-3284	35
Sodium Bicarbonate	6 1/2
Infusorial Earth	1 3/4
Polyvinyl Acetate	9

The plastic binder is added as a solution in acetone. Most of the solvent evaporates during mixing and granulation. This is not a standardized formula but similar "castable" mixtures have been reported by others. Unsaturated polyester with added styrene has also been tried for this purpose.

FORMULA 140

Insecticide Dispersal

	%
DNPT*	44
Pyrethrum Flowers	50
Piperonyl Butoxide	6

The mixture can be used in powdered form, enclosed in small polyethylene bags. It is ignited with a nitrate-impregnated paper strip.

FORMULA 141

Tear Gas Grenade (CS) Filling

	%
Potassium Chlorate	40
Sugar	28
Magnesium Carbonate	32

100 parts granulated with 8 parts of NC (JAN-N-244 Grade D) in 92 parts of acetone. Approximately 150 g of this fuel are mixed with 115 g of CS.

FORMULA 142

Cloud-Seeding with Prepared Crystals

	%
Silver Iodide or Lead Iodide	40—60
Ammonium Perchlorate	24—45
Iditol (synthetic resin)	10—25
Graphite or Oil	1 1/2—2

* Dinotrosopentamethylenetetramine $C_5H_{10}O_2N_6$

FORMULA 143
Cloud-Seeding with Chemically Formed Agent

	%
Powdered Lead	20—25
Ammonium Iodide	25—34
Ammonium Perchlorate	20—30
Iditol (synthetic resin)	10—20

Formulas 142 and 143 are from Shidlovsky, 3rd ed.

FORMULAS 144 AND 145
American Cloud-Seeding Mixtures

	144 %	**145** %
Silver Iodate, $AgIO_3$	75	—
Lead Iodate, $Pb(IO_3)_2$	—	75
Magnesium 25 μ	15	10
Laminac (no hardener)	10	15

Prior to addition of the resin, the components were moistened with acetone and the mixture pressed into grains several feet long and four inches in diameter.

FORMULAS 146, 147, AND 148
Black (Gun) Powder

	146 %	**147** %	**148** %
Charcoal JAN-C-178A	15.6	—	16
Semibituminous Coal	—	14	—
Sulfur JAN-S-487	10.4	16	12
Potassium Nitrate MIL-P-156B	74.0	70	—
Sodium Nitrate MIL-S-322B	—	—	72

The standard military black powder (#146) and the slow-burning powder for fuzes (#147) carry a $\pm 1\%$ tolerance for all components; #148 a tolerance of $\pm 2\%$. For glazing, a small quantity of finely powdered graphite (JAN-G-155) is added during the finishing process.

FORMULA 149
Celluloid

	149 %	**149′** %
Nitrocellulose (ca. 11% N) 40-60 sec viscosity	72—78	71

Camphor	28—22	26
Triacetin	—	3

Formula 149 is according to MIL-C-15567 (1950), Cellulose Nitrate Plastic (Celluloid or Pyroxylin Type), for use in ammunition; also called Composition B. Up to 5% of residual solvent is permitted. Also covered in MIL-C-15567 is a Composition A for sheets and other shapes of "approximately 3 parts to 1 part" NC to camphor with 3/4% urea added.

FORMULAS 150—154

High Nitrogen Type Gas Producers

	150	151	152	153	154
	%	%	%	%	%
Ammonium Nitrate	78.5	72	68.4	—	—
Potassium Nitrate	9.0	—	—	—	—
Sodium Nitrate	—	16	—	—	—
Guanidine Nitrate	—	—	—	94.5	—
Nitroguanidine	—	—	13.7	—	80
Ammonium Chloride	—	4	—	—	—
Dicyandiamide	—	—	8.5	—	—
Ammonium Oxalate (anhy.)	6.9	—	—	—	—
Ammonium Dichromate	5.6	8	9.4	—	—
Potassium Chromate	—	—	—	—	20
Other Additives	*			†	

Formulas 151 and 152 are castable after melting at 105 to 120°C.

FORMULA 155

Railroad Torpedo

	%
Potassium Chlorate	40
Sulfur	16
Sand (—60 mesh)	37
Binder	5
Neutralizer	2

FORMULAS 156—160

Whistling Compositions

	156	157	158	159	160
	%	%	%	%	%
Potassium Chlorate	73	—	—	—	—
Potassium Perchlorate	—	—	70	—	721/2

* 0.7-2.5 pts. China Clay to 100 pts

† 0.5% V_2O_5 and 5.0% Cu_2O

Potassium Nitrate	—	50	—	30	—
Red Gum	3	—	—	—	—
Gallic Acid	24	—	—	—	—
Potassium Picrate	—	50	—	—	—
Potassium Benzoate	—	—	30	—	—
Potassium Dinitrophenate	—	—	—	70	—
Sodium Salicylate	—	—	—	—	27 1/2

Formula 156 is the mixture used in the *M119 Whistling Booby Trap Simulator* (the proportions may be changed to 77:4:19). Formula 157 is supposedly an alternate but the picrate should be avoided as a more hazardous material. Formulas 158 and 159 are from the quoted article by Maxwell.[392] Formula 160 is an interesting experimental mixture of very slow B.T.

Shidlovsky[3] mentions variants of Formula 157 with as low as 15% potassium nitrate and from Weingart's description of the subject[7] it appears that the potassium picrate is effective by itself.

Formula 161
Starter Mixture VI (W/B)*

	%
Potassium Chlorate	43.2
Sulfur	16.8
Sodium Bicarbonate	30.0
Cornstarch	10.0

60 parts of this mixture are mixed with 40 parts of a 4% nitrocellulose solution in acetone. Built obviously in imitation of an actual colored smoke mixture, but without dye and hotter burning, this mixture, though prescribed for some colored smokes, has several undesirable features.

Formula 162
Starter Mixture

	%
Potassium Perchlorate	49
Wood Flour (Hardwood)	7
Charcoal	6
Red Iron Oxide	7
Finely Powdered Glass	31

To 100 parts, add 37 parts of a 25% solution of 1/2-sec nitrocellulose in ethyl acetate. This W. W. II formula was used for initiation of colored smokes.

* With binder

FORMULA 163
Starter Mixture XII (W/B)

	%
Potassium Nitrate, MIL-P-156B, Class 2 99.9%—60 mesh; 40%—100 mesh	70.5
Charcoal, JAN-C-178, Class D 98%—60 mesh; 80%—140 mesh; 50% min., 88% max.—325 mesh	29.5

50 parts dry mixed with 50 parts of 4% nitrocellulose, Grade D (MIL-N-244) in 90% acetone.

The mixture of nitrate and charcoal without the binder is also called "sulfurless black powder."

FORMULA 164
Ignition Mixture

	%
Sodium Nitrate MIL-S-00322A, Grade A, Class 1	47
Sugar, JJJ-S-791 Type 1, Grade A	47
Charcoal, JAN-C-178, Class D	6

This loose powder mixture, sealed in a plastic bag together with a transfer line such as the end of a length of delay fuse (the latter provided with some adhering first-fire mixture at its flash end) is a safe, cool-burning starter for such items as mixtures of paper and nitrate for document destruction.

FORMULA 165
Flare Ignition and Ejection Disk (BuOrd 1619437)

	%
Potassium Perchlorate, JAN-P-217 Grade D	74.0
Ammonium Dichromate, Fed. O-A-498	10.5
Red Iron Oxide, JAN-I-706 Purity 69.2-70.2% Fe; 100% —325 mesh	5.0
Nitrocellulose, JAN-N-244 Grade A, Type II	10.5*

* Dissolved in butylacetate and methanol

FORMULA 166

Starter Mixture V for Smokes (W/B)

	%
Silicon, MIL-S-230, Grade II, Class C 98% —170 mesh; 90% —230 mesh	40.0
Potassium Nitrate, MIL-P-156B 100% —14 mesh; 55% —60 mesh; 20% max. —170 mesh	54.0
Charcoal, JAN-C-178 98% —60 mesh; 80% —140 mesh; 50% min., 80% max. —325 mesh	6.0

70 parts of this mixture added to 30 parts of a solution of 4 parts NC, MIL-N-244, Grade D, in 96 parts acetone, O-A-51. This formula goes back to (at least) 1942 where it is presented in a Chemical Corps specification for starting an HC smoke mixture as follows:

Silicon 40; Potassium Nitrate 40; *Sulfurless Meal Powder* 20 (See Formula 163).

FORMULA 167

Starter Mixture XXV for HC Smokes (W/B)

	%
Silicon	26
Potassium Nitrate	35
Black Iron Oxide, MIL-I-275, Class C	22
Charcoal	4
Aluminum, MIL-A-512A Type II, Grade C, Class 4	13

83.3 parts of above mixed with 16.7 parts of a 6% NC solution in acetone.

FORMULA 168

First Fire for Illuminating Flares

	%
Silicon, MIL-S-230, Class A, Grade 1	20
Zirconium Hydride, MIL-Z-21353, $5 \pm 3\ \mu$	15
TNC, MIL-T-13723, technical grade	10
Barium Nitrate, JAN-B-162, Class A, Gran. 1	50
Binder*	5

This efficient, pressed-on first fire is the final igniting compound in many illuminating flares of the magnesium/sodium nitrate/resin binder-type such as the *Mk 24 Parachute Flare*, and various illumi-

* 98.6% unsaturated polyester resin (Laminac 4110 or 4116) and 1.4% catalyst (Lupersol DDM).

nating candles such as for *Shell M-314* for the *105 mm Howitzer* or for the 5″/38 projectile.

FORMULA 169

First Fire X (W/B)

	%
Silicon	25
Red Lead, Pb_3O_4	50
Titanium*	25

A maximum of 0.6 parts (per 100 parts dry) of graphite, MIL-G-155, and 1.8 NC/Camphor (celluloid) of MIL-B-10854, dry base (but as 8-10% solution in acetone, Spec. O-A-51) may be added if so desired.

This is an efficient, pressed-on first fire for delay trains of related composition as in the *M201 A1* and similar delay fuzes for chemical grenades; also for thermate.

FORMULA 170

First Fire (B2-56)

	%
Silicon	25
Red Lead	25
Titanium	25
Red Iron Oxide	25

In loose powder form, this mixture can serve as a "gap-bridging" primer flash acceptor on top of pressed-on #166.

FORMULAS 171 AND 172

Ignition Powder and Starter Composition

	171 %	**172** %
Silicon	33 1/3	50
Lead Dioxide JAN-L-376, Grade B	33 1/3	20
Cuprous Oxide MIL-C-15169	33 1/3	—
Cupric Oxide	—	30

Formula 171 is used for phosphorus candles of float or drift signals (Formula 41). Formula 172 is for the *Mk* 25 *Mod 0* and *Mod* 2 *Flare* (according to BuWeps Dwg. #2150073, Rev. B of LD #282857).

* Titanium with B.T. according to MIL-T-13405

FORMULAS 173—176
Artillery Tracer Ignition

	173 %	174 %	175 %	176 %
Magnesium, Type III, Gran. 12 JAN-M-382	20	17.5	25	14.0
Barium Peroxide, JAN-B-153	78	80.5	4	79.0
Strontium Peroxide, JAN-S-612 Grade 8	—	—	70	—
Asphaltum	2	—	—	—
Calcium Resinate	—	2.0	—	—
Zinc Stearate	—	—	1	0.9
Toluidine Red Toner	—	—	—	0.5
Parlon (Chlorinated Rubber) FA-PD-FED-1657	—	—	—	5.6

Formula 173 is known as igniter composition "K"; #174 is a similar one for a tracer for 40-mm ammunition. Formulas 175 and 176 are known under the Ordnance Corps designations I-237 and I-508. I-237 with the addition of 2% of Dechlorane (Hooker Chemical Co.) is coded I-531. As an intentionally intensely bright "Headlight Tracer," #175 has been used for 0.50 cal. munition.

FORMULAS 177 AND 178
Thermite Ignition

	177 %	178 %
Atomized Aluminum, fine	40	—
Magnesium Powder	—	9
Black Iron Oxide	29	—
Barium Peroxide	31	91

FORMULA 179
Ignition Mixture III (W/B)

	%
Zirconium, JAN-Z-399A	17.5
Titanium, 95% —200 mesh	32.5
Iron Oxide, TT-P-375 Indian Red and Bright Red, Fe_2O_3	50.0

Also known as Chemical Corps B2-50 or "Boom Powder," where the requirement for Ti is 100% —80 and 100% +325 mesh. 1.8% NC/Camphor is added in solution, and the mixture is granulated.

FORMULA 180
Ignition Mixture

	%
Amorphous Boron	10
Barium Chromate	90

A "gasless" mixture that is relatively safe to process and to handle. It will burn under greatly diminished atmospheric pressure. Used with NC binder and added powdered glass, it adheres to the substratum and its ash continues to adhere for efficient fire transfer. Its efficacy can be increased by an increase in the percentage of boron and replacement of barium chromate by calcium chromate.

FORMULA 181
Ignition Mixture "A1A"

	%
Zirconium (120A Foote Mineral Co. or eq.)	65
Red Iron Oxide, JAN-I-706 (with exceptions)	25
Superfloss*	10

A rather violently burning, hazardous mixture of very high static sensitivity. Originally dubbed XD-8A, later A1A, detailed processing and testing procedures are given on BuOrd Dwg. 1170731 (1955). Its specification number is MIL-P-22264, Amendment 2 (1962). A1A has an excess of fuel and a heat output of 550 cal. Its gas release at STP is given as 25 ml/g.

A related mixture F-33B with 41% Zr and 49% Fe_2O_3 is less violent but also more limited in ignitibility and fire transfer qualities. It is recommended for the ignition of D-16 delay powders.

FORMULA 182
Ignition Pellet, OS 9833a ("Alclo") and Powder, OS 9878

	%
Aluminum	35
Potassium perchlorate	64
Vegetable Oil	1

Heat of explosion 2.2 to 2.5 kcal/g; in pellet form this material at d=2.45 develops 6 kcal/cc. While extremely high in heat output, it

* Superfloss is a tradename for a type of finely ground and calcined diatomaceous earth.

tends to irregular burning, which can be rectified by metal additives.[638]

FORMULA 183
"Tichloral" Rocket Igniter Mixture

	%
Aluminum (Reynolds I-511)	20
Titanium	13
Potassium Perchlorate	63
Vistanex (polyisobutylene)	4

FORMULA 184
Rocket Igniter Pellet

	%
Amorphous Boron	23.7 ± 2
Potassium Nitrate $<15\ \mu$	70.7 ± 2
Laminac 4116 (containing 2% Lupersol catalyst)	5.6 ± 0.5

This relatively high caloric mixture (1.5 kcal/g) is described in BuOrd SK 445983. It is used most often in pellet or tablet form.

FORMULA 185
Delay Mixture V (W/B)

	%
Silicon	20 ± 2
Red Lead, Pb_3O_4, TT-R-191, Type I, Grade 85, 99% —325 mesh	80 ± 2

3 to 7 parts (per 100 parts dry mixture) of diatomaceous earth, MIL-D-20550, Grade A, may be added for adjustment of BT; add celluloid/camphor and graphite as for First Fire X (Formula 169); the latter is used for its first fire in hand grenade fuzes.

Mixture X was formerly (1950) designated B2-53 but "Fuller's Earth" was specified in lieu of diatomaceous earth. A similar formula was B2-54 with 15/85 ratio of Si to Pb_3O_4.

FORMULAS 186 AND 187
Boron Type Delays

	186	**187**
	%	%
Amorphous Boron	5	10
Barium Chromate	95	90

These mixtures are suitable for use in photoflash cartridge fuzes for 1 and 2-sec delay time. The large change in BT with slight changes in fuel (186: 2 1/2 sec/in.; 187: 1/2 sec/in.) is a disadvantage as is the tendency of all boron mixtures to reach their "final" stable BT after some aging. Formula 187 will reliably function at diminished pressure and both will not self-ignite until about 700°C is reached.

FORMULA 188

Zr/Ni Alloy Type Delays

	a %	b %	c %	d %	e %	f %
Zr/Ni Alloy, MIL-Z-11410A	54	—	—	26	9	3
Type I (70% Zr)						
(50% Zr)	—	54	—	—	—	—
Type II (30% Zr)	—	—	54	—	17	23
Barium Chromate	31	31	31	60	60	60
Potassium Perchlorate	15	15	15	14	14	14
Rated BT sec/in. (approx.)	2 1/2	5	8	2	5	12

Formulas d, e, and f according to specification MIL-C-13739 (Ord) (1954), called Types I, II, and III, respectively.

The Zr/Ni alloy is treated with sodium dichromate according to U.S. Patent 2,696,429[416] to promote stability of delay time on prolonged storage of finished items.

FORMULA 189

D-16, Manganese-Type Delays

	a %	b %	c %	d %
Manganese, JAN-M-476A, Grade I	44	39	37	33
Barium Chromate, JAN-B-550, Class A (with exceptions)	3	14	20	31
Lead Chromate, 5 μ JAN-L-488	53	47	43	36
Rated BT sec/in.	3.7	5.8	8.7	13.5

A very reliable mixture that must be carefully prepared according to NavOrd 9360 (1st Rev. 1957). However, recent investigations[601] have disclosed that the tedious protective treatment required by the original procedure[600] is actually undesirable.

D-16 carries now the designation MIL-M-21383 (NOrd) (1958) superseding NavOrd OS 5445D.

Note. It is obvious that in a three-component system where all three materials are chemically active, different formulations may produce identical burning times. Thus, the burning times of Formulas 189c and 189d have also been reported for the ratios of 33/30/37 and 33/37/30 respectively for $Mn/BaCrO_4/PbCrO_4$.

FORMULA 190
Slow-Burning Delays with Tungsten

	a	b	c	d	e	f	g	h
	%	%	%	%	%	%	%	%
Tungsten (Wolfram) 7-10 μ MIL-T-13827 (1961)	27	30	33	34	49	63	80	58*
Barium Chromate	58	56	52	52	41	22	12	32
Potassium Perchlorate	10	9	10	9	5	5	5	5
Superfloss	5	5	5	5	5	10	3	5
Rated BT sec/in. (approx.)	40	32	29	18	10	3 1/2	1 1/2	1

Formula 190a is probably the only reliably functioning gasless delay powder with a burning time as long as 40 sec/in. in unobturated state. Burning times may vary widely with particle size and manufacturing procedure of the metal powder and only certain grades (selected by trial and error) will produce ignitible and well-burning columns at all.

FORMULA 191
Delays Based on Chromium

	a	b	c	d	e
	%	%	%	%	%
Chromium	80	70	62	50	40
Barium Chromate	10	20	25	25	50
Potassium Perchlorate	10	10	13	25	10
Rated BT sec/in.	1.5	2.5	3.8	7.0	7.7

These delay formulas of a U.S. patent[410] seem not to have received the attention they deserve.

FORMULA 192
Delays Based on Molybdenum†

	a	b††	c	d	e
	%	%	%	%	%
Molybdenum, 4 μ	80—89	80	55	35	30

* Particle size of metal about 2 μ

† 5 parts of diatomaceous earth per 100 improved performance, especially in the slower burning formulas.

†† This formula has been completely misprinted in the cited patent under "Example IV."

Barium Chromate	—	10	40	55	65
Potassium Perchlorate	20—11	10	5	10	5
Rated BT sec/in.	0.01—0.04	0.1	2	6	18

Molybdenum delays, also of the above U.S. patent,[410] extend over the seconds as well as the millisecond range with proper selection of the metal powder qualities. In loose powder form, the fast mixtures burn very violently.

FORMULAS 193, 194, AND 195

Delays for Detonators

	193 %	**194** %	**195** %
Barium Peroxide	80—84	40	70
Selenium	20—16	20	—
Tellurium	—	40	30
Rated BT sec/in.	3.6—4.0	9.0	4.4

These mixtures described in a U.S. patent[422] are used in combination with Pb/Se or Pb/Te ignition mixtures. The composition of the ignition mixtures and their arrangement play a part in accuracy of the delay proper.

FORMULAS 196 AND 197

Thermate

	196 %	**197** %
Aluminum, granulated JAN-A-512, Grade II 100%—12 mesh; max. 3%—140 mesh	16	19
Aluminum, grained JAN-A-512, Grade I 100%—50 mesh; max. 97%—200 mesh	9	3
Black Iron Oxide MIL-I-275A, Grade B 10% max. —140 mesh	44	51
Barium Nitrate MIL-B-162, Class 5 99%—12 mesh; 5% max. —200 mesh	29	22
Sulfur JAN-S-487, Grade E	2	—
Resin, Low Pressure Laminating MIL-R-7575, QPL 7575 Alkyd-Styrene	—	5

The granulated aluminum and the black iron oxide in both formulas can be added combined as *Thermite, Plain Incendiary*, MIL-T-407A (1952). The resin in Formula 197 is usually the commercial resin

Laminac 4116 with catalyst added.

FORMULA 198
Plain Incendiary Thermite

	%
Aluminum JAN-A-512, Grade II	25—28
Iron Oxide, Magnetic JAN-I-275, Class B min. 99.9% —12 mesh; min. 95%—25 mesh; max. 70% —50 mesh	75—72

The specification calls for a calorific output of 825 cal/g; theoretical value is 870 cal.

FORMULAS 199 AND 200
Special Thermites for Welding Wires

	199 %	200 %
Magnesium	20—30	4½
Aluminum	2—7	18
Iron/Manganese (50%/50%)	—	5
Black Iron Oxide	67—78	72½

Formula 199 is a nonliquefying modified thermite mixture said to be useful to weld aluminum cables. Formula 200 is claimed to be suited for welding of grounding wires to metallic structures.

FORMULA 201
Oxygen Candle

	%
Sodium Chlorate	92
Barium Peroxide	4
Steel Wool, Grade 2 or 00	4

This formula is said to be lower in chlorine content (<0.5 ppm) than older formulas with iron powder and glass fiber.

FORMULA 202
Oxygen Candle

	%
Lithium Perchlorate	84.82
Lithium Peroxide	4.24
Manganese Metal Powder	10.94

A development for the Air Force by the Foote Mineral Company.[496]

Part XII

The Language of Pyrotechnics

chapter 48

On Style and Words

> Dictionaries are like watches; the worst is better than none, and the best cannot be expected to go quite true.
>
> Samuel Johnson[639]

If pyrotechnics were merely what the word implies, a *technique*, it would suffice for purposes of communication to add to the spoken and written language the relatively few special terms not normally or ever encountered outside of it. This is achieved by a Glossary (from the Greek *glossa*, the tongue; hence, language, and a word needing explanation) that follows this chapter.

The word pyrotechnics, according to its modern meaning, is somewhat of a misnomer, since it is fixed in people's mind as dealing with fireworks and firecrackers. No better and perhaps more dignified word has been found, though the author has attempted to speak of *pyrochemistry* and *pyrochemical* reactions wherever it seemed proper.

To say, on the other hand, that pyrotechnics has become a science is going too far. It is true that the term science is being downgraded while pyrotechnics is being upgraded, but this does not mean that the two have met. Science, so-called, is riding high, heavily endowed by industry and the military. It has been all but forgotten that the term was once synonymous with *natural philosophy*. When the natural sciences arose as disciplines of their own, the four faculties of the university (theology, jurisprudence, medicine, and philosophy) tagged them onto philosophy, hence the Ph.D.—*philosophiae doctor*.

Perhaps the reader may think that this little excursion is to exhort the pyrotechnician to assume what is called a "philosophical attitude" in the face of the problems that plague him. Actually, all the writer wishes to say is that modern pyrotechnics is respectable enough and is developing an extensive literature so as to deserve a language that is not only clear and unambiguous but also presents the material in a *style* that makes it a pleasure to read, rather than a bore and a chore. No style manuals, of which there are quite a few, can give more than some guidelines and admonitions. Style comes out of love for the written word, which is nurtured by copiously reading well-written and preferably nonscientific literature. It is marred by grammatical illiteracies (*this* data, *this* phenomena) even though these may parade as "modern usage." Furthermore, the unusual and special word has its place in a special field of knowledge and its use is not an unnecessary show of learnedness or deliberate obscurity, though such excesses may occur.

The glossary that follows is a collection of words appearing in the text (and a few not occuring there) that may be unfamiliar to some readers and also to otherwise competent and scientifically schooled secretaries. A more comprehensive glossary of ordnance terms is *Ordnance Technical Terminology*[311] and its short supplement from the AMC Military Pyrotechnics Series.[31] The author has been guided by his own preferences, aiming at the pitfalls rather than at comprehensiveness. This should save some leafings through a standard dictionary, but for more help with strictly chemical terms, at least two good special dictionaries are available.[113,640]

Glossary

Abbreviations—For names of government installations and a few related terms, see list at end of Glossary.

Ablation—The wearing away of surface material by hot gases, resulting in protection of the material (nose cone) below it.

Absorption—Act of "soaking up" and retaining a gas in liquid or a liquid in a solid; also retention (opposite—reflection) of radiant energy.

Accuracy—Closeness to true value; not to be used indiscriminately as a synonym for *precision,* which means reproducibility of measurement.

Actuator (*explosive*)—A self-contained power transmitting device to convert chemical energy into mechanical force. Removers (retractors), thrusters, dimple motors, etc. fall into this category.

Additive—Something added to a basic composition (mostly in small proportions) to accomplish some specific subordinate result.

Adiabatic—Occurring without (external) gain or loss of heat; used *flame temperature* as a calculated value, often extremely high but only theoretical.

Adsorption—Taking up a gas or liquid by a solid due to surface energy.

Aerial—Of or pertaining to operations in the air; items used in the air or delivered from aircraft.

Aerosol—A system of extremely fine droplets or solid particles in a gas (air); popularly and commercially, a cloud of fairly coarse liquid droplets out of a pressure can.

Ambient—Surrounding on all sides; used mainly for the existing "normal" conditions of temperature and pressure in terrestrial environment. Its use implies (if not stated otherwise) that the object has reached the ambient condition.

Amine—An organic compound containing the NH_2 group, such as ethylenediamine.

Ammine—A part of an inorganic coordination compound containing the NH_3 molecule, such as the compound hexammine cobalt (III) nitrate.

Amorce (*French*)—A toy cap; mainly used in Europe.

Anode—The positive *electrode* in electroplating toward which the negative *ions* move; in electrolysis the pole toward which negatively charged *anions* move, sometimes dissolving the anode (anodic corrosion) or changing chemically its surface (anodizing). Opposite—*cathode*.

Art—The sum total of skill and knowledge of a specific technological branch.

Atomizing—An unfortunate term referring to the conversion of molten metal, by a blast from air or inert gases, into tiny spherical droplets, which form a powder on solidification.

Attenuate—Outside of electronics, meaning to reduce, weaken, mitigate.

Autoxidation—Fairly rapid combination with oxygen of the air under ambient (q.v.) conditions.

Auxiliary—Thus, though frequently misspelled "ll."

Average—Strangely derived from the French word *havarie* or shipwreck, indicating the early relation of ships' insurance (underwriting) to statistics. Here, the word "mean" (value) is preferred.

Ballistics—Applying to the motion and behavior characteristics of missiles. Interior ballistics—within the bore of the weapon; exterior ballistics—in flight.

Bandoleer—A wide ribbon of fabric or plastic provided with numerous adjoining pockets to accommodate small-arms ammunition or similarly shaped items.

Batch—See *Lot*.

Belleville spring—A curved, springy piece of steel or plastic, mostly circular, sometimes with a center hole; used as a spring tension washer or, arranged to "flip over," as a driving force on the percussion primer of an AP (antipersonnel) mine or for other uses of mechanical force.

Billion—Representing the magnitude 10^9 in the U.S. but 10^{12} in several other countries and to be shunned for this reason in scientific terminology.

Black (*grey*) *body*—An ideal concept of a solid radiating substance.

Black powder (*gun powder*)—An "old fashioned" mixture of sulfur, charcoal, and a nitrate, becoming obsolete, but still used to a limited extent in pyrotechnics, especially in civilian fireworks.

Blasting cap—A narrow, short tube filled with explosives that detonate from an electric bridgewire or from safety fuse for the purpose of initiating a high explosive.

Booby trap—A concealed AP explosive charge, set off unexpectedly by a trip wire, pressure, opening of a door, etc. If actuated by being stepped on, it is called an AP *mine.*

Bouchon (*French*: *cork, stopper*)—A closure of a grenade, etc.; hence, *bouchon fuze*, a pressure release type fuze so installed on a hand grenade.

Brazing—Welding with coalescence by heating, using nonferrous filler metal; welding, i.e. joining by heat, is the generic term.

Brisance (*adj*: *brisant*)—Ability of an explosive to shatter its solid confines or surroundings.

Burning—Applied in a broad sense to propellants and pyrotechnic mixtures, though the correct term would be "reacting," since no external air is required to "burn" the material.

Burning rate (*BR*)/ *Burning time* (*BT*)—The linear regression of the reaction zone measured in in./sec or in other units is the burning rate; the reciprocal term sec/in. should not be called burning *rate* but rather burning *time* or inverse burning rate. "Rate" is common for propellants, "time" for pyrotechnics such as flares and delay columns.

Caloric (*adj.*)—Pertaining to heat, but mostly used in lieu of the word calorific, which means heat-furnishing; (as a noun)—a historic and obsolete term for "heat-substance."

Calorimeter—A device for measuring *heat of combustion* (under compressed oxygen or with sodium peroxide) or *heat of explosion* (under an inert gas such as argon): The latter is used for propellants, explosives, and pyrochemical mixtures, which react without outside oxygen. Not to be mistaken for *colorimeter*, an apparatus to measure color.

Canister—A metal can for hand-thrown chemical munitions, mostly the size of a beer can. The second most misspelled word in pyrotechnics (—nn—).

Catalyst—A promoting agent for a chemical reaction that itself is not chemically changed; in plastics technology sometimes loosely used for an additive that becomes part of the final product through chemical interaction.

Celsius (*°C*)—The original name for *centigrade.* Adopted officially again in the U.S. in 1948 by N.B.S.

Chemical spelling—Small differences may mean entirely different compounds. Example: chloride, chlorite, chlorate, perchlorate. Mistaking one for the other may have fatal consequences! Sulfur, sulfide, sulfuric acid, etc. have yielded to modernized spelling: f in lieu of ph.

Chemiluminescence—Light emission from chemical action, not ascribable to incandescence (q.v.)

Cigarette-type burning—A linear progression (or perhaps better, regression) of the flame or glow front of a column of solid reactants, the action resembling the way a burning cigarette is consumed.

Clandestine—In secrecy, such as of sabotage and espionage, before these activities became respectable and fashionable.

Classification—Arranging information (verbal, documentary, objects) into groups of restricted availability with the intention of safeguarding national advantages. In ascending order of severity, the present U.S. labels are: confidential, secret, top secret. Formerly the lowest label was "restricted," but this is now just a general term again.

Colored smoke—An aerosol of special dyestuffs or chemical reactants dispersed by pyrochemical heat or by explosion. Used for signaling and spotting.

Comestible—Pertaining to foodstuff; rare, but found in a pyrotechnic patent.

Comminute (*noun*: *comminution*)—Pulverize, triturate, break up into small particles.

Compatibility—Absence of unwanted interaction of the components in chemical mixtures or of such mixtures or compounds with adjacent metals, plastics, etc. (Not -ability!).

Composite propellant—A solid propellant that contains an inorganic oxidizer salt and a plastic binder, which is also the fuel.

Composition—Specifically, a castable or moldable explosive containing more than one ingredient, such as Composition B, Composition C-4; also a pyrotechnic mixture. A composition may be compounded (by mixing) but the term *a compound* must be reserved to chemical individuals. "A mix" is slang.

Critical material—Something not available in sufficient quantity in war time.

Decimals—Caution! 7 gram is not the same as 7.0, which again is different from 7.000. And if one adds 7.0 and 7.055, the result is 14.0, *not* 14.055.

Deflagration—The brisk "burning" of a compound or mixture without detonation (q. v.). Some confusion exists on the use of the word deflagration to characterize nondetonating explosives. It would perhaps be preferable to restrict the word deflagration to fast exothermic reactions that are *not* regarded as explosions.

Delay (*train*)—A pyrotechnic item, mostly of relatively slow reaction time (0.1 to 30 sec/in.), suitable for timing sequences of events. "Delays" may also be mechanical (clocks), electronic, physicochemical (actuation by corrosion, or by softening of a plastic), or physical (release of compressed air).

Deliquescent—A hygroscopic (q.v.) solid material that liquefies when enough water has been attracted.

Destruct system (*to destruct*)—Modern "back-formations" from the word *destruction* signifying the means to destroy a military object in order to make it harmless or unrecognizable to an enemy.

Detent—A mechanism that locks and unlocks a movement such as a flip-over lever on certain booby trap release designs; the "catch" itself is called a *sear.*

Deterrent (*coating*)—A surface treatment of a solid propellant, which reduces initial burning rate.

Detonation—A type of explosion of highest rate of progression, accompanied by a shock wave. The word detonation should not be used indiscriminately. A deafening "bang" does not signify detonation because of its loudness. Detonation is also called "high order" explosion in contradistinction to "low order" explosion by mere expansion (as of black powder).

Detonator—A somewhat more general term than *blasting cap* (q.v.) for an explosive initiating device, which in turn is initiated by a primer (q.v.) or other forms of heat input.

Diatomaceous earth—A bulky light-weight mineral deposit (and commercial product) consisting of silica "skeletons" of tiny marine organisms. Also called *infusorial earth* and (rarely heard) *diatomite. Fuller's Earth* and *Bentonite* have somewhat similar properties, but are chemically related to clay.

Diluent—Specifically (in lacquer technology) a liquid compatible with a solution, though not a solvent by itself. Also an inert, dry, admixed powder.

Discrete; *Discreet*—One must use discretion in keeping apart the meaning, as in "*discrete* (i.e. separate) particles," or in "being *discreet* in talking to strangers about military work."

Disk—Preferred spelling, except as a botanical or an anatomical term, where *disc* is used.

Divers; *diverse*—Divers (i.e. several) opinions may be expressed about diverse (i.e. different, distinct, dissimilar) ideas.

Double-base propellant—A propellant containing two active ingredients, specifically nitrocellulose and nitroglycerine; there are *single*-base and *triple*-base propellants.

Drift signal—See *float signal.*

DTA; *DDTA*—Differential thermal analysis; derivative differential thermal analysis. Test methods for pyrotechnic mixtures by gradual heating.

Dummy—A nonexplosive item resembling in outside dimensions a bomb or the like, to be used for demonstration, camouflage, training. Different from an *inert* object, which is an actual, usable item that does not contain any "hot" materials such as active primers, pyrotechnic mixtures, or explosives.

Endothermic (*reaction*)—Proceeding under absorption of heat. An endothermic compound, however, is (theoretically) formed under heat absorption and decomposes into the elements under release of heat. See *exothermic.*

Enthalpy—In thermodynamics, a term meaning total heat energy.

Envelope—Used to signify a container; also, the overall dimensions of an item.

Environmental testing—Tests referring to exposure of items to climatic, mechanical, and other external stresses.

Equivalent weight—The amount of an element (or compound) that combines with one (gram) equivalent, i.e. 8 g of oxygen.

Error—Not the same as a *mistake*; error pertains to deviations from a true value and may be unavoidable.

Eutectic—Of solid mixtures of elements or compounds, the one (or several) having minimum melting points.

Exothermic—Occurring with release of heat; of compounds having been formed from the elements with release of heat.

Expendable—Something that is consumed, lost, abandoned, or destroyed in the action; opposite—reloadable, reusable.

Explosion—A very fast chemical reaction, not necessarily as fast as a detonation.

Explosives—*High explosives* are detonating explosives when initiated by a *primary* explosive. TNT, RDX, HMX, MOX are such (secondary) high explosives. Black powder is a nondetonating (low) explosive. Tetryl is an intermediary *booster explosive.*

Explosive switch—A self-contained electrically initiated small unit that causes one or more electric circuits to be opened and/or closed by "explosive" (actually, propulsive) action.

Feasibility study—A much misused and over-used term to denote the determination of the practicability, advisability, or adaptability of an item or technique for an intended purpose. Not just any kind of initial effort.

Field expendient—A material or a device such as an explosive or a weapon that can be made by simple means from available substances, especially behind the lines or in guerilla warfare.

Firing device—Any item designed to initiate, mostly by mechanical means, a blasting cap or an igniter.

Flechette—An aerial dart or a small fin-stabilized missile.

Float signal (*drift signal*)—Light- and smoke-producing day and night markers, floating on water and dropped from airplanes as reference points. (Both terms obsolete as official designations.)

Fluorine (*and derivative words*)—Another candidate for top honors in a misspelling contest (-ou-).

Frangible—A solid material that breaks up into a powder or small fragments. It may be quite strong, which makes it somewhat different from one that is *friable.*

Fuel—In pyrotechnics, anything combustible or acting as a chemical reducing agent, such as sulfur, aluminum powder, iron powder, charcoal, magnesium, gums, and plastic binder. Opposite—*oxidizer.*

Fuse; *Fuze*; *Fusee*; *Fuzee*—Fuse: An igniting or explosive device in the form of a string or tube that contains pyrochemical mixture. Slow burning fuse is called *safety fuse*; other types: quickmatch, firecracker fuse, detonating cord (Primacord). Fuze: A hardware item for the same purpose, but its function may be essentially mechanical or electronic. *Fusee* its a warning (R. R.) flare or a special type of strongly burning match, sometimes spelled *fuzee.* Neither word is related to the word (to) *fuse* (from Latin *fundere*, *fusus*, pour out) but derives from the Latin *fusus*, a spindle, by way of the French word (*la*) *fusée*, a spindle of tow. *Fusée* has become, in English, *fusee.*

Gage—Thus, also gauge, but not guage.

Gasless—To denote a pyrotechnic material that reacts with the release of very small amounts of volatile matter. It stands, mostly without express qualification, for "virtually gasless."

Grain, rocket propellant—A preformed item of a solid propellant regardless of size.

Gram—Metric unit of weight for which, unfortunately, no generally adopted abbreviation exists; g or gm is most frequently used, also g. (Chemical Abstracts) and Gm. (medical); gr stands for grain. 1/1000 g is one milligram (mg); 1/1000 mg is called one microgram.

Grenade—A small explosive or chemical (pyrotechnic) missile, thrown by hand or propelled from a rifle or special launcher.

Hardware—Actual units in contradistinction to chemical components; also, loosely, opposed to experimental effort (". . . the usual clamor for hardware before the developmental phase is completed").

Hermetic (seal)—A type of closure that lets neither gas nor liquid enter or escape.

Homing—Of a missile, one which guides itself toward a target.

Howitzer—A weapon firing slower than a gun and faster, but at lower angle, than a *mortar*.

Hydrophilic—(not -illi-). Having a strong affinity to water, swelling, or dissolving in it. Example, animal glue. (Opposite—*hydrophobic*).

Hygroscopic—Eagerly absorbing water in vapor form by adsorption or chemical bond under heat evolution. (not hydro-).

Hypergolic—Of propellants, one that ignites spontaneously upon contact with an oxidizer. A coined word derived from a German (W. W. II) code word "gola." Similar coinages such as "propergol," "monergol," "katergol," and "lithergol" for propellant, monopropellant, catalytically self-ignited monopropellant, and solid/liquid (or gaseous) bipropellant are used in Germany and occasionally in this country.

Igniter—A capsule, etc. containing ignition material, mostly electrically initiated. (not ignitor).

Ignitible (-able)—Ignitible and ignitable are both correct, but uniformity should be preserved throughout a script—a matter of personal preference; -ible in this book except when quoted verbatim as -able.

Illuminant—A pyrotechnic mixture formulated to give light.

Implosion—Sounds like a joke, but is a proper expression. The opposite of explosion; an inward burst of particles, fragments, etc. due to reduced pressure or from surrounding explosive.

Incandescence—Light emission from a substance because of its high temperature.

Incendiary—(Noun or adjective). Substance used for destructive ignition.

Inert—Not chemically reactive under specific circumstances.

Infrared (*IR*)—Invisible radiation of longer wavelength than visible light; heat radiation. Invisible radiation of shorter wavelength in the range adjoining the visible (0.40-0.78 μ) is called ultraviolet (UV).

Initiator—The first element of an explosive or pyrotechnic train normally containing a small amount of sensitive prime explosive.

Ion—An electrically charged particle such as emitted by an "ion flare."

In situ (*Latin*)—"In place" e.g. (in pyrotechnics) produced right at the spot where it is used.

Intense; *Intensive*—The latter is often used where the meaning of the first should correctly apply; intense heat but intensive treatment.

Iso—Equal, in terms such as isobaric, isothermal, and in unusual formations such as isochronal, meaning of equal time.

Jell [*gel* (*preferred*)]—Something gelatinized (like "Jello") but spelled "jellied gasoline," (gasoline thickened with "Napalm").

Kerosene; *Kerosine*—The former spelling is the usual one but Department of Defense Cataloging Handbook H6-1 dictates the latter.

Laminac—Proprietary name for a plastic binder material, a polymerizable unsaturated polyester.

Lanyard—A cord or cable for setting off a firing mechanism from a distance.

Leidenfrost phenomenon—The observation that a drop of water on a red-hot metal plate will not evaporate at once, since it is protected by its own layer of steam.

Logistics—Aspects of a (military) operation that deal with providing material and services and comprising determination of needs, planning, and implementation.

Lot—A quantity of material or of manufactured items produced under identical conditions (as far as possible) though not necessarily in one *batch*. A batch is a uniform product assumed to be uniform because of the nature of its processing. Several lots (or batches) of powder and the like can be *homogeneously blended* to produce a *homogeneous lot*. This method is suitable for powders or liquids but undesirable for discrete items.

Mark (*Mk*)—Followed by a number, designates Navy items; an added "Mod" means modifications starting with *zero* (pronounced "oh," i.e. none) to 1, 2, ...; M stands for ordnance items in general.

Match—Something resembling a commercial match such as an electric match; a magnesium matchhead. Also a type of fuse—quickmatch.

Materiel; *Material*—The first word is more generic (and used only in singular) for "all kinds of materials," but sometimes restricted to military items other than of the common "household type," such as uniforms, food, etc.

Medium—(Noun). A liquid in which solid particles are suspended; also called a *vehicle*.

Memory—A material, such as a piece of plastic, is said to have memory if it tends to return to the shape and size into which it was originally cast or molded.

Metathetical—Referring to "double decomposition" such as in a reaction of the type $AB + CD \longrightarrow AC + BD$.

Micron (μ)—1/1000 of one millimeter (mm), not to be confused with 1/1000 of one *inch*, which is called one *mil*; 1/1000 μ is a millimicron (mμ).

Missile—Be sure *not* to spell missle! A missive is mostly a written message, but may have the meaning of missile.

Misznay-Schardin—A special type of shaped charge (q.v.). As formidable in action as it is to spell!

Mockup—A model (often crude) for study or training.

Moment (*of inertia*)—Has nothing to do with time.

Napalm—Several materials that convert combustible liquids (gasoline, Stoddard solvent) in the congealed state. By extension, the gel itself is now often called Napalm.

Nascent—In the state of "being born," such as a compound or element at the moment it is chemically created; hence, very active. Chemists, who formerly learned more Latin than is customary now, used to speak of hydrogen *in statu nascendi*—a powerful reducing agent.

Obsolete; *Obsolescent*—The second word means in the process of becoming obsolete.

Obturate—To seal-in a delay element, etc. to prevent escape of gas.

Ordnance—Military material such as weapons and ammunition, and accessory items such as vehicles and repair tools.

Packaging; *Packing*—The first word means the wrapping, cushioning, etc. of the individual item; the second is the process of putting the packages in a shipping container.

PAD—Stands for *p*ropellant *a*ctuated *d*evice; older term *CAD*—cartridge actuated device.

Percussion primer—A small capsule containing prime explosive, ignited by a blunt firing pin.

*Phlegmatize**—To slow down, to make less active; a word showing our cultural heritage from the ancient Greek.

Phosphorus—Uncontestedly, the most misspelled word in pyrotechnics (-ous).

Photoflash (*cartridge, bomb*)—Photoflash powder is a loose pyrotechnic mixture that yields a very large amount of light for a small fraction of a second on exploding.

Prime; *Primary Ignition*—The first step in a pyrotechnic (or explosive) system that raises the temperature by release of pyrochemical energy. A *primer* is a small capsule containing prime (primary) explosive, which is easily exploded by mechanical or electrical energy.

Propellant; *Propellent*—Propellant is the noun; propellent the adjective—the latter rarely used in the U.S., but in British usage the accepted spelling for noun and adjective both is propell*ent*.

Pyrochemical—To be used (with discretion) in lieu of pyrotechnical, especially in regard to the chemical reactions of pyrotechnics.

Pyrophoric—Igniting spontaneously on exposure to air; sometimes used of alloys such as lighter "flints" that on abrasion yield a profusion of sparks. A noun, *the pyrophorus*, is now obsolete.

Pyrotechnic; *Pyrotechnical*—No difference in meaning, the longer word is scoffed at by Fowler but quite proper.

(*the*) *Pyrotechnic*—This noun, paralleling the word "the explosive," sounds odd but seems to fill a need for the layman, who uses it most frequently. This author has never needed it.

Qualitative; *Quantitative*—Terms from analytical chemistry pertaining to determination of either the nature or the amount of ingredients or elements in a compound or mixture.

Rarefaction; *Rarefied*—Pertaining to the atmosphere at high altitude.

---

* It is most surprising that neither Webster's Unabridged Dictionary nor the large OED carry this useful word or the corresponding *phlegmatization*. Davis[8] uses the verb in describing the effect of castor oil on unmanageably sensitive chlorate explosives; the noun is given to indicate the two references in the book's index. The just published (October 1966) Random House Dictionary[640a] has "phlegmatized" and its proper explanation.

Redundant—Over-abundant and hence superfluous, but in military parlance referring to a multiplication of a single auxiliary device (such as an ignition system) in order to increase probability of success.

Refractory—Very resistant, especially to high temperature; *refractive* pertains to optics.

Respirable—Suitable for being breathed, e.g. uncontaminated oxygen gas.

Retrofit(ted)—A partial change in older equipment.

Rifted—From solid particles, those that contain many deep cracks and cleavages and thus have a much increased true surface area.

Riser—In foundry practice (also in plastics molding), a vertical channel on the mold for escape of air and as a receptacle for excess material.

Safety fuse—See *fuse*.

Safety pin—Something similar to a bobby pin that locks movable parts of a fuze during transportation but is speedily removed by means of a pull-ring either by hand or with a lanyard (q.v.).

Scavenge(r)—Sometimes used to indicate removal of (gaseous) material by energetic chemical action.

Scoria—Slag or dross, occasionally used for the "chimney" forming on some flares during burning.

Seek; *Seeker*—Refers to instrumentation for the guiding of missiles to the target.

Setback—Sudden rearward motion due to inertia, as in launching of a projectile.

Shaped charge—An explosive charge of special configuration that produces very high penetrations, also called *lined cavity charge*. Do not write "shape" charge.

Shelf item—Something made and put in storage for prospective customers before an order is received. Hardly ever attempted in pyrotechnics, except for standardized commercial items.

Shelf life—Length of time in storage during which an item will remain in serviceable condition. Same as *storage life*.

Silicon; *Silicone*; *Silica*; *Siliceous*—*Silicon* is a chemical element; *silicones* are plastic organic substances containing the element silicon; *silica* is a general term for minerals (such as quartz sand) consisting of an oxide of silicon. *Siliceous material* includes more complex inorganic compounds such as glass, asbestos, bentonite.

Simulator—An item that more or less harmlessly simulates a hazardous device for training purposes; also an imitation setup for teaching complex sequences of operations.

Smoke—In pyrotechnics, an aerosol (q.v.) of white or colored chemical particles (droplets) for signaling or for obscuring objects ("screening").

Specification—A more or less successful attempt to describe an item and its expected functioning so that anyone skilled in "the art" can produce it. The bane of the Government contractor.

Spontaneous ignition—Occurring without external energy input from mere exposure to air (see also *autoxidation*).

Squib—An electrically fired igniter.

Star—A pyrotechnic light signal of short duration (but not a flash) that burns as a spot of light in the air.

Stoichiometry; *Stoichiometric*—Referring to exact balance of weights of materials in a reacting mixture, based on the stipulations of the *Daltonide* compounds in which each molecule contains atoms in proportions of small whole numbers. Opposite—nonstoichiometric or *Berthollide* compounds.

Stores—Items or supplies for an army, a ship, or a station; *naval stores* are certain resinous products.

Strategic material—Needed for the industrial support of a war effort (cf. *critical material*).

Sublime; *Sublimate*—A solid *sublimes* when, after passing from the solid to the gaseous state without liquefaction, it condenses again directly to the solid state; the person doing it *sublimates* the substance, but *to sublimate* is also used intransitively meaning the same as *to sublime.*

Subsieve size—Of a fineness in particle size too small to be separated and characterized by so-called standard sieves. Below about 44 μ particle diameter; in pyrotechnics, generally in the 1-10 μ average diameter class.

Surveillance—Originally merely a term for watchfulness and observation, it has assumed the meaning of the observation of survival or breakdown of ordnance items under severe test conditions.

Temperature coefficient—The relative change of some measurable quantity (e.g. burning time per unit length) with change of temperature, mostly expressed as mean change per degree in percent of mean temperature within a certain range.

Thermate—Incendiary filler consisting of thermite with additives.

Thermite—A mixture that produces very high temperature; derived from the coined and proprietary word Thermit. Do not use "thermit."

TGA—Thermal gravimetric analysis, a test method by means of heating and determining loss of weight.

Topical—Relating to a place, more often used in medicine (a topical application of an ointment).

Vehicle—A liquid in which solid particles are dispersed or soaked.

Very Pistol—A popular but obsolete term for a pistol-shaped firing device for pyrotechnic cartridges, now better called a *Pyrotechnic Pistol.*

Viscosity; *Consistency*—Related but quite different rheological (pertaining to flow) terms. A metal ball will penetrate a *viscous* substance (asphaltum) no matter how long it takes, but may stay on top of a weak but highly *consistent* semisolid (starch paste) for an indefinite period.

Vitiated (*air*)—Partially "used up" in burning, i.e. of reduced oxygen content.

Abbreviations

The following abbreviations are given as an aid in identifying letter combinations and acronyms pertaining to Government installations that have some connection with pyrotechnics and a few related terms appearing in the text and references.

AEC	Atomic Energy Commission
AFOSR	Air Force Office of Scientific Research
AMC	Army Material Command
AOA	American Ordnance Association
ARDC	Air Research & Development Center
ARPA	Advanced Research Projects Agency
ASTIA	Armed Services Technical Information Agency (obsolete)
ASTM	American Society for Testing Materials
BIOS	British Intelligence Objectives Sub-Committee
BM	Bureau of Mines
BuOrd	Bureau of Ordnance (USN) (obsolete)
BuWeps	Bureau of Naval Weapons (USN)

CC or CmlC	Chemical Corps (obsolete)
CBR	Chemical, Biological, Radiological
CIOS	Combined Intelligence Objectives Sub-Committee
CPIA	Chemical Propulsion Information Agency (formerly SPIA), Silver Spring, Md.
CRDL	Chemical Research and Development Laboratories, Edgewood Arsenal, Md.
CWS	Chemical Warfare Service (obsolete)
DDC	Defense Documentation Center, Alexandria, Va.
DOFL	Diamond Ordnance Fuze Laboratories (obsolete, see HDL)
DOD	Department of Defense
FA	Frankford Arsenal, Philadelphia, Pa.
HDL	Harry Diamond Laboratories, Washington, D. C.
JANAF	Joint Army-Navy-Air Force
NAD	Naval Ammunition Depot
NASA	National Aeronautics & Space Administration
NDRC	National Defense Research Committee (W. W. II)
NOL	Naval Ordnance Laboratory, White Oak, Md. (also Corona, Calif.)
NOTS	Naval Ordnance Test Station, China Lake, Calif.
NRL	Naval Research Laboratory, Washington, D. C.
NWL	Naval Weapons Laboratories, Dahlgren, Va.
ONR	Office of Naval Research
OSRD	Office of Scientific Research and Development (W. W. II)
PA	Picatinny Arsenal, Dover, N. J.
USAERDL	U.S. Army Engineer Research and Development Laboratories, Ft. Belvoir, Va.
USNPP	U.S. Naval Propellant Plant, Indian Head, Md. (formerly "Powder Factory")
WADC	Wright Air Development Center, Ohio

Part XIII

Aftermath ; Afterthoughts

chapter 50

Aftermath

Worthwhile new material comes to one's attention while a manuscript takes shape. This is a seemingly never ending invitation to making changes. However, it is less bothersome to the author than his becoming aware that he has omitted certain facts that in the course of writing either did not fit into the scheme of things or emerged into his consciousness too late to be integrated in the text.

Aftermath is defined in the dictionary as a second-growth crop, and the following notes are just that—a selection from the crop of information that came to the author's knowledge after completion of the respective chapters. In addition, to stay within the metaphor, gleanings from the harvested field of older information have been added, but only those that seem to fulfill a useful purpose.

To Part I

With Mellor's Treatise of the years 1922 to 1937 being only slowly supplemented, and Gmelin's *Handbuch* being forbiddingly expensive to many, a French encyclopedia of chemistry may fill the gap.[641] Twenty volumes were scheduled to be issued, most of them priced around NF 125—or about $25 per volume.

A newcomer to the field of explosives is a translation of a book from the Polish language.[642]

The area of thermodynamic behavior of the oxides of the groups II—VI of the periodic systems, including the pyrotechnically important and little known facts on their evaporation and existence of lower oxides, is part of the subject of a translation from the Russian language.[643] It was not available at the time of completion of this book.

To Part II

The list of spontaneously igniting compounds becomes steadily longer. Add two newcomers: The compound SiH_3PH_2 (B.p. 12.7°C),[644] a close relative of $P(SiH_3)_3$;[80] the other is the compound lithium dimethylamide, Li $N(CH_3)_2$, a white solid, self-igniting with a "steady, subdued flame."[645]

To Part III

A phenomenon cited by Gmelin that deserves further scrutiny is the ignition of aluminum foil in oxgen under glass by exposure to photoflash (*Blitzlicht*) but not sun or arc light. The original article is by W. Zimmermann.[646]

It was pointed out in the text that with flares for highest amount of infrared emission, the preferences of white-light production do not apply. As long as no practical means exist for the shifting of radiant energy into the desired IR ranges, a maximum of grey-body emission must be achieved. This occurs with magnesium flares in which the oxidizer is *not* a salt but rather an organic fluorine compound. Pressed magnesium/teflon compositions were formerly used, but now it is known that advantages are gained by liquid or plastic processing, perhaps because of the benefits from greater intimacy of fuel and oxidizer.

One such flare, described in some advertising literature,[647] claims the use of fluoroalkyl-ester monomers as oxidizer and (after casting and curing) as binder. Heat of reaction is up to 2 kcal/g, and an auto-ignition temperature above 470°C has been reported.

This subject, explored since the mid-fifties, has been treated in the restricted literature, which is the reason why it was formerly excluded from open discussion.

In this context, a few additional remarks on *mixing* may not be amiss. The subject applies, of course, not only to light sources but to all compositions of pyrotechnics.

When nonporous metal powders and crystalline oxidizer salts or oxides are combined, an intermingling of the particles to the point of greatest homogeneity may be all that can be achieved, as with flash powders or thermite. A much more thorough permeation occurs with black powder where a porous fuel, charcoal, and a plastic-deformable fuel—sulfur—are subjected to mixing under heavy rollers.[8] In other cases, where liquid or semiliquid components are involved, various degrees of intimacy of mixture are possible. One of the benefits of a liquid or plastic component entering pores and interstices of granules seems to be the improvement of behavior under high-altitude conditions. However, even where only dry powders are concerned, they can be mixed under a suitable liquid vehicle and at proper pH (an adjustable state of "active" acidity not too far from the point of neutrality) to undergo a process called *agglomerate mixing*.[648] Here, the homogeneously dispersed particles are forcefully re-agglomerated so that after removal of the liquid a superior state of intimacy and of proper distribution of particles is achieved.

A liquid vehicle is also involved in a production process dubbed "Quickmix."[649,650] It was originally devised for composite propellants, but is applicable to such mixtures as colored smokes. It not only provides intimate mixing but it is automated to bring together the ingredients in exact proportions. The use of a liquid vehicle is a safety factor, especially valuable where large quantities of fuel-oxidizer mixtures must be processed.

What might be called the opposite approach to the foregoing is reported in a U.S. patent.[651] Here, the liquid binder—a plasticized polyvinyl dispersion—is frozen by dry ice, ground and mixed with ammonium perchlorate, for the compounding of a propellant formulation.

To Part IV

Small and inexpensive white HC-smoke cartridges for civilian uses are marketed by the Superior Signal Company under the name Sportsman Smoke Signal. The 3-in.—long item delivers a good white smoke, purportedly amounting to 10,000 ft^3, for 75 sec.

Small smoke cartridges claimed to be nontoxic and suitable for indoor use are sold commercially for the purpose of detecting leaks in air ducts or for following visually the air flow from an air conditioning or heating system. Such smoke generators burn from 1-5 min and are claimed to emit from 8,000-100,000 ft^3 of smoke.

Colored smokes for use on low-spin, folding fin aircraft rockets of 20-30 sec/in. BT (2-3 in./min BR) have recently (1966) been formulated using conventional formulation with addition of small amounts of polyvinyl alcohol/acetate (VAAR) resin.[651a] By use of high loading pressures, the volume of composition can be significantly decreased and burning time prolonged. Figures are given of 12% increase in BT and 20% of reduction of volume when compacting pressure was raised from 9,000 to 45,000 lb/in^2

One of the queerest facets of pyrotechnics is the use of potassium chlorate with pipe tobacco as a substitute for hashish (*Cannabis sativa*), a drug similar to marijuana. Since the publication of *Modern Pyrotechnics*, the author had occasion to study the original article.[652] The description of the symptoms experienced by the students and hospital personnel who underwent clinical tests confirms the efficacy of the substitute "drug." The side effects seemed to have been so unpleasant that the spread of such usage is unlikely. Nothing is said about the reasons for the physiological behavior of the smoke except that no CO-hemoglobin or methemoglobin was found in the blood of the users.

The containment and dispersion of unruly or aggressive crowds has made it necessary to provide weapons that are not only nonlethal but also do not cause injuries or lasting after effects. Neither the burning type nor the exploding tear gas grenades are suitable weapons in mob control.

By converting CN or CS agent into talc-size dustlike ("micronized") particles and expelling it from a nonfragmenting grenade by the action of a piston, an effective dust cloud of highly irritating material that adheres to clothes and moist skin areas can be created. A similar principle is involved in an item called by the tradename Chemical Mace in which the offensive material is dispersed in a liquid solution.[652a]

To Part V

The difficulties encountered with *inhibition* of propellant powders have been mentioned, specifically of case-bonding, i.e. the attachment of cylindrical powder grains to the envelope in order to achieve cigarette-type or internal-surface burning. The subject applies also to flare candles.

A patent,[653] "Pyrotechnical Devices and Methods of Making Same" (the bad English is not the inventor's) claims the immersion of the grain in a semisolid silicone grease so that the external wall surfaces of grains or pellets are effectively inhibited in a nonrigid manner that offers several advantages. One of these is the use of the technique of pressing and assembling pellets rather than pressing the material into a flare casing. There are also disadvantages in the method, but altogether it is a departure from conventional practice that at least experimentally has proved its usefulness.

To Part VII

Sulfur in combination with aluminum also appears obliquely in Bebie's manual[19] under the heading "Thermit," where it is mentioned as a binder: "When sulfur is used, the resulting Thermit (sic!) is called *Daisite.*" The author has been unable to locate any other reference to Daisite or to ascertain if sulfur is regarded as a major active ingredient (as oxidizer) in such a mixture.

An accumulation of soot in chimneys and intermediary passages has long been combated by aiding the burning off of the carbon, which, however, stubbornly resists such efforts. A handful of salt thrown on the hot coal in the furnace often achieves this, perhaps by a kind of hot-spot or catalytic effect of salt particles deposited on the soot. A French patent[654] does the same with a mixture of 50% NaCl, 23% $NaNO_3$, 25% S, and 2% Zr. This can be put directly in the chimney, since on ignition it furnishes its own heat for dispersion of alkali salts.

Formulas 82-86 in the author's earlier book *Modern Pyrotechnics* give the composition of delay mixtures using the metals niobium (columbium) and tantalum. These formulations were actually made and tested in the author's laboratory, but the limited investigation was not further pursued. At that time, the newly developed delay mixtures with chromium, molybdenum, and tungsten (wolfram) were still all or in part classified information. Hence, the reference to those "exotic" delay mixtures was merely a way of introducing the subject of reactive, high density metals without disclosing what was then restricted information. In the meantime, Shidlovsky[3] as well as Fedoroff and Sheffield in the third volume of the Encyclopedia of Explosives[49] have cited the formulas with niobium and tantalum. What the writer wanted to emphasize was that these were not established delay mixtures at the time they were quoted, and to his knowledge the subject has also not been treated further by others.

An interesting variant of delay action by means of corrosion is the controlled removal of metal by anodic corrosion until a certain mechanical effect such as a spring release has taken place. The device requires a source of direct current that can be furnished by a so-called sea-water battery, as in the case of arming a mine following its placing it in the ocean, claimed by F. F. Farnsworth et al.[655]

A time delay switch based on pneumatic principles whereby a small air space is put under compression and the escape of the air at variable rates terminates in a mechanical action is the subject of another U.S. patent.[656]

A subject that ties in with the topic of destruction by flame and heat is that of the combustible cartridge case. In World War II, the Germans prepared such cases from nitrocellulose[48] and this type, with modifications, is still in use or under consideration[657] and subject of a recent U.S. patent.[657a] Unfortunately, most of the work concerning the other approaches is classified. An unclassified Frankford Arsenal report[658] mentions the use of the molded solid propellant charge itself as the casing. A Picatinny Arsenal report[659] discusses various synthetic resins combined with oxygen-containing substances, an approach perhaps closest to pyrotechnics proper.

In all these instances, a cartridge may either be consumed in its entirety or for ballistic reasons a metal base used, leading to a "partially combustible case." The advantages of a combustible case are obvious. Enormous amounts of strategic metal would be released for other purposes and there is no question of collection and reuse of such metal. Special benefits are derived for the gunner within a tank: The hot used metal cases are an encumbrance in the close quarters, and residual toxic gases from burned propellant powder remain in the conventional metal cases.

To Part VIII

An example of the high-temperature thermite process that seems to be little known and hence had completely escaped the author's attention earlier is the production of tungsten (wolfram) metal from the trioxide (WO_3) with a large excess of zinc dust. The powder mixture tamped in a crucible and covered with zinc dust will start to react at about 500°C (dark red heat) and the process will be completed in a few minutes. It yields a fairly pure metal after chemical removal of other matter. The process has been described by Brauer[97] as a preparative method and is also mentioned by Mellor.[67]

It is remarkable that the element zinc thus becomes reactive in a thermitic reaction. Heat output is 0.11 kcal/g or 0.81 kcal/cm^3, which is quite low; this explains the need for incipient heat input. The process is applicable to the metal molybdenum.

To Part IX

Because of the great susceptibility of finely powdered zirconium metal to electrostatic charges, efforts have been made to reduce such sensitivity without impairing the usefulness of the powders. Two methods have been reported: According to P. Karlowicz,[660] treatment with a 1% hydrogen fluoride solution fulfills the purpose; this method is also claimed in a U.S. patent.[661] M. F. Murphy and B. F. Larrick claim[662] a desensitization of various metal powders, especially of zirconium, by ball-milling the powder with a dispersion of dimethylpolysiloxane or methyl phenylsiloxane. They cite as a result an increase in the 50% ignition point from 125 ergs for the untreated zirconium to 50,000 ergs for the treated one, and no ignition of the latter was found to occur at 5000 ergs when the powders were tested for susceptibility to ignition by electrostatic charges.

To Part X

A good example of the surprises that one may encounter in the field of new oxidizers is an Italian reference.[663] The double salt $Ba(ClO_2)_2.5Ba(NO_3)_2.12H_2O$ is described as extremely stable except for losing water at 60°C and becoming completely dehydrated at 130°C without decomposition. Note that this is a *chlorite*/nitrate combination. There are no corresponding calcium or strontium compounds, according to the reference.

chapter 51

Afterthoughts

When in my youthful fifties I wrote *Modern Pyrotechnics*, I was full of crusading spirit while being racked by frustration when contemplating the State of the Art—the sad state of the upgraded fireworks trade called military pyrotechnics.

Now, calmer or just resigned, let me reiterate what I said: The need for a planned effort at all levels of investigation is still as great as ever, even though substantial work is being done that does not apply merely to the exigencies of immediate military requirements. However, the output along lines of strictly scientific and theoretical endeavor in pyrochemistry is still minute and the gathering of empirical basic facts, divorced from creation of "hardware," is sporadic and seemingly without overall planning, capriciously depending on uncommitted funds, available personnel, and time schedules at the government arsenals and laboratories as well as at their contracting universities and industrial establishments. As to civilian pyrotechnics, it is economically such a limited field that only a developmental effort within the narrow confines of product improvements and development of minor novel items can be expected from industry.

The path appears to be just as thorny along lines of other subordinate concerted efforts. Wastefulness could be reduced by *standardization* and by reducing in numbers the minor components such as squibs and igniters. *Specifications* of materials should be thoroughly revised, upgraded as to the information they contain, and combined wherever it is suitable. All major *accidents* in government installations and by contractors should be critically evaluated and the information made available to everyone in the field.

The state of the art of pyrotechnics has not much chance to advance in a manner comparable to the spectacular technical progress in other fields. As to the basic studies concerning solid state reactions, one must look forward with whatever hope one can muster to a revival of the spirit of and love for mere *inquiry* divorced from commerce and war.

References

References

1. Henry B. Faber. Military Pyrotechnics, *The History and Development of Military Pyrotechnics*, 3 Vol., Government Printing Office, Washington, 1919.
2. A. St. H. Brock, *A History of Fireworks*, George G. Harrap & Co., Ltd., London (1949).

3.* A. A. Shidlovsky, *Osnovy Pirotekhniki* ("Fundamentals of Pyrotechnics"), Government Publication of the Defense Industry, 1st Ed., Moscow, 1943 (The Library of Congress, Card No. UF 860.S5); 2nd Ed., Moscow, 1954 (Translation as "Foundation of Pyrotechnics" by Wright Patternson Air Force Base, Ohio, AD 602687 30 April 1964); 3rd Ed., Moscow, 1964 (Translated by U.S. Joint Publication Research Service, Feltman Research Laboratories, Picatinny Arsenal, Dover, N. J. as Technical Memo 1615, "Fundamentals of Pyrotechnics," May 1965).

3a. A. F. Belyayev and L. D. Komkova, *Zavisimost' skorosti goreniya termitov ot davleniya* ("Dependence of Speed of Burning of Thermites on Pressure"), J. Phys. Chem. U.S.S.R. **24** [11], 1302 (1950).

3b. A. F. Belyayev and A. B. Nalbandyan, *K voprosy o vzryvchatykh svoystvakh bezgazovykh sistem* ("To the Question of Explosive Qualities of Gasless Systems"), Doklady Akad. Nauk S.S.S.R. **46** [3], 113 (1945).

3c. P. F. Bubnov and I. P. Sykhov, *Sredstva initsiirovaniya* ("Means of Initiation"), Oborongiz, Moscow, 1945.

3d. I. V. Bystrov, *Kratky kurs pirotekhni* ("Short Course in Pyrotechnics"), Oborongiz, Moscow, 1940.

3e. G. P. Bystrov, *Spichechnoye proizvodstvo* ("Match Production"), Goslesbymizdam, Moscow & Leningard, 1950.

3f. P. G. Demidov, *Osnovy goreniya veshchestv* ("Primer of Fuel Materials"), RSFSR, Moscow, 1951.

3g. A. N. Demidov, *Vvedeniye v pirotekhniku* ("Introduction to

* Transliteration of References 3a-3ee according to Gregory Razran, Transliteration of Russian, *Science*, **129**, 1111 (1959).

Pyrotechnics"), Voyenizdat, Moscow, 1939.

3h. A. A. Freyman, *Kratky kurs pirotekhniki*, ("Short Course in Pyrotechnics"), Oborongiz, Moscow, 1940.

3i. I. A. Fuks, *Uspekhi v izuchenii aerozoley i prakticheskiye dostizheniya v etoy oblasti* ("Progress & Studies of Aerosols & Practical Achievements in this Field"), Usp. Khim. **19,** [2], 1950.

3j. A. P. Gorlov, *Zazhigatel'nye sredstva, ikh primeneniye i boryba c nimi,* ("Incendiary Articles, Their Application & Wartime Uses"), 1943, Izd. Narkomkhoza RSFSR, Moscow & Leningrad, 1943.

3k. A. G. Gorst, *Porokha i vzryvchatye veshchestva* ("Powders & Explosives"), Oborongiz, Moscow, 1949.

3l. M. Karasev, *Termit i termitnaya svarka rel'sov* ("Thermite & Thermite Welding of Rails"), Gostransizdat, Moscow, 1936.

3m. P. P. Karpov, *Sredstva initsiirovaniya* ("Means of Initiation"), Oborongiz, Moscow, 1945.

3n. M. D. Konshin, *Aerofototopografiya* ("Aerial Photographic Topography"), Geodezizdat, Moscow, 1949.

3o. A. N. Kukin, *Novye vidy termitnoy svarki* ("New Means of Thermite Welding"), Transzheldorizdat, Moscow, 1951.

3p. I. B. Levitin, *Vidimost' i maskirovka korabley*, ("Visibility & Camouflage of Ships"), Voyenizdat Moscow, 1949.

3q. V. A. Likhachev, *Pirotekhnika v kino* ("Pyrotechnics in Moving Pictures"), Goskinoizdat, Moscow, 1944.

3r. B. Ya. Mikhaylov, *Fotografiya i aerofotografiya* ("Photography and Aerial Photography"), Geodezizdat, Moscow, 1952.

3s. M. B. Ravich, *Poverkhnostnoye besplamennoye goreniye* ("Flameless Surface Burning"), Izd. AN. SSSR. Moscow, 1949.

3t. N. P. Rozhdestvin, *Aerofotografiya* ("Aerial Photography") Voyenizdat, Moscow, 1947.

3u. L. T. Safronov, *Nochnoye vozdushnoye fotografirovaniye*, ("Aerial Night Photography") Voyenizdat, Moscow, 1947.

3v. N. A. Shilling,, *Kurs dymnykh porokhov* ("Course in Smoke Compositions"), Oborongiz, Moscow, 1940.

3w. V. P. Sivkov *Primeneniye dymovykh zaves tankovymi podrazdeleniyami* ("Application of Smoke Screens by Tank Subdivisions") Voyenizdat, Moscow, 1946.

3x. V. Ya. Smirnov, *Pirotekhnicheskiye materialy* ("Pyrotechnical Materials"), Oborongiz, Moscow, 1939.

3y. V. Ya. Smirnov, *Vvedeniye v tekhnologiyu piroteknicheskikh proizvodstv* ("Introduction in the Technology of Pyrotechnical Production"), Oborongiz, Moscow, 1940.

3z. B. M. Solodovnikov, *Pirotekhnika* (*proizvodstvo i szhiganie feyerverka*) ("Pyrotechnics [Production & Combustion of Fireworks]") Oborongiz, Moscow, 1938.

3aa. I. P. Tolmachev, *Proizvodstvo alyuminiyevo poroshka, alyuminiyevoy pudry i termita* ("Production of Aluminum Powders & Thermite"), GNTI, Moscow & Leningrad, 1938.

3bb. Yu. I. Veytser i G. P. Luchinsky, *Maskiruyushchiye dymy* ("Screening Smokes"), Goskhimizdat, Moscow & Leningrad, 1947.

3cc. G. M. Zhabrova, *Teplo bez plameni* ("Heat Without Flame"), Izd. AN SSSR, Moscow, 1945.

3dd. O. A. Yesin and P. V. Gel'd, *Fizicheskaya khimiya pirometallurgicheskikh protsessov, chast' I* ("Physical Chemistry of Pyro-Metallurgical Processes, Part I") Metallurgizdam, Moscow, 1950.

3ee. N. F. Zhirov, *Svecheniye pirotekhnicheskovo plameni* ("Light Output of the Pyrotechnical Flame"), Oborongiz, Moscow, 1939.

4. Herbert Ellern, *Modern Pyrotechnics*, Chemical Publishing Co., Inc., N. Y., 1961.

5. *Aerospace Ordnance Handbook*, (Frank B. Pollard and Jack H. Arnold, Jr., Editors) Prentice-Hall, Inc., Englewood Cliffs, N. J., 1966.

6. James Taylor, *Solid Propellent and Exothermic Compositions*, Gge. Newnes Ltd., London, 1959.

7. G. W. Weingart, *Pyrotechnics*, 2nd Ed., Chemical Publishing Co., Brooklyn, N. Y., 1947. Reprinted 1968.

8. T. L. Davis, *The Chemistry of Powder & Explosives*, John Wiley & Sons, Inc., N. Y., 1941.

9. Colonnello Dott. Attilio Izzo, *Pirotecnia e Fuochi Artificiali* (Ulrico Hoepli, Editor) Milano, 1950.

10. Julius Hutstein and Prof. Dr. Martin Websky, *Lustfeuerwerkkunst*, Leipzig, 1878, at the Library of the University of Washington, Seattle.

11. Amédée Denisse, *Traité Pratique Complet des Feux d'Artifice*, Paris, 1882, at the New York Public Library (not for loan).

12. Domenico Antoni, *Trattao-Pratico Di Pirotechnia Civile*, Trieste, 1893, not located in the U.S.

13. Vannoccio Biringuccio, *The Pirotecnia*, Basic Books, Inc. (Translation), N. Y. (re-issued) 1959; also paperback edition, MIT Press, Cambridge, Mass., 1966.

14. J. R. Partington, *A History of Greek Fire & Gunpowder*, W. Heffer & Sons, Ltd., Cambridge, 1960.
15. R. Connor et. al., *Chemistry, Science in World War II OSRD.* (W. A. Noyes, Jr., Editor) Little, Brown & Co., Boston, 1948.
16. Dr. C. Beyling and Dr. K. Drekopf, *Sprengstoffe & Zündmittel* ("Explosives & Ignition Materials"), Julius Springer Verlag, Berlin, Germany, 1936.
17. Dr. A. Stettbacher, *Spreng- und Schiesstoffe, Atomzerfallselemente und ihre Entladungserscheinungen* ("Explosives & Ammunition, Nuclear Fission & Its Phenomena"), Rascher Verlag, Zürich, Switzerland, 1948.
18. L. F. Fieser, *The Scientific Method*, Reinhold Publishing Corp., N. Y., 1964.
19. J. Bebie, *Manual of Explosives, Military Pyrotechnics & Chemical Warfare Agents*, The MacMillan Co., N. Y., 1943.
20. M. F. Crass, Jr., "A History of the Match Industry," *J. Chem. Educ.* **18,** [3, 6, 7, 8, 9] (1941).
21. M. F. Crass, Jr., "The Match Industry: Raw Materials Employed," *Chem. Ind.*, April, May 1941.
22. F. Ullmann, *Encyklopädie der technischen Chemie* ("Encyclopedia of Technological Chemistry"), 3rd Edition issued by Dr. Wm. Foerst, München, 1951 to—.
23. *Encyclopedia of Chemical Technology*, Vol. 11, "Pyrotechnics (Military)," Interscience Publishers, Inc., N. Y., 1953.
24. *Kirk-Othmer Encyclopedia of Chemical Technology* (Anthony Standen, Executive Editor) 2nd Ed., Interscience Publishers, Inc., N. Y., 1963 to—.
25. *The Encyclopedia Americana*, Encyclopedia Americana Corp., N. Y. and Chicago, 1965.
26. *Encyclopaedia Britannica*, Encyclopaedia Britannica, Inc., Chicago, 1965.
27. *Collier's Encyclopedia*, P. F. Collier & Son Corp., N. Y., 1950.
28. *Chamber's Encyclopaedia*, Oxford University Press, N. Y., 1950.
29. F. A. Brockhaus, *Brockhaus Konversationslexikon* (" 'Brockhaus' Encyclopedia"), Wiesbaden, 1954.
30. AMC Pamphlet, *Military Pyrotechnics Series, Part One* [Pamphlet 706-185, issued 1967].
31. AMC Pamphlet 706-186, *Engineering Design Handbook, Military Pyrotechnics Series*; *Part Two, Safety, Procedures & Glossary*, Hq. U.S. Army Material Command, October 1963.

32. AMC Pamphlet 706-187, *Engineering Design Handbook, Military Pyrotechnics Series, Part Three, Properties of Materials Used in Pyrotechnic Compositions*, Hq. U.S. Army Materiel Command, October 1963.

32a. AMC Pamphlet 706-189, *Engineering Design Handbook, Military Pyrotechnic Series*, Part Five, Bibliography, October 1966.

33-42. Department of the Army, Washington, D. C. Technical Manuals (TM) (cf. Ref. 47b. 47c, and 387):

33. *Military Pyrotechnics*, TM9-1370-200 (1958).

34. *Military Explosives*, TM9-1910 (1955).

35. *Military Chemistry & Chemical Agents*, TM3-215 (1963).

36. *Ammunition, General*, TM9-1900 (1956).

37. *Foreign Mine Warfare Equipment*, TM 5-223 (1957).

38. *Ground Chemical Munitions*, TM3-300 (1956).

39. *Use and Installation of Boobytraps*, FM5-31 (1956).

40. *Explosives & Demolitions*, FM5-25 (1954).

41. *Artillery Ammunition*, TM9-1901 (1950).

42. *Bombs for Aircraft*, TM9-1980 (1950).

42a. *Flame Thrower and Fire Bomb Fuels*, TM-3 (1958).

43. *Pyrotechnics and Miscellaneous Explosive Items*, Dept. of the Navy OP 2213, Bureau of Ordnance Publication (1957); *Pyrotechnic, Screening, & Dye Marking Devices*, NavWeps OP 2213, 1st rev.

44. *Aircraft Bombs, Fuzes, & Associated Components*, Vol. 1, Dept. of the Navy, NavWeps OP 2216, 1960.

45. *Ordnance Explosive Train Designers' Handbook*, NOLR 1111 (1952), declassified April 1964, Naval Ordnance Laboratory.

46. *Research & Development of Materiel, Engineering Design Handbook, Propellant Actuated Devices*, AMC Pamphlet 706-270, U.S. Army Materiel Command, August 1963.

47. *Power Cartridge Handbook*, 2nd Edition, U.S. Naval Weapons Lab., Dahlgren, Va., NavWeps Report 7836 (1961).

47a. *Cartridges & Cartridge Actuated Devices for Aircraft & Associated Equipment*, Fourth Revision, U.S. Naval Weapons Lab., Dahlgren, Va., NavWeps OP 2606 (15 July 1964).

47b. *Unconventional Warfare Devices & Techniques, References*, Department of Army, Technical Manual, TM31-200-1 (1966).

47c. *Unconventional Warfare Devices & Techniques, Incendiaries*, Department of Army, Technical Manual, TM31-201-1 (1966).

47d. *Improvised Munitions Handbook*, Frankford Arsenal, Philadelaphia, Pa.

48. B. T. Fedoroff et al., *Dictionary of Explosives, Ammunition & Weapons* (*German Section*), Picatinny Arsenal TR #2510, Dover, N. J., 1958.

49. *Encyclopedia of Explosives & Related Items*, Vol. 1, A to Azoxy (1960), Fedoroff et al.; Vol. 2, B (Explosif) to Chloric Acid (1962), Fedoroff & Sheffield; Vol. 3, C (Chlorides) to Detonating Relays (1966), Fedoroff & Sheffield, Picatinny Arsenal, Dover, N.J.

50. *Toxic Hazards Associated with Pyrotechnic Items*, AD 436880, Bureau of Naval Weapons, NavWeps OP 2793 (1963).

51. *General Safety Manual*, AMCR 385-224, U.S. Army Materiel Command, (formerly ORDM 7-224; 1951—supplemented by ORDM 7-225, 1952).

52. *Safety Handbook*, AD 233149, Picatinny Arsenal, Dover, N. J., Explosives and Propellants Laboratory, Explosives Research Section, (1959).

53. *Explosives, Propellants & Pyrotechnic Safety*, Naval Ordnance Laboratory, NOLTR 61-138, R. McGill, AD 272424, (1961).

54. *Ordnance Safety Precautions*, OP-1014 (First Review 1955).

55. D. T. Smith, "Safety in the Chemical Laboratory," *J. Chem. Educ.* **41,** A520 (1964).

56. C. L. Knapp, "High Energy Fuels, Safe Handling," *Ind. Eng. Chem.*, **55,** 25 (1963).

57. F. W. Sears, *Optics*, Addison-Wesley Press, Inc., Cambridge, Mass., 1949.

58. Arthur C. Hardy and Fred H. Perrin, *The Principles of Optics*, 1st Ed., McGraw-Hill Book Co., N. Y., 1932.

59. F. Ephraim, *Inorganic Chemistry*, 6th English Ed. (P. C. L. Thorne and E. R. Roberts, Editors), Interscience Publishers, Inc., N. Y., 1954.

60. H. Remy, *Treatise on Inorganic Chemistry*, 2 Vol., Elsevier Publishing Co., N. Y., 1956; reprinted 1965.

61. Dr. Karl A. Hofmann, *Anorganische Chemie* ("Inorganic Chemistry") [Prof. Dr. Ulrich Hofmann and Prof. Dr. Walter Rüdorff, Editors] 17th Ed., 1963; 18th Ed., 1965.

62. T. Moeller, *Inorganic Chemistry, An Advanced Textbook*, John Wiley & Sons, Inc., N. Y., 1952.

63. F. D. Rossini et al., *Selected Values of Chemical Thermodynamic Properties*, Circular of the National Bureau of Standards 500, U.S. Government Printing Office, Washington, D. in two parts, C., (reprinted 1961).

64. O. Kubaschewski & E. Ll. Evans, *Metallurgical Thermochemistry*, John Wiley & Sons, N. Y., 1956.

65. N. A. Lange, *Handbook of Chemistry*, 10th Ed., Handbook Publishers, Sandusky, Ohio, 1961. Revised Edition 1967.

66. *Gmelin's Handbuch der anorganischen Chemie* ("Encyclopedia of Inorganic Chemistry"), 8th Ed., 71 System-Nummern ("Parts"), Verlag Chemie GMBH, Weinheim, Bergstrasse (Germany), 1922 to—.

67. J. W. Mellor, *A Comprehensive Treatise of Inorganic & Theoretical Chemistry*, 16 Vol. (1927), New Impression (1952); also Supplements I (1956), II (1961), III (1963) to Vol. II, Supplement I (1964) to Vol. III, Supplement I (Part I) to Volume VIII, Longmans Green & Co., Ltd., London; the last 3 volumes sold in the U.S. through John Wiley & Sons, Inc., N. Y.

68. *A Textbook of Inorganic Chemistry*, (J. Newton Friend, Editor), Chas. Griffin & Co., Ltd., London, 1928.

69. F. Beilstein, *Handbuch der organischen Chemie*, ("Encyclopedia of Organic Chemistry"), 4th Ed. by F. Richter, Julius Springer Verlag, Berlin, Germany; supplemental series of volumes are continually issued, series III being the latest.

70. *Chemistry of the Solid State*, (W. E. Garner, Editor), Butterworths Scientific Publications, London, 1955; U.S. edition, Academic Press, Inc., N. Y.

71. K. Hauffe, *Reaktionen in und an festen Stoffen* ("Reactions in and upon Solids"). Springer Verlag, Berlin, 1955.

72. E. F. Van Dersarl, U.S. Patent 3,110,259 (1961).

73. A. Stock and C. Somieski, *Ber.* **49,** 111 (1916); *ibid.* **50,** 169 (1917); *ibid.* **55,** 3961 (1922).

74. R. Schwarz, *Z. Anorg. Chem.*, **215,** 288 (1933).

75. C. S. Herrick et al., "Borane Pilot Plants," *Ind. Eng. Chem.* **52,** 105 (1960).

76. *Tech. Bull.* 507*C and* 508*A*, Metal Hydrides, Inc., Beverly, Mass.

77. Sneed-Brasted, *Comprehensive Inorganic Chemistry*, Vol. 6, "The Alkali Metals" by John F. Suttle, D. Van Nostrand Co., Inc., N. Y., 1957.

78. W. Freundlich and B. Claudel, *Bull. Soc. Chem. France*, 967-70 (1956).

79. V. I. Mikheeva, *Hydrides of the Transition Metals* (1960), Translation Series AEC-TR-5224 (October 1962).

80. E. Amberger & H. Boeters, *Angew. Chemie*, **74,** 32-3 (1962); *C. A.* **56,** 12522 (1962).

81. A. W. Stewart and C. L. Wilson, *Recent Advances in Physical*

& Inorganic Chemistry, 7th Ed. Longmans, Green & Co., London, 1946.

82. L. Spialter and C. A. MacKenzie, U.S. Pat. 2,614,906 (1952).
83. *Angew. Chem.* **75** (24), 1205 (1963); *C. A.* **60,** 6692 (1964).
84. G. E. Coates, *Organo-Metallic Compounds*, Methuen & Co., Ltd., London; J. Wiley & Sons, Inc., N. Y., 1956.
85. J. E. Knap et al., "Safe Handling of Alkylaluminum Compounds," *Ind. Eng. Chem.* **49,** 874 (1957).
86. Heck, Jr. & Johnson, "Aluminum Alkyls Safe Handling," *Ind. Eng. Chem.* **54,** 35 (1962).
87. *Henley's 20th Century Book of Formulas, Processes & Trade Secrets*, N. W. Henley Publishing Co., N. Y., 1947.
88. *Chem. Ind.* (London), **52,** 2132 (1964); *C. A.* **62,** 7577 (1965).
89. W. Holt and W. E. Sims, *J. Chem. Soc.* **65,** 432 (1894).
90. C. B. Jackson, *Liquid Metals Handbook*, Supplement T.I.D. 5227, U.S.A.E.C., 3rd Ed., Washington, D. C., 1955.
91. H. J. Emeléus & J. S. Anderson, *Modern Aspects of Inorganic Chemistry*, 2nd Ed., D. Van Nostrand Co., Inc., N. Y., 1952.
92. *C. A.* **50,** 656 (1956).
93. N. B. Pilling and R. E. Bedworth, *J. Inst. Metals,* **29,** 529 (1923).
94. R. N. Lyon, *Liquid Metals Handbook*, NAVEXOS P-733 (Rev.), U.S. Gov. Printing Office, Washington, D. C., 1952.
95. *Handling & Uses of the Alkali Metals*, A. E. C. Publication, May 1957.
96. *Handling Metallic Sodium on a Plant Scale*, U.S. Industrial Chemicals Co., 3rd Ed., N. Y., 1959.
97. Georg Brauer, *Handbuch der präparativen anorganischen Chemie* ("Handbook of Inorganic Synthesis"), Ferd. Enke Verlag, Stuttgart, Germany, 1954.
98. E. Tiede, *Ber.* **49,** 1742, 1744 (1916).
99. M. Faraday, *A Course of Six Lectures on the Chemical History of a Candle*, Chautauqua Press, N. Y., Re-issued, "The Chemical History of a Candle," Viking Press, N. Y., 1960.
100. L. M. Kefeli and S. L. Lel'chuk, *Doklady Akad. Nauk*, S.S.S.R. **83,** 697-9 (1952).
101. F. Pawlek, *Z. Metallk.* **41,** 451-3 (1950); *C. A.* **48,** 4271 (1954).
102. *C. A.* **60,** 5045 (1964).
103. Hartman, Nagy and Jacobson, *U.S. Bureau of Mines* (*Pittsburgh, Pa.*) *Rpt. Invest.* #4835 (1951).
104. J. J. Katz & E. Rabinowitch, *The Chemistry of Uranium*,

Part. I, McGraw-Hill Book Co., Inc., N. Y., 1951.
105. "Ultrafine Particles," *Symposium Papers*, Sponsored by the Electrothermics & Metallurgy Division of the Electrochemical Society, Indianapolis, J. Wiley & Sons, N. Y., 1963.
106. M. Ya. Gen et al., *Doklady Akad. Nauk*, S.S.S.R. **127,** 366-8 (1959); *C. A.* **54,** 23603 (1960).
107. *I. & E. C. Prod. Res. & Dev.* **2,** 212 (September 1963).
108. E. Raub & M. Engels, *Metallforschung*, **2,** 115-19 (1947).
109. R. Fricke and S. Rihl, *Z. Anorg. Allgem. Chem.* **251,** 414-21 (1943).
110. *C. A.* **60,** 5058 (1965).
111. U.S.A.E.C., *SRIA*-29 (1960); *C. A.* **55,** 8230 (1961).
112. R. E. Mitchell and H. Bradley, British Patent 666,141.
113. *The Van Nostrand Chemist's Dictionary*, D. Van Nostrand Co., Inc., N. Y., 1953.
114. P. P. Alexander, U.S. Patent 2,611,316 (1952).
115. C. Balke, D. Hill, and W. Graff, U.S. Patent 2,819,163 (1958).
116. H. S. Cooper, U.S. Patent 1,562,540 (1925).
117. H. L. Mencken, *A New Dictionary of Quotations*, A. A. Knopf, N. Y., 1942.
117a. Georgius Agricola, *De Re Metallica*, translated from the Latin Edition of 1556 by Herbert C. and Lou H. Hoover, Dover Publications, Inc., N. Y., 1950.
118. "Table of Materials Subject to Spontaneous Heating," National Fire Protection Association, Boston (1947).
119. *Factory Mutual Eng. Div. Loss Prev. Bull.* #36.10, Boston (1950).
120. L. Morris and A. J. W. Headlee, *J. Chem. Educ.* **10,** 637 (1933).
121. *J. Res. Natl. Bur. Std.* **61,** 413-17, Res. Paper 2909 (1958); *C. A.* **53,** 8629 (1959).
122. Anon., *Textil-Rundschau,* **12,** 273 (1957).
123. M. Kehren, *Textil-Rundschau,* **3,** 409-17 (1948); *Chem. Zentr.* (Russian Zone Ed.), **1,** 943 (1949).
124. W. D. Maclay, Dir., Northern Div., U.S. Dept. of Agriculture, Agricultural Research Service, Peoria, Ill., *Personal Communication.*
125. Bolland and Gee, *Trans. Farad. Soc.* **42,** 244 (1946).
126. Drinberg and Shepelov, *J. Gen. Chem. USSR*, **10,** 2049 (1940).
127. Bjorn Holmgren, *Svensk Papperstid.* **51,** 230-2.
128. N. D. Mitchell, "Self Ignition," NBS, reprint from *Nat. Fire Prot. Assoc. Quarterly* (Oct. 1951), Boston.

129. K. N. Smith, "Self-heating of Wood Fiber," *Tappi*.**42,** 869-872 (1959); Fire Research Abstracts and Reviews, **2,** 49 (1960).
130. H. P. Rothbaum, *J. Appl. Chem.* (London) **13,** 291-302 (1963); *C. A.* **59,** 7310 (1963).
131. M. B. Cunningham, *Tappi.* **44** [3], 194A-198A (1961); *C. A.* **55,** 14911 (1961).
132. *Special Interest Bulletin* #214 American Insurance Association (Feb. 1965).
133. *Fire Protection Handbook*, 12th Ed., Chapter V, "Fire Hazard Properties of Dusts," National Fire Protection Association, Boston, 1962.
134. A. C. Fieldner and W. E. Rice, *U.S. Bur. of Mines Inf. Circ.* 7241 (1943).
135. I. Hartmann and H. P. Greenwald, *Mining & Met.* **26,** 331-5 (1945); *C. A.* **40,** 2629 (1946).
136. H. S. Kalish, *Pyrophoricity & Toxicity of Metal Powders*, Speech before Powder Metallurgy Committee, A.I.M.E., N. Y. Section, Nov. 1959.
137. *Fire Research Abstracts*, Vol. 3, #1 (Jan. 1961); H. Selle and J. Zehr, *VDI*, *Berichte*, **19** 73-87 (1957).
138. N. P. Setchkin, *Modern Plastics*, **38,** 119 (1961).
139. *U.S. Bureau of Mines Report Invest.* #6516 (9), (1964); *C. A.* **61,** 13117 (1964).
140. *U.S. Bureau of Mines Report Invest.* #6543 (10), (1964); *C. A.* **62,** 394 (1965).
141. "When Explosion Threatens, Every Millisecond Counts," *Occupational Hazards*, p. 52-54, June 1965.
142. Shin-Ichi Aoyama and Eizo Kanda, *Bull. Chem. Soc.* (Japan), **12,** 521-4 (1937); *C. A.* **32,** 4843 (1938).
143. *Product Information Technical Bulletin TA* 8532-1 of Baker & Adamson Products, Gen. Chem. Div., Allied Chemical & Dye Corp., N. Y.
144. *American Rocket Society J.* **32** [3], 384-7 (1962); *C. A.* **61,** 4138 (1964).
145. H. N. Gilbert, "Some Unique Properties of Sodium & Potassium," *Chem. Eng. News*, **26,** 2604 (1943).
146. *Angew. Chem.* **75** [24], 1025 (1963); *C. A.* **60,** 6692 (1964).
147. E. W. Lindeijer, *Chemical Weekblad*, **46,** 571 (1950); *C. A.* **46,** 7769 (1952).
148. J. Van Hinte, *Veiligheid* (*Amsterdam*), **28,** 121 (1952); *C. A.* **47,** 10152 (1953).
149. *Chem. Eng. News*, **36,** 40 (1958).

150. *J. Chem. Ed.* **40,** 633 (1963).
151. H. A. Smith, *The Compleat Practical Joker*, Pocket Books, Inc., N. Y., 1956.
152. *Tech. Bulletin 401D.* Metal Hydrides, Inc., Beverly, Mass.
153. *Tech. Bulletin 402C.* Metal Hydrides, Inc., Beverly, Mass.
154. *Tech. Bulletin 502F.* Metal Hydrides, Inc., Beverly, Mass.
155. Anon., Pamphlet, *Nitroparaffine Symposium* sponsored by Commercial Solvents Corp., (1956).
156. E. Masdupuy and F. Gallais, *Compt. rend. 78th Congr. socs. savants Paris et dépt., Sect. sci.* 369 (1953).
157. *Fire Research Abstracts*, **4** 150 (1962); *C. A.* **55** 9878 (1961).
158. It. Patent 446,010 (1949); *C. A.* **45,** 1770 (1951).
159. J. Amiel, *Compt. Rend.* **198,** 1033-5 (1934); *ibid.* **199,** 787-9 (1934).
160. F. Taradoire, *Compt. Rend.* **199,** 603-5 (1934).
161. E. C. Kirkpatrick, Rpt. OSRD #3507 (declassified), *Chemical Ignition of Flame Throwers*, 1944.
162. A. Engelbrecht and A. V. Grosse, *J. Am. Chem. Soc.* **64,** 5262-4 (1942).
163. A. V. Grosse, U.S. Patent 2,684,284 (1954).
164. M. Schmeisser and D. Luetzow, *Angew. Chem.* **66,** 230 (1954).
165. A. Engelbrecht and A. V. Grosse, *J. Am. Chem. Soc.* **76,** 2042-5 (1954).
166. *Chemical Reviews*, **62,** 1 (1962).
167. K. A. Hofmann and A. v. Zedtwitz, *Ber.* **42,** 2021 (1909).
168. Dr. A. Stettbacher, *Die chemische Zündung und ihre Anwendungsmöglichkeiten* ("Chemical Ignition & Its Uses"), Nitrozellulose, **9,** S. 75-7, 100-1, 138-41 (1938).
169. Dr. A. Stettbacher, *Schweiz. Chem. Z.* S. 27-37 (1944).
170. J. C. Schumacher, Editor, *Perchlorates*, ACS Monograph 146, Reinhold Publishing Co., N. Y., 1960.
171. *Chemical Safety Data Sheet* (*SD*-11), Manufacturing Chemists' Association, Inc., Washington, D. C. (1947).
172. H. Rathsburg and H. Gawlick, *Chem. Z.* **65,** 426-7 (1941).
173. *Webster's Unabridged* Dictionary, G. & C. Merriam Co., Springfield, Mass., 1963.
174. *The Oxford English Dictionary*, At the Clarendon Press (1933), reprinted 1961.
175. *German Methods of Production of Amorces & Sundry Pyrotechnical Stores*, BIOS, *Final Rept.* #1313, *Item* 17, 37, London—HM Stationery Office, 38-46 Cadogan Sq., London SWI, (1947).
176. M. Mauder, *Almanach für Scheidekünstler*, **1,** 36 (1780.)

177. *Evaluation Test Results on Service & Experimental Squibs*, U.S. Navy Report, NavOrd Rept. 6061 by Peet and Gowen, 1958.
178. *Engineering Design Handbook, Explosives Series, Explosive Trains*, AMC Pamphlet 706-179, AD 462254, Hq. U.S. Army Materiel Command, March 1965.
179. *Percussion and Electric Primers*, 3rd Ed. Olin Mathieson Chemical Corp., Winchester-Western Div., 1962.
180. *Explosive Specialities Manual*, E. I. du Pont deNemours, Wilmington 98, Del.
181. Levi, *Atti Linc.* (5), **31** I, 214 (1922); *ibid.* **32** I, 166 (1923).
182. Millon, *Ann. Chim. Phys.* **3,** 329 (1843).
183. Schiel, *Lieb. Ann.* **109,** 321 (1859).
184. E. von Herz, *Zeit. ges. Schiess- und Sprengstoffw.* **11,** 365 (1916).
185. J. D. McNutt and S. D. Ehrlich, U.S. Patent 2,292,956 (1942).
186. J. F. Kenney, U.S. Patent 2,689,788 (1954) and 2,702,746 (1955).
187. W. B. Woodring and H. T. McAdams, U.S. Patent 2,970,900 (1961); *C. A.* **55,** 11850 (1961).
188. R. A. Carboni, U.S. Patent 2,904,544 (1959).
188a. L. B. Steinmark and L. Saffian, *Automated Manufacturing of Electric Detonators*, PA-TM-1490, AD 448451 Picatinny Arsenal, Dover, N. J., (1964).
189. Atlas Match M-100 etc., *Data Sheets*, Wilmington, Del., 1957.
190. *Datalog*, Unidynamics/Phoenix Div., UMC Industries, Inc., Phoenix, Ariz., 1966.
191. *Explosive Ordnance Technical Data Book*, McCormick-Selph Assoc., Hollister, Calif.
192. O. A. Pickett, U.S. Patent 2,008,366 (1935).
193. E. E. Kilmer, U.S. Patent 3,150,020 (1964).
194. F. R. Seavey and N. J. Wilkaitis, U.S. Patent 3,142,253 (1964).
195. Wm. E. Schulz, U.S. Patent 2,953,447 (1960).
196. *Pulse-Sensitive Electro-Explosive Devices*, AD 406346, Aerojet-General Corp., Sacramento, Calif.
197. Dr. D. D. Taylor, *Electric Initiators of the Conductive Mix Type*, American Ordnance Association, Loading Section Meeting, May 1961.
198. Th. Q. Ciccone, U.S. Patent 2,970,047 (1961).
199. G. W. Peet and L. F. Gowen, U.S. Patent 3,090,310 (1963).
200. F. M. Correll, *Development of RF-Protected Electro-Explosive Devices*, PA-TM-1258, AD 600944, Picatinny

Arsenal, Dover, N. J.
201. A. McLellan Yuill, U.S. Patent 3,096,714 (1963).
202. M. M. J. Sutherland, *A Textbook of Inorganic Chemistry*, Vol. X, "The Metal Ammines," Chas. Griffin & Co., Ltd., London, 1928.
203. J. C. Bailar, Jr., *The Chemistry of Coordination Compounds*, ACS Monograph Series, Reinhold Publishing Co., N. Y., 1956.
204. H. Ellern and D. E. Olander, *J. Chem. Educ.* **32,** 24 (1955).
205. J. Amiel, *Compt. Rend.* **199,** 201-3 (1934); *C. A.* **28,** 5774 (1934).
206. W. R. Tomlinson, K. G. Ottoson, and L. F. Audrieth, *J. Am. Chem. Soc.* **71,** 375-6 (1949).
207. L. Médard and J. Barlot, *Mém. Poudres*, **34,** 159 (1952).
208. Dr. Taylor Abegg and Wm. Meikle, *C. A.* **63,** 2839 (1965).
209. R. Salvadori, *Gazz. Chim. Ital.* **40,** II, 12 (1910); **42,** I, 458 (1912).
210. W. Friederich and P, Vervoorst, *Z. ges. Schiess- und Sprengstoffw.* **21,** 49-52, 65-69, 84-87, 103-105, 143-146 (1926).
211. Maissen and Schwarzenbach, *Helv. Chim. Acta*, **34,** 2084 (1951).
212. N. I. Lobanov, *C. A.* **52,** 1832 (1958); *C. A.* **53,** 21354 (1959); *C. A.* **55,** 23155 (1961); *C. A.* **59,** 8344 (1963); *Zh. Neorgan. Khim.* **2,** 1035-9 (1957); **4,** 344-5 (1959); **6,** 870-3 (1961); **8** (5), 1112-15 (1963).
213. Fustel de Coulanges, *The Ancient City*, Doubleday Anchor Books, Doubleday & Co., Inc., Garden City, N. Y., 1956.
214. P. M. Woolley, "A Hundred Years of Making Matches at Bryant & May, Ltd.," *Chemistry and Industry*, *Soc. of Chem. Industry*, p. 374 f. (25 March 1961).
215. Kenneth Lunny, "Matchmaking Past and Present," *Canadian Chem. Processing*, p. 54-55, Toronto, Ont., (20 June 1952).
216. Charles Darwin, *The Voyage of the Beagle*, Everyman Paperback Edition, E. P. Dutton & Co., Inc., N. Y. (1961); Doubleday Anchor Book, Doubleday & Co., Inc., Garden City, N. Y. (1962).
217. A. Osol and G. E. Farrar, *The Dispensatory of the United States*, J. B. Lippincott Co., Phila., 1950.
218. Henri Sévène and Emile David Cahen, U.S. Patent 614,350 (1898).
219. U.S. Department of Commerce, *Census of Manufacturers* (1958).

220. Herbert Ellern, U.S. Patent 2,495,575 (1950); Canadian Patent 526,012 (1956); U.S. Patent 2,647,048 (1953).

221. Zoltán Földi and Rezsö König, Hungarian Patent 108,056 1934), 108,188 (1934); British Patent 409,291 (1934); *C. A.* **28,** 6314 (1934); U.S. Patent 2,093,516 (1937); *C. A.* **31,** 8202 (1937); K. E. Olson, Swedish Patent 83,904; *C. A.* **31,** 2824 (1937); Jos. Ellenbogen, British Patent 590,036; U.S. Patent 2,545,076 (1951); Benjamin Edelberg, U.S. Patent 2,566,560 (1951); John B. Tigrett, Belgian Patent 625,124 (1963); French Patent 1,366,748; French Patent 1,366,705 (1964).

222. *Blasters' Handbook*, 14th Ed., E. I. du Pont deNemours, Wilmington 98, Del., 1963; 15th Ed., 1966.

223. *Dictionary of Americanisms on Historical Principles*, (M. M. Mathews, Editor) U. of Chicago Press, 1951.

224. F. de Capitani, British Patents 746,435 (1954); 758,858 (1953); 800,596 (1955) [Group VI].

224a. F. De Capitani, U.S. Patent 3,262,456 (1966); *C. A.* **65** 13452 (1966).

225. M. S. Silverstein et al., U.S. Patents 2,635,953 (1953); 2,664,343 (1953); 2,664,344 (1953); 2,645,571 (1953); 2,440,303 (1948).

226. *Letter Circular* 423, U.S. Dept. of Commerce, NBS, Wash., D. C. (1934).

227. I. Kowarsky, U.S. Patent 2,722,484 (1955).

228. René Arditti, Hubert Gaudry, and Yvon Laure, *Compt. Rend.* **226,** 1179-80 (1948); *C. A.* **42,** 7044 (1948).

229. Federal Specification EE-M-101H (1964), "Match, Safety and Match, Non-Safety (Strike Anywhere)."

230. *The Merck Manual of Diagnosis & Therapy*, 10th Ed., Merck & Co., Inc., Rahway, N. J. (1961).*

231. N. J. Sax, *Dangerous Properties of Industrial Materials*, Reinhold Publishing Co., N. Y., 1957; 2nd Ed., 1963.

232. Federal Specification MIL-M-1513 (October 1949), cancelled 1963 (Originally QMC Tent, JCQD #1007, April 1944, and #1007A, Sept. 1944).

233. *Modern Plastics Encyclopedia*, McGraw-Hill, Inc., N. Y., 1965.

234. *Materials in Design Engineering*, Materials Selector Issue, Reinhold Publishing Co., N. Y. (Oct. 1964).

* The misleading statement mentioned in the text does not appear in the 11th edition (1966).

234a. Alexander Lebovits, *Modern Plastics*, **43,** 139 ff (March 1966).

235. NASA Report SP 55, Air Force ML-TDR-64-159, *Symposium* on "Thermal Radiation of Solids," San Francisco (1964).

236. R. A. Smith et al., *The Detection & Measurement of Infrared Radiation*, Oxford at the Clarendon Press, 1957.

237. H. L. Hackforth, *Infrared Radiation*, McGraw-Hill Book Co., N. Y., 1960.

238. *Radiation Calculator GEN.* 15-*B* 9-56 (2500) by Optics & Color Engineering Lab., General Electric Lab., Schenectady, N. Y.

239. E. L. Nichols and H. L. Howes, *Phys. Rev.* **19,** 300-18 (1922); *C. A.* **16,** 2265 (1922).

240. A. G. Gaydon and H. G. Wolfhard, *Flames, Their Structure, Radiation & Temperature*, Chapman & Hall, 2nd Ed., London, 1960.

241. *Optika v Voyennom Dele* ("Optics in Military Affairs"), Collection of Articles Edited by S. I. Vavilov, AN SSSR, 1945.

242. *Missiles and Rockets*, **14,** 9 (October 5, 1964).

243. *Missiles and Rockets*, **14,** 31 (October 5, 1964).

244. W. E. Knowles Middleton, *Vision Through the Atmosphere*, University of Toronto Press, 1952, reprint (1963) in U.S.

245. "Temperature," *Symposium of the American Inst. of Physics*, in N. Y. City in 1939, Reinhold Publ. Corp., N. Y. (1941); *ibid.* Article by A. G. Worthing, "Temperature Radiation Emissivities & Emittances"; *ibid.* W. F. Roeser and H. T. Wensel (NBS), Tables 16, 17, 18, 19 (in Appendix).

246. Dr. Robert W. Evans, *Comments Concerning the Production of Light by Visual Flares*, Speech at American Ordnance Association, Military Pyrotechnics Section Meeting, November 1965, Monterey, Calif.

247. D. Hart and H. S. Eppig, *Long Range Research on Pyrotechnics*; *Burning Characteristics of Binary Mixtures*, TR-1669, DDC-ATI 66289, Picatinny Arsenal, Dover, N. J. October 1947.

248. *C. A.* **55,** 24014 (1961).

249. *C. A.* **60,** 6689 (1964).

250. Z. W. Flagg* and R. Friedman, 'Solid Powders as Sources of Cesium Plasma," *Raketnaya tekhnika* (Rocket Engineering), N. **1,** 122 (1961).

* Thus, in the translation; Fagg, in the original.

251. B. Douda, NAD, Crane, Ind., *Colored Flare Ingredient Synthesis Program*, RDTR No. 43, 10 July 1964; also U.S. Patent 3,296,045 (1967).

252. B. Douda, NAD, Crane, Ind., *Theory of Colored Flame Production*, RDTN No. 71, 20 March 1964.

253. B. Douda, NAD, Crane, Ind., *Relationships Observed in Colored Flames*, RDTR No. 45, 25 Sept. 1964.

254. B. E. Douda, *J. Opt. Soc.* **55** No. 7, pp. 787-793, (July 1965).

255. Dr. David Hart, Picatinny Arsenal, Dover, N. J., *Pyrotechnic Illumination of Bombing Targets*, Speech at American Ordnance Association, Military Pyrotechnics Section Meeting at NOL, March 1958.

256. H. N. Cohen and G. F. Kottler, *The Optimum Height of a Burning Flare*, Samuel Feltman Labs., Picatinny Arsenal, Dover, N. J., TR 2081, (Oct. 1954.)

257. Maj. T. C. Ohart, *Elements of Ammunition*, J. Wiley & Sons, Inc., N. Y., 1952.

258. Naval Airborne Ordnance, *NavPers* 10826-*A*, 1st Ed. 1958, Reprinted with minor changes 1961, U.S. Gov. Printing Office.

259. P. S. Strauss and G. R. DeTogni, *Personnel Target Acquisition Under Flare Illumination*, Report #3012, Picatinny Arsenal, Dover, N. J., July 1962.

259a. *Pamphlet*, "Underwater Flare," Special Devices, Inc., Newhall, Calif.

260. E. W. Rolle, U.S. Patent 3,107,614 (1963).

261. P. L. Harrison, *The Combustion of Titanium & Zirconium*, *Seventh Symposium* (*International*) *on Combustion* (1958), Butterworths Scientific Publications, London (1959).

262. V. J. Korneyev and I. I. Vernidub, *Issled. po Zharoproch. Splavam*, *Akad. Nauk SSSR*, *Inst. Met. A. A. Baikova* ("Investigations of Heat Resistant Alloys, Academy of Sciences of the USSR, A. A. Baikov Institute of Metallurgy"), **7**, 309-16 (1961); *C. A.* **56** 8428 (1962).

263. K. P. Coffin, "Some Physical Aspects of the Combustion of Magnesium Ribbons," *Fifth Symposium on Combustion* (1954), Reinhold Publ. Co., N. Y., 1955.

264. Henry C. Lottes, NAD, Crane, Ind., *Flare Performance Investigation*, NavWeps Report 8250 (2 Nov. 1962).

265. G. J. Schladt, U.S. Patents 2,035,509 (1936) and 2,149,314 (1939).

266. *German Illuminating Flares*, CIOS, Target No. 3a/164 and 17/34.

267. *German Tracer Compositions*, CIOS, by H. Eppig, Item No. 3, File No. XXXII-20, (4 Sept. 1945).

268. *Pyrotechnic Anti-Pathfinder Devices*, CIOS, by H. Eppig, Item No. 3 and 17, File No. XXXII-56, (June-July 1945).

269. *German Pyrotechnic Factories*, BIOS, Final Report No. 477, Item No. 17.

270. *Technical Investigation of "Exploding" Flare Fatality*, Res. and Develop. Dept., NAD, Crane, Ind., RDTR No. 39, (1963).

271. *From My Life, The Memoirs of Richard Willstätter*, translated from German by Lilli S. Hornig, W. A. Benjamin, Inc., N. Y., 1965.

272. A. Taschler and Seymour M. Kaye, Feltman Res. Labs., Picatinny Arsenal, Dover, N. J., *Effects of Magnesium Content, Case Material, and Case Coating on Burning Characteristics of a Flare System*, April 1960.

273. Ralph Chipman, NAD, Crane, Ind., *Experimental High Intensity Flare Systems Data Reduction and Analysis*, RDTR No. 57 (11 May 1965).

273a. D. E. Middlebrooks and S. M. Kaye, Picatinny Arsenal Dover, N. J., *The Effects of Processing on Pyrotechnics Compositions*, Part I, PA-TR-3252, AD 472872 (Sept. 1965).

273b. W. J. Nolan and H. Dorfman, Picatinny Arsenal, Dover, N.J., PA-TR-2027 (May 1954).

274. James Swinson, NAD, Crane, Ind., *Computer Program for the Analysis of Visible Spectrometric Data: Chromaticity*, RDTR No. 59 (26 Aug. 1965).

275. French Patent 828,664.

276. *Chem. Eng. News*, p. 50 (12 Oct. 1964).

277. L. D. Lockwood, U.S. Patent 2,263,179 (1941).

278. M. Pipkin, U.S. Patent 2,285,125 (1942).

279. R. E. Vollrath, U.S. Patent 2,149,694 (1939).

280. *Airborne Photographic Equipment*, Air Force Systems Command, Wright Patterson AFB, Ohio, Reconnaissance Application Branch Report, 1964.

281. *Investigation of Current Techniques of Low Altitude Pyrotechnic Flash Night Aerial Reconnaissance Photography*, U.S. Naval Air Development Center, Johnsville, Pa., Report No. NADC-6012, AD 257359 (1960).

281a. F. P. Clark, *Special Effects in Motion Pictures*, The Society of Motion Picture & Television Engineers, N. Y., 1966.

281b. *Special Effects Materials & Prices*, Special Devices, Inc., Newhall, Calif., Leaflet.

282. D. D. Collins, Missile Firing Lab., Cape Canaveral, Fla., Speech, *Pyrotechnic Flashes & Flares in Guided Missile Tracking*, at Aberdeen Proving Ground, Md. (Oct. 1959), Meeting of Military Pyrotechnics Section, AOA.

283. B. Jackson, S. M. Kaye, & G. Weingarten, Picatinny Arsenal, Dover, N. J. *Development of Substitute Compositions for High Altitude Flash Systems Containing Elemental Calcium*, TR 3068, AD 457878; *C. A.* **65,** 12056 (1966).

283a. S. Lopatin and David Hart, U.S. Patent 3,261,731 (1966).

284. S. M. Kaye and J. Harris, Picatinny Arsenal, Dover, N. J., *Effect of Fuel & Oxidant Particle Size on the Performance Characteristics of* 60/40 *Potassium Perchlorate/Aluminum Flash Composition*, AD 266486; *C. A.* **58,** 5443 (1963).

285. Hershkowitz, Dalrymple and H. Dorfman, Picatinny Arsenal, Dover, N. J., TR 3098, AD 415926.

286. S. Lopatin, Picatinny Arsenal, Dover, N. J., *Sea Level & High Altitude Performance of Experimental Photoflash Compositions*, Tech. Rept. FRL-TR-29, AD 266213 (Oct. 1961).

287. E. D. Crane, B. Werbel, and G. Weingarten, Picatinny Arsenal, Dover, N. J., *Development of Pyrotechnic Spotting & Ejection Charges for Use in the Davy Crocket* 37 *mm Spotting Round*, PA-TM-1431, AD 460333 (March 1965).

288. J. G. Holmes, "Recognition of Colored Light Signals," *Trans. Illumination Eng. Soc.* **6,** #1, pp. 71-97 (1941); *Nature* (London), **147,** 423-4 (1941).

289. R. F. Barrow and E. F. Caldin, *Proc. Phys. Soc.* (London), **62B,** 32-9 (1949).

290. A. Chapanis, "Color Names for Color Space," *The American Scientist*, **53,** 327-345 (1965).

291. G. F. Smith and E. G. Koch, *Z. anorg. allgem. Chem.* **223** 17 (1935).

291a. B. E. Douda, U.S. Patent 3,258,373 (1966).

292. Ross M. Hedrick and Edward H. Mottus, U.S. Patent 3,094,444 (1963).

293. W. M. Fredericks, U.S. Patent 2,538,360 (1951).

294. *Specification for Red Railroad Fusees or Red Highway Fusees*, Bureau of Explosives, Revised May 1, 1959.

295. I. Niditch, U.S. Patent 954,330 (1910).

296. L. S. Ross, U.S. Patent 1,019,190 (1912).

297. I. Niditch, U.S. Patent 1,029,884 (1912).

298. E. C. Pfeil, U.S. Patent 1,441,878 (1923).

299. L. A. Sherman, U.S. Patent 2,120,580 (1938).

300. J. Vinton, U.S. Patent 2,628,897 (1953).

301. H. C. Clauser and R. S. Long, U.S. Patent 2,640,771 (1953).
302. *Evaluation of Pocket Pen Flare,* Arctic Aeromedical Lab., Fort Wainwright, Alaska, Tech. Note 63-3, AD 411357, (1963).
303. *Operational Test & Evaluation Aerial Signal Flare,* Tactical Air Command, Langley AFB, Va., AD 613006, (March 1965).
304. R. O. Stefan and A. G. Lang, U.S. Patent 3,102,477 (1963) and 3,044,360 (1962).
305. T. Stevenson and E. R. Rechel, Frankford Arsenal, Phila., Pa., *Delay Action Tracer Ammunition,* R-44 (1940).
306. T. Stevenson, Frankford, Arsenal, Phila., Pa., *Delay Action & Dim Igniters for Small Arms Tracer Bullets,* R-708 (1946).
307. Tech. Rept. #1335, Picatinny Arsenal, Dover, N. J. (6 Sept. 1943), declassified.
308. S. P. Meek, U.S. Patent 1,708,186 (1929).
309. C. H. Pritham, U.S. Patent 2,123,201 (1938).
310. "Fireworks: No Boom," *Chemical Week,* **87,** p. 55 (2 July 1960).
311. *Ordnance Technical Terminology,* ST 9-152, U.S. Army Ordnance School, Aberdeen Proving Ground, Md. (June 1962), superseding June 1959 Preliminary Edition, *Glossary of Ordnance Terms.*
312. German Patent 565,250 and 646,129.
313. H. Bennett, *The Chemical Formulary,* Vols. I through XIII, Chemical Publ. Co., Inc., N. Y. (1933-1966).
314. Max Wullschleger-Hirschi, Swiss Patent 319,241 (1957); *C. A.* **51,** 15132 f. (1957).
315. *The Chemical Compositions of German Pyrotechnic Colored Signal Lights,* CIOS, by H. Eppig, Target No. 3a/163; 7/222; 17/33.
316. Leo Finklestein, U.S. Army Edgewood Arsenal, Md., *History of Research and Development of the Chemical Warfare Service in W. W. II,* CRDL Spec. Publ. 1-42, AD 461129 (June 1964).
317. Maj. Gen. W. M. Creasy, "Smoke to Protect from A-Bomb Heat," *Armed Forces Chem. J.* (May/June 1955).
318. E. H. Engquist, "Attenuation of Thermal Radiation," *Armed Forces Chem. J.* (Jan./Feb. 1958).
318a. Final Engineering Report, Phase II, Preparedness Study, for Improved HC Smoke Mixture, Universal Match Corp., Ferguson, Mo. to Pine Bluff Arsenal, Ark. (1954-declassified 1961).

318b. J. H. McLain and S. Mayer, Correlation of the Properties of HC-Type E with the Density of the Constituent Zinc Oxide, TDMR 750 (1943).
319. J. O. M. Brock, U.S. Patent 2,939,779 and 2,939,780 (1960).
320. R. P. Teele and H. K. Hammond, *Development of a Laboratory Test for Evaluation of the Effectiveness of Smoke Signals*, NBS Rept. 4792 (1956), U.S. Dept. of Commerce, Washington, D. C.
321. M. H. Weasner and J. Carlock, Picatinny Arsenal, Dover, N. J., *Smoke Marker Detection and Identification Study*, PA-TR-3259 (1965).
322. *Speech*, Leo Frey, Picatinny Arsenal, Dover, N. J., at American Ordnance Association, Military Pyrotechnics Section Meeting, December 1964.
323. J. S. Hatcher, *Textbook of Pistols and Revolvers*, Small Arms Technical Publ. Co., Plantersville, S. C., 1935.
324. T. L. Davis, U.S. Patent 2,409,111 (1946).
325. R. L. Tuve, U.S. Patent 2,469,421 (1949).
326. S. J. Magram, U.S. Patent 2,775,515 (1956).
327. Wm. G. Finnegan, Rex L. Smith, and L. R. Burkardt, U.S. Patent 3,046,168 (1962); *C. A.* **57,** 14046 (1962).
328. G. Graff, U.S. Patents 1,920,254 (1933); 1,975,099 (1934); 1,975,785 (1934); 2,091,977 (1937).
329. A. F. Tatyrek, Picatinny Arsenal, Dover, N. J., *The Production of Colored Smokes from Highly Reactive Hydrolyzable Metal Chlorides*, PA-TM-1644, (1965).
330. German Patent 1,188,490 (1965); *C. A.* **62,** 12970 (1965).
331. E. Crane, B. Werbel, and G. Weingarten, Picatinny Arsenal, Dover, N. J., *Improved Green, Red, Yellow and Violet Smoke Compositions for Rocket-Type Parachute Ground Signals*, TM 1033 (1963).
331a. E. Crane, B. Werbel, and G. Weingarten, Picatinny Arsenal, Dover, N. J., *Development of Miniature Smoke Signal Package for Inclusion in Survival Kits*, TR-3294, October 1966.
332. J. W. Van Karner, U.S. Patent 2,411,070 (1946).
333. J. W. Orelup, U.S. Patents 2,419,851 (1947) and 2,478,418 (1949).
334. *Munsell Book of Color*, Munsell Color Company, 10 E. Franklin St., Baltimore, Md.
335. A. F. Tatyrek, Picatinny Arsenal, Dover, N. J., *The Health Hazards of Certain Smoke Dyes in Current Use*, PA-TM-1674 (1965).
336. N. Anson and P. Parent, Army Medical Center, Md., Toxi-

cological Information Center, *Toxicity of Nine Dyes, A Literature Study*, Tech. Memo 47-6 (1959).
337. B. B. Blackford et al., U.S. Patent 2,546,964 (1951).
338. Leonard Jans, U.S. Patent 3,103,171 (1963).
339. Lt. Col. Rex Applegate, "New Riot Control Weapons," *Ordnance*, **49,** 67 (July/August 1964).
340. I. Nimeroff and S. W. Wilson, U.S. Patent 2,802,390 (1957).
341. C. Katzenberger, U.S. Patent 1,207,766 (1916); W. McLain et al., U.S. Patent 2,385,636 (1945); J. Dinsdale, P. Holmes, and P. Martin, U.S. Patent 2,557,814 (1951); E. Wheelwright et al., U.S. Patent 2,557,815 (1951); E. Bateman et al., U.S. Patent 2,606,095 (1952); J. Gillies and W. Cunningham, U.S. Patent 2,606,858 (1952); Australian Patents 142,197 (1951) and 152,786 (1953).
342. H. L. Green and W. R. Lane, *Particulate Clouds, Dusts, Smokes, and Mists*, 2nd Ed., D. Van Nostrand Co., Inc., Princeton, N. J., 1964.
343. German Patent 1,058,895 (1959); *Ber.* **21,** 2737 (1888); *Ber.* **21,** 2883 (1888).
344. Hans G. Bremer, "Pyrethrum Reaction Aerosol," *J. Econ. Entomol.* **57,** 62-67 (1964).
345. *Zentr. Biol. Aerosol Forsch.* **11,** 114-31 (1963).
346. *Aerozoli v Sel'skom Khozyaystve* ("Aerosols in Agriculture"), Collection of Articles Edited by A. G. Amelin, Sel'khozgiz ("State Agricultural Publishing House), 1956.
347. A. I. Sidorov, *Zashchita Rasteniy ot Vrediteley i Bolezney,* ("Aerosol Smoke Pots"), No. 2, **8,** 38-39 (1960).
348. *Kratkaya Khimicheskaya Entsiklopediya* ("Brief Chemical Encyclopedia KKhE"), Vols. 1, 2, and 3 (publ. to be continued), 1961-1964.
349. D. Marke and C. Lilly, "The Formation of Insecticide Smokes," *J. Sci. Food Agr.* **2,** 56 (1951).
350. C. R. Weinert, U.S. Patent 3,085,047 (1963).
351. I. I. Vernidub et al., *Izv. Akad. Nauk SSSR, Ser. Geofiz.* ("Bulletin of the Academy of Sciences of the USSR, Geophysical Series"), **9,** 1286-1293 (1962), "Investigation of the Ice-Forming Properties of Lead Iodide."
352. B. D. Meyson, *Fizika Oblakov* ("Physics of Clouds"), Gidrometeoizdat (Hydrometeorological Publ. House), 1961; V. Ya. Nikandrov, *Iskusstvennyye Vozdeystviya na Oblaka i Tumany* ("Artificial Effects on Clouds and Fogs"), Gidrometeoizdat, Leningrad (1959).
353. *Science News Letter*, 17 August 1963, Ref. to *Nature*, **199,**

475 (1963).
354. *Science News Letter*, 22 February 1964, Ref. to *J. Atmospheric Sciences*, **20,** 563 (1964).
355. R. M. Stillman and E. R. LaChapelle, "The Control of Snow Avalanches," *Scientific American*, **214,** 99-101 (1966),
355a. N. W. Rosenberg, "Chemical Releases at High Altitudes," *Science*, Vol. 152, No. 3725 (20 May 1966).
356. *The Diary of Samuel Pepys, Selections*, Harper Torchbooks, The Academy Library, Harper & Bros., N. Y., 1960.
357. *Johnson's Dictionary*, Pantheon Books, Div. of Random House, N. Y., 1963.
358. S. Hedden, J. Massey, Jr., and P. Stamoulas, Naval Weapons Lab., Dahlgren, Va., *An Evaluation of the High-Low Ballistic Principle for CAD*, Rept. #1964 (1965).
359. Frankford Arsenal, Phila., Pa., *Propellant Actuated Devices, Engineering Manual*, IEP 65-6370-8, (April 1965), loose-leaf with revisions.
360. G. P. Sutton, *Rocket Propulsion Elements*, 3rd Ed., J. Wiley & Sons, N. Y., 1963.
361. R. N. Wimpress, *Internal Ballistics of Solid Fuel Rockets*, McGraw Hill Book Co., Inc., N. Y., 1950.
362. F. A. Warren, *Rocket Propellants*, Reinhold Publ. Corp., N. Y., 1958.
363. B. Kit and D. S. Evered, *Rocket Propellant Handbook*, The MacMillan Co., N. Y., 1960.
364. R. H. Bleikamp, E. R. Lake, and D. R. McGovern, McDonnell Aircraft Corp., St. Louis, Mo., *Investigation of PAD for Use in Emergency Crew Escape Systems for Advanced Aerospace Vehicles*, AFFDL-TR-65-26, (1964).
365. Henry Semat, *Fundamentals of Physics*, 3rd Ed., Rinehart & Co., Inc., N. Y., 1957.
366. W. H. Rinkenbach and V. C. Allison, U.S. Patent 2,415,848 (1947).
367. J. W. Burns, *Black Powder Manual*, Sec. H, Div. 3, NDRC Report (unclassified).
368. F. P. Bowden and J. D. Blackwood, Brit. Patent 712,765.
369. F. P. Bowden and J. D. Blackwood, Brit. Patent 715,827.
370. F. P. Bowden and J. D. Blackwood, Brit. Patent 715,828.
371. F. P. Bowden and J. D. Blackwood, Brit. Patent 715,829.
372. A. Douillet, *Mém. Poudres*, **37,** 167-96 (1955); *C. A.* **51,** p. 717 (1957).
373. A. T. Camp, U.S. Patent 3,088,858 (1963); *C. A.* **51,** 6191 (1963).

374. R. S. Jessup and E. J. Prosen, *J. Res. Natl. Bur. Standards*, **44**, 387, (1950).

375. E. Ott and H. M. Spurlin, *High Polymers*, Vol. V, Part II, Interscience Publ. Co., N. Y., 1951.

376. Frank Douglas Miles, *Cellulose Nitrate*, Interscience Publ. Co., N. Y., 1955.

377. Phokion Naoúm *Nitroglycerine and Nitroglycerine Explosive* (translated by E. M. Symmes, Hercules Powder Co.) The Williams & Wilkins Co., 1928.

378. J. Taylor and G. P. Sillito, "Catalytics of Decomposition of Ammonium Nitrate by Chromates," Paper #73, *3rd Symposium on Combustion and Flame & Explosion Phenomena*, The Williams & Wilkins Co., Baltimore, Md., 1949.

379. J. Taylor, U. K. Patents 570,075 and 570,211.

380. T. A. Burgwald et al., U.S. Patent 2,942,964 (1960).

381. J. R. Eiszner and Wm. G. Stanley, U.S. Patent 2,942,961 (1960); *C. A.* **54,** 25830 (1960).

382. Paul O. Marti, Jr., U.S. Patent 3,013,871 (1961).

383. J. Taylor and A. C. Hutchison, U.S. Patents 2,604,391 (1952) and 2,653,086 (1953).

384. A. C. Hutchison, U.S. Patents 2,682,461 (1954) and 2,710,-793 (1955).

385. David P. Moore, U.S. Patent 3,135,634 (1964).

386. A. M. Halstead, Frankford Arsenal, Phila., Pa., *A Study of Selected Propellant Actuated Devices to Eliminate or Minimize Toxic Propellant Gases*, M65-19-1, SEG TR 65-27 (May 1965).

387. Dept. of Army Technical Manual, *Demolition Materials*, TM9-1946 (1955).

388. Bureau of Explosives, Specifications for Standard Track Torpedoes, Revised February 1, 1949.

389. R. A. Hunter, U.S. Patent 2,189,398 (1940); F. Dutcher, U.S. Patent 1,421,187 (1922).

390. L. A. Sherman, U.S. Patent 2,061,854 (1936).

391. Francis B. Paca, U.S. Army Engineer R & D Labs., Fort Belvoir, Va., *Reduction of Noise Arising from Demolition Activities*, Rept. 1703-TR (20 December 1961).

392. W. R. Maxwell, "Pyrotechnic Whistles," *Fourth Symposium on Combustion*, Paper No. 111, p. 906, The Williams & Wilkins Co., Baltimore, Md. (1953).

393. Duane M. Johnson, NAD, Crane, Ind., *Ignition Theory: Application to the Design of New Ignition Systems*, RDTR No. 56, AD 627257 (24 Nov. 1965).

394. Wm. Ripley, NAD, Crane, Ind., *Investigation of Mk* 25 *Mod* 2 *Starter Composition*, Rept. No. 36, AD 436599 (March 1964), cf. Ref. 623.

395. *Pyrotechnic Starter Composition*, NAD, Crane, Ind., AD 288746 (20 Feb. 1963).

396. Wm. Ripley, NAD. Crane, Ind., *Chemical Analysis of a Typical* 6-6-8 *Pyrotechnic Starter*, RDTR No. 27 (13 July 1962).

397. R. H. Heiskell, U.S. Patents 2,716,599 (1955); 2,726,943 (1955); 2,726,944 (1955).

398. F. B. Clay and R. A. Sahlin, U.S. Patent 2,709,129 (1955).

399. G. C. Hale, U.S. Patent 1,805,214 (1931).

400. S. J. Magram, U.S. Patent 2,421,029 (1947).

401. H. Zenftman, U.S. Patent 2,497,387 (1950).

402. S. J. Magram and J. J. Blissel, U.S. Patent 2,640,770 (1953); *C. A.* **47,** 7779 (1953).

403. J. H. McLain and T. A. Ruble, U.S. Patent 2,643,946 (1953); *C. A.* **47,** 10851 (1953).

404. D. Hart, U.S. Patent 3,030,243 (1962); *C. A.* **57,** 2489 (1962).

405. H. Williams and W. A. Gey, U.S. Patent 2,900,242 (1959).

406. Jean Piccard, U.S. Patent 1,971,502 (1934).

407. H. Ellern and D. E. Olander, U.S. Patent 2,954,735 (1960); *Pyroclok* Trademark Certificate No. 688260 (Nov. 1959).

408. Shoji Nakahara, *Kogyo Kayaku Kyokaishi*, **21,** 363-74 (1960); *C. A.* **55,** 20433 (1961), "Theory of Burning Rate of Delay Powders."

409. A. Broido and S. B. Martin, U.S. Naval Radiological Defense Lab., San Francisco, Calif., *Ignition of Cellulose*, AD 268729; *C. A.* **58,** 1627 (1963).

409a. "Tobacco," Lloyd Peterson, *Science & Technology*, No. 31, p. 46 ff., (July 1964).

410. D. E. Olander, U.S. Patent 3,028,229 (1962); *C. A.* **57,** 10096-7 (1962).

411. G. C. Hale and D. Hart, U.S. Patent 2,468,061 (1949).

412. L. B. Johnson, Jr., Research Lab for Engineering Sciences, U. of Va., Charlottesville, Va., *Ind. Eng. Chem.* **52,** 868 (1960).

413. G. C. Hale and D. Hart, U.S. Patents 2,461,544 (1949) and 2,467,334 (1949).

414. O. G. Bennett and J. Dubin, U.S. Patent 2,457,860 (1949).

415. G. C. Hale, U.S. Patent 2,450,892 (1948).

416. D. Hart, U.S. Patent 2,696,429 (1954).

417. E. M. Patterson, U.S. Patent 2,478,501 (1949).
418. H. W. Coleman, U.S. Patent 2,725,821 (1955).
419. H. E. Nash, U.S. Patent 1,960,591 (1934).
420. D. T. Zebree, U.S. Patent 2,892,695 (1959); *C. A.* **53,** 17514 (1959).
421. D. T. Zebree, U.S. Patent 3,094,933 (1963).
422. D. T. Zebree, U.S. Patent 3,113,519 (1963).
423. I. Kabik et al., U.S. Patent 3,088,006 (1963).
424. R. H. Comyn and R. E. McIntyre, *Influence of Body on Delay Time*, NASA Doc. N63-20944 (1964); *C. A.* **60,** 11840 (1964).
425. James Cohn, U.S. Patent 2,911,504 (1959).
426. "Pyrofuze," pamphlet by Pyrofuze Corp., Mt. Vernon, N. Y. (1962); Engineering Data #66 Series, (1966).
427. W. R. Peterson, Frankford Arsenal, Phila., Pa., *Investigation of a Close Tolerance Pyrotechnic Metallic Delay Element*, Rept. R-1693, ASD-TDR-63-563, (Sept. 1963).
428. M. Palmieri and S. D. Ehrlich, U.S. Patent 2,103,014 (1937).
429. M. Manus, *Underwater Saboteur*, Wm. Kimber & Co., Ltd., London, 1956.
430. Ira Marcus, Diamond Ordnance Fuze Laboratories, Washington, D. C., *Electronic Long-Delay Timer*, Report TR-995, (12 Dec. 1961).
431. *Development and Qualification of Delay Elements for Propellant Actuated Devices*, Frankford Arsenal, Phila., Pa., AD 451654, (1964).
432. W. Katz and R. Huxster, Picatinny Arsenal, Dover, N. J., *Development of a Temperature Stabilized Ordnance—Binary Timer*, Tech. Memo #1017, AD 292923.
433. W. J. Youden, *Statistical Methods for Chemists*, J. Wiley & Sons, Inc., N. Y., 1951.
434. K. A. Brownlee, *Industrial Experimentation*, Chemical Publishing Co., Inc., N. Y., 1952.
435. M. J. Monroney, *Facts from Figures*, Penguin Books, Inc., Baltimore, Md., 1957.
436. *Statistical Methods in Research and Production*, O. L. Davies, Editor, Hafner Publishing Co., N. Y., 1957.
437. Nolan D. C. Lewis and Helen Yarnell, *Pathological Firesetting* (*Pyromania*), Dept. of Psychiatry, Columbia Univ., Nervous and Mental Disease Monographs, N. Y., 1951.
438. E. Josué, U.S. Patent 1,534,962 (1925).
439. L. F. Fieser, U.S. Patent 2,606,107 (1952).
440. Belgian Patent 644,290 (1964); *C. A.* **63,** 9736 (1965).

441. L. J. Novak, U.S. Patent 2,921,846 (1960).
442. P. C. L. Thorne and C. G. Smith, "Solidified Alcohol" in *Colloid Chemistry*, Vol. IV, The Chemical Catalog Co., Inc., N. Y., 1931.
443. John E. Schultze, U.S. Patent 2,721,120 (1955).
444. K. A. Loftman, Cabot Corporation, Boston, Paper on *Theoretical and Practical Aspects of the Rheological Properties of Pyrogenic Silica in Aqueous and Non-Aqueous Media*, April 1960.
445. H. Kartluke et al., Technidyne, Inc., West Chester, Pa., *Gelling of Liquid Hydrogen*, Final Report RR 64-47, NASA CR-54-55 (31 July 1964).
446. E. E. Bauer and G. Broughton, U.S. Patent 2,922,703 (1960).
447. R. E. Schaad, U.S. Patent 2,891,852 (1959); *C. A.* **53,** 17512 (1959).
448. Frank N. Vannucci, U.S. Patent 3,126,259 (1964).
449. C. G. Long, U.S. Patent 3,035,950 (1962); *C. A.* **57,** 7510 (1962).
450. G. J. B. Fisher, *Incendiary Warfare*, McGraw Hill Book Co., Inc., N. Y., 1946.
451. Federal Specification R-R-M-31A (1960).
452. N. J. Thompson, U.S. Patent 2,791,178 (1957).
453. D. L. Woodberry, W. W. Howerton, and A. Dunbar, U.S. Patent 2,452,091 (1948).
454. David Irving, *The Destruction of Dresden*, Ballantine Books, N. Y. (1965).
455. Thomas Stevenson, U.S. Patent 2,951,752 (1960); *C. A.* **55,** 2108 (1961).
456. Italian Patent 448,101 and 460,630.
457. British Patent 126,394.
458. Edmund Ritter von Herz, German Patent 1,114,419 (applied for 1960).
459. Harry K. Linzell, U.S. Patent 2,424,937 (1947).
460. *Weapons Technology*, "The Importance of Pyrotechnics in Modern Weaponry," American Ordnance Association, Wash., D. C., Technical Division Report, November, 1960.
461. "Operation School Burning," National Fire Protection Association, Boston, 1959.
462. W. A. Caldwell and J. Gillies, *Ind. Chemist*, **26,** 301 (1950).
463. M. Bamberger and F. Boeck, U.S. Patent 802,256.
464. W. H. Foster, U.S. Patent 2,157,169 (1939).
465. W. M. Bruner, U.S. Patent 2,261,221 (1941).
466. French Patents 808,510 and 820,530.

467. British Patents 22,242; 474,249; 475,411; 479,671; 531,250.
468. *Chem. Eng. News*, 5 Feb. 1962.
469. British Patent 936,310.
470. O. G. Bennett, U.S. Patents 2,500,790 (1950) and 2,531,548 (1950).
471. *Quik-Shot*, Advertising Leaflet, The Kemode Manufacturing Co., Inc., N. Y.
472. *Thermal Battery Technical Discussion*, Vol. 1, Eureka-Williams Co., Bloomington, Ill.
473. J. P. Schrodt et al., U.S. Patent 2,442,380 (1948).
474. British Pat. 891,035 (1962); *C. A.* **56,** 12615 (1962).
475. J. H. Deppeler, Jr., Thermex Metallurgical Co., Lakehurst, N.J., *Personal Communication*, 26 August 1965.
476. Albert T. Camp and Gerald L. MacKenzie, U.S. Patent 3,156,595 (1964).
477. William A. Gey and Robert W. Van Dolah, U.S. Patent 3,110,640 (1963).
478. Swiss Patent, 268,855, Klasse 39c, (1950).
479. *Missiles and Rockets*, **13,** 48 (1963).
480. Anon., "Putting on the Heat," *Chem. Eng. News*, Vol. 34, p. 3442 (1956).
481. J. B. Conway and A. V. Grosse, *Powdered-Metal Flames*, Third Technical Report, High Temperature Project Contr. N9-onr-87301, Research Institute of Temple Univ., August 1953.
482. C. S. Stokes et al., *Ind. Eng. Chem.* **52,** 75 (1960).
483. A. V. Grosse, *Chem. Eng. News*, **29,** 1704 (1951) and **31,** 2160 (1953).
484. A. V. Grosse and J. B. Conway, *Ind. Eng. Chem.* **50,** 663 (1958).
485. J. B. Conway, R. H. Wilson, Jr., and A. V. Grosse, *J. Am. Chem. Soc.* **75,** 499 (1953).
486. A. V. Grosse and A. D. Kirshenbaum, *J. Am. Chem. Soc.* **77,** 5012 (1955).
487. J. B. Conway, et al., *J. Am. Chem. Soc.* **77,** 2026 (1955).
488. A. D. Kirshenbaum and A. V. Grosse, *J. Am. Chem. Soc.* **78,** 2020 (1956).
489. R. A. Baker and F. M. Strong, *Ind. Eng. Chem.* **22,** 788 (1930).
490. B. Lewis, "High Temperatures: Flame," *Sci. American*, special issue on Heat, **191,** No. 3, p. 84 (Sept. 1954).
491. L. Brewer and A. W. Searcy, *J. Am. Chem. Soc.* **73,** 5308 (1951).

492. W. H. Schechter, U.S. Patent 2,469,414 (1949).
493. J. Moni and Ph. LeChartier de Sedony, French Patent 1,142, 360; *C. A.* **54,** 784 (1960).
494. P. R. Gustafson, S. H. Smith, Jr., and R. R. Miller, Naval Research Lab., Washington, D. C., *Chlorate-Candle Fabrication by Hot Pressing*, NRL 5732, AD 272580, (January 1962).
495. C. B. Jackson and R. M. Bovard, U.S. Patent 2,558,756 (1951).
496. M. M. Markowitz and Eugene W. Dezmelyk, Foote Mineral Co., Exton, Pa., *A Study of the Application of Lithium Chemicals to Air Regeneration Techniques in Manned, Sealed Environments*, Tech. Doc. Rept. No. AMRL-TDR-64-1, AD 435815, (Feb. 1964).
497. B. M. Fabuss and A. S. Borsanyi, Monsanto Research Corp., Everett, Mass., *Self-Contained Generator for Shelter Use*, AD 616639 (1964).
498. C. B. Jackson and R. C. Werner, Mine Safety Appliances Co., Callery, Pa., *Manufacture and Use of Potassium Superoxide* (cf. Ref. 95).
499. Mine Safety Appliances Co., Callery, Pa., Final Report, *Solid Sources of Gaseous Oxygen* (1962).
500. *Missiles and Rockets*, p. 27, (17 May 1965).
501. H. Ellern, I. Kowarsky, and D. E. Olander, U.S. Patent 2,842,477 (1958).
502. G. F. Jaubert, French Patent 406,930.
503. M. M. Markowitz et al., Foote Mineral Co., Exton, Pa., *J. Phys. Chem.* **68** (8), 2282-9 (1964); *C. A.* **61,** 7930 (1964).
504. G. G. Marvin and L. B. Woolaver, *Ind. Eng. Chem.* Vol. 17, 474-6 (1945).
505. S. Patai and E. Hoffmann, "Pre-ignition Reactions of Some Combustible Solids with Solid Oxidants," *J. Appl. Chem.* (London), **2,** 8 (1952).
506. L. L. Bircumshaw and T. R. Phillips, *J. Chem. Soc.* p. 703-7 (1953).
507. A. E. Harvey et al., *J. Am. Chem. Soc.* **76,** 3270 (1954); (cf. Ref. 170).
508. F. A. Warren et al., Southwest Research Institute, San Antonio, Texas, *Chlorates and Perchlorates, Their Manufacture, Properties, & Uses*, Vol. I, NavOrd Report #7147, AD 242192 (1960).
509. Senze et al., *Jahresb. Chem.-Tech. Reichsanstalt*, **8,** 1-3 (1931).

510. C. Dufraisse and M. German, *Compt. Rend.*, **207,** 1221-4 (1938).
511. Dr. Hans Goldschmidt, German Patent 96,317 (1895).
512. *Ceram. Ind.* **57** [2], 54-5, 57-8, 100, (1951).
513. K. A. Kühne, British Patent 11,606 (1907); *C. A.* **3,** 166 (1909).
514. Cueilleron and Pascand, *Compt. Rend.* **233,** 745-7, (1951); *C. A.* **46,** 2983, (1952).
515. Weiss and Aichel, *Liebigs Annalen*, **337,** 380 (1904).
516. C. A. Hampel, *Rare Metals Handbook*, Reinhold Publishing Corp., N. Y., 1956, (2nd Ed., 1961).
517. British Patent 675,933 (1952); *C. A.* **46,** 11092, (1952).
518. W. Rutkowski, *Prace Inst. Min. Hutniczych*, **6,** 176-83, (1954).
519. *C. A.* **47,** 7413, (1953); *ibid.* p. 12195.
520. D. Gardner, U.S. Patent 2,556,912 (1951).
521. Thermex Metallurgical Co., Lakehurst, N. J., *Brochure*, 1965.
522. F. and S. S. Singer, *Industrial Ceramics*, Chapman & Hall, Ltd., London, 1963; U.S. Edition (1964), Chemical Publishing, N. Y.
523. J. D. Walton, Jr. and N. E. Poulos, *J. Am. Ceram. Soc.* **42,** 40-49, (1959).
524. British Patent 892,048 (1962); *C. A.* **56,** 15172, (1962).
525. Hans Béller, U.S. Patent 2,900,245 (1959).
526. Gerh. Jangg et al., *Z. Metallk.* **50,** 460-5, (1959); *C. A.* **53,** 21530, (1959).
527. O. G. Bennett and Joseph C. W. Frazer, U.S. Patent 1,893,-879 (1933).
528. H. A. Wilhelm and P. Chiotti, U.S. Patent 2,635,956 (1953).
529. *Metallurgy at Metal Hydrides*, Metal Hydrides, Inc., Beverly, Mass., pamphlet.
530. Riyad R. Irani, *Intern. J. of Powder Metallurgy*, **1,** (4), 22-27, (1965).
531. C. Orr, Jr. and J. M. Dallavalle, *Fine Particle Measurement*, The Macmillan Co., N. Y., 1959.
532. H. E. Rose, *The Measurement of Particle Size in Very Fine Powders*, Chemical Publishing Co., Inc., N. Y., 1954.
533. R. D. Cadle, *Particle Size Determination*, Interscience Publishers, Inc., N. Y., 1955; 2nd Rev. Ed., 1960.
534. G. Herdan, *Small Particle Statistics*, The Elsevier Press, N. Y., 1953; 2nd Ed., 1960.
535. Metals Disintegrating Co., Elizabeth, N. J., *Particle Size*

Analysis of Metal Powders, 1946.

536. S. M. Kaye, D. E. Middlebrooks, and G. Weingarten, Picatinny Arsenal, Dover, N. J., *Evaluation of Sharples Micromerograph for Particle Size Distribution Analysis*, Tech. Report, FRL-TR-54.

537. R. H. Comyn, Diamond Ordnance Fuze Laboratories, Washington, D. C., *Measurement of Particle Size of Components of Gasless Mixtures*, Tech. Rept. #636 (1958).

538. R. H. Berg, "Electronic Size Analysis of Sub-sieve Particles Through a Small Liquid Resistor," *Am. Soc. Testing Mater.* Special Tech. Publ. #234 (1958).

539. Cl. G. Goetzel, *Treatise on Powder Metallurgy*, Interscience Publishers, Inc., N. Y., 1949.

540. R. D. Cadle, Nat. Center for Atmos. Research, Boulder, Colo., *Particle Size: Theory and Industrial Application*, Reinhold Book Div., N. Y., 1965.

541. Fansteel Metallurgical Corp., N. Chicago, Ill., *Data Sheet.*

542. L. Brewer, "The Thermodynamic Properties of the Oxides and Their Vaporization Processes," *Chem. Rev.* **52,** 1 (1953).

543. W. D. Kingery, *Property Measurements at High Temperatures*, J. Wiley & Sons, Inc., N. Y., 1959.

544. I. E. Campbell (Ed.-in-Chief), *High Temperature Technology*, *J.* Wiley & Sons, Inc., N. Y., 1956.

545. *High Temperature Technology*, Proceedings of an International Symposium, McGraw-Hill Book Co., Inc., N. Y., 1960.

546. P. T. B. Shaffer, *Handbook of High Temperature Materials* #1, *Materials Index*, Plenum Press, N. Y., 1964.

547. G. V. Samsonov, *Handbook of High Temperature Materials* #2, *Properties Index*, Plenum Press, N. Y., 1964.

548. W. D. Wood, H. W. Deen, and C. F. Lucks, *Handbook of High Temperature Materials* #3, *Thermal Radiative Properties*, Plenum Press, N. Y., 1964.

549. M. A. Bramson, *Handbook of High Temperature Materials* #4, *Infrared Radiation from Hot Bodies*, Plenum Press, N. Y., 1964.

550. Advertising Pamphlet, *Materials Technology*, Carborundum Co., 1963.

551. Samuel Glasstone, *Textbook of Physical Chemistry*, 2nd Ed., D. Van Nostrand Co., Inc., N. Y., 1946.

552. L. L. Weil, U.S. Patent 3,106,497 (1963).

553. Foote Mineral Co., Exton, Pa., *Report TD*-105, October 1959.

554. A. P. Vitoria, *Anales Real Soc. Espan. Fis. Quim.* **27,** 787 (1929).
555. V. D. Hogan and S. Gordon, Picatinny Arsenal, Dover, N. J., PB Report 143903 (1959); *C. A.* **55,** 16249 (1961).
556. J. A. Fitzpatrick, U.S. Patent 2,885,277 (1959).
557. Diamond Ordnance Fuze Laboratories, Wash., D. C., *The Measurement of Heat Evolved by Thermite Mixtures,* Part I, TR-488; Part II, TR-494; Part III, TR-576; Part IV, TR-844 (March 1957).
558. R. H. Comyn and I. R. Marcus, Diamond Ordnance Fuze Laboratories, Wash., D. C., *Summary of Heat-Powder Calorimetry,* TR-862 (1 Aug. 1960).
559. Sheldon G. Levin and Robert E. McIntyre, Harry Diamond Labs., Wash., D. C., *Statistical Evaluation of Calorimetric Measurements on Heat-Source Materials,* TR-1028 (March 1963).
560. Frederick D. Rossini (Editor), *Experimental Thermochemistry, Measurements of Heats of Reaction,* Interscience Publishers, Inc., N. Y., 1956.
561. W. Eggersgluess et al., *Trans. Faraday Soc.* **45,** 661 (1949).
562. G. V. Samsonov et al., *Bor, yego soyedineniya i splavy* ("Boron, Its Compounds and Alloys"), AN SSSR, Kiev. 1960.
563. Harry F. Rizzo, "Oxidation of Boron at Temperatures Between 400 and 1300°C in Air," in *Boron, Synthesis, Structure, and Properties,* Plenum Press, Inc., N. Y., 1960.
564. *Engineering Data Sheet,* No. 66-24, Pyrofuze Corp., Mt. Vernon, N. Y. (cf. Ref. 426).
565. *Angew. Chem.* **75** (18), 859 (1963); *C. A.* **59,** 14657 (1963).
566. Metal Hydrides, Inc., Beverly, Mass., *Tech. Bulletin* 201-D.
567. A. Sieverts and A. Gotta, *Z. Anorg. Allgem. Chem.* **199,** 384 (1931).
568. R. A. Rhein, Jet Propulsion Lab., California Institute of Technology, Pasadena, Calif., *The Ignition of Powdered Metals in Nitrogen and in Carbon Dioxide,* Tech. Rept. 32-679 (Sept. 1964).
564a. O. Kubaschewski and J. A. Catterall, *Thermochemical Data of Alloys,* Pergamon Press, London-New York (1956).
569. E. Raisen, S. Katz, and K. Franson, IITRI (formerly Armour Research Institute), Chicago, Ill., *Survey of Thermochemistry of High Energy Reactions,* ASD-TDR-63-846 (August 1963).
569a. R. E. McIntyre and M. L. Couch, Diamond Ordnance Fuze

Labs., Wash., D. C., *Specification of Zirconium for Use in Gasless Mixtures*, TR-821 (9 March 1960).

570. H. B. Alexander, U.S. Patent 1,989,729 (1935).
571. G. H. Smith, U.S. Patent 2,189,301 (1940).
572. L. B. Johnson, Jr., Univ. of Va., *Ind. Eng. Chem.* **52,** 241 (1960).
573. NASA Doc. N63-11332 (1962); *C. A.* **60,** 14627 (1964).
574. J. E. Spice and L. A. K. Stavely, *J. Soc. Chem. Ind.* Part I, **68,** 313 (1949); Part II, **68,** 348 (1949).
575. R. A. Hill, L. E. Sutton, R. B. Temple, and A. White, *Research* (London), **3,** 569 (1950).
576. R. A. Hill, *Nature*, **170,** 29 (1952); *Proc. Roy. Soc.* (London), **A226,** 455 (1954); *ibid.* **A476,** 239 (1957).
577. R. A. Hill, L. E. Sutton, and T. L. Cotrell, "Studies of Combustion Waves in Solids," *Fourth Symposium on Combustion*, The Williams & Wilkins Co., Baltimore, Md., 1953.
578. F. Booth, *Trans. Faraday Soc.* **49,** Pt. 3,272 (1953).
579. B. L. Hicks, *J. Chem. Phys.* **22,** 414 (1954).
580. W. J. Smothers and Yao Chiang, *Differential Thermal Analysis*, Chemical Publishing Co., Inc., N. Y., 1958; new edition (1966), *Handbook of Differential Thermal Analysis.*
581. C. B. Murphy, *Anal. Chem.* **32** [5], pp. 1R-292R (April 1960).
582. C. Campbell and G. Weingarten, Picatinny Arsenal, Dover, N. J., "A Thermoanalytical Study of the Ignition and Combustion Reactions of Black Powder," Monograph 43; *Trans. Farad. Soc.* **55,** Part 12 (1959).
583. Cl. Duval, *Inorganic Thermogravimetric Analysis*, 2nd Rev. Ed., Elsevier Publishing Co., N. Y., 1963.
584. E. S. Freeman and D. Edelman, *Anal. Chem.* **31,** 624 (1959).
585. Cl. Campbell, S. Gordon, and Cl. Smith, *Anal. Chem.* **31,** 1188 (1959).

585a. Wesley Wm. Wendlandt, *Thermal Methods of Analysis*, Interscience Publishers, Div. of John Wiley & Sons, N. Y., 1964.

586. Henkin and McGill, *Ind. Eng. Chem.* **44,** 1391 (1952).
587. Dr. James E. Sinclair, The Autoignition of Various Pyrotechnics, Speech at Military Pyrotechnics Section Meeting, AOA, November 1965.
588. David J. Edelman, Picatinny Arsenal, Dover, N. J., *Electrostatic Sensitivity of Pyrotechnic and Explosive Compositions: First Report, PL IR* 8.
589. R. W. Brown, D. J. Kusler, and F. C. Gibson, *Bureau of*

Mines Report of Investigation #5002, "Sensitivity of Explosives to Initiation by Electrostatic Discharges," (September 1953).

590. A. J. Clear, Picatinny Arsenal, Dover, N. J., *Standard Laboratory Procedures for Sensitivity, Brisance, and Stability of Explosives*, Tech. Rept. FRL-TR-25, January 1961.

591. T. Stevenson, Frankford Arsenal, Philadelphia, Pa., "Tracer Bullet Ignition," *Weapons Technology*, American Ordnance Ass'n., Technical Div. Report, November 1960.

592. J. H. McIvor, Picatinny Arsenal, Dover, N. J., *Friction Pendulum*, Manual 7-1 (May 1950).

593. Joseph Kristal and S. M. Kaye, Picatinny Arsenal, Dover, N. J., *Survey of Sensitivity Characteristics of Typical Delay, Igniter, Flash, and Signal Type Pyrotechnic Compositions*, Tech. Memo 1316 (April 1964).

594. C. R. Grande, Picatinny Arsenal, Dover, N. J., *New Methods for the Measurement of Relative Ignitibility & Ignition Efficiency*, Tech. Rept. 2469 (Feb. 1958).

595. D. Costa et al., *Chim. Ind.* (Milan), **34,** 645-54 (1952).

596. NASA report, Univ. of Va., *C. A.* **61,** 13118 (1964).

597. *Light Metals*, p. 40-41 (December 1964).

598. A. A. Shidlovsky, *J. Appl. Chem.* U.S.S.R., **19,** 371-8 (1946).

599. H. F. Payne, *Organic Coating Technology*, Vol. II, J. Wiley & Sons, N. Y., 1961.

600. R. H. Comyn, U.S. Patent 2,832,704 (1958).

601. R. H. Comyn, M. L. Couch, and R. E. McIntyre, Diamond Ordnance Fuze Labs., Wash., D.C., *Stability of Manganese Delay Mixtures*, AD 268079 (1961); *C. A.* **61,** 526 (1964).

602. Stig Yngve E. K. and John A. D. Eldh, U.S. Patent 2,894,-864 (1959).

603. H. A. Barbian, "Finish Systems on Magnesium," *Materials and Methods*, **1,** 102 (1954).

604. Frederick W. Fink, "Metal Corrosion in Saline Water," *Battelle Technical Review*, September 1963.

605. Air Force Report, *The Role of Bacteria in Electrochemical Corrosion of Steel in Sea Water*, AD 602332.

606. Naval Research Laboratory, *Corrosion of Metals in Tropical Environments*—Part 6, AD 609618.

607. Jos. H. McLain and Donald V. Lewis, U.S. Army Edgewood Arsenal, Md., *Effect of Phase Change in Solid-Solid Reactions*, Report. No. WCDC-6465, Annual Report, October 1965.

608. Robert E. Johnson, Rock Island Arsenal Lab., Rock Island,

Ill., *The Use of Volatile Corrosion Inhibitors as a Preservative Medium for Long Term Storage of Ordnance Material—Addendum VII—Results After Ten Years of Exposure*, RIA Lab Rept. 61-544, (9 Feb. 1961).

609. W. L. Porter, Headquarters, QM R & D Command, Natick, Mass., *Occurrence of High Temperatures in Standing Boxcars*, Tech. Rept. EP-27, (1956).

610. Glenn A. Greathouse and Carl J. Wessel, *Deterioration of Materials*, Reinhold Publishing Corp., N. Y., 1954.

611. E. S. Freeman and D. A. Anderson, Picatinny Arsenal, Dover, N. J., PA-TR-2673.

612. B. R. Phillips and Duncan Taylor, *J. Chem. Soc.* 5583-90, (1963); *C. A.* **60,** 2537, (1964).

613. J. H. DeBoer et al., *Reactivity of Solids*, Elsevier Publishing Co., London, 1961.

614. Clyde H. Teesdale, *Modern Glues and Glue Handling*, The Periodical Publ. Co., Grand Rapids, Mich., 1922.

615. *Animal Glue in Industry*, prepared and published by National Association of Glue Manufacturers, Inc., N. Y., 1951.

616. L. K. J. Tong and W. O. Kenyon, *J. Am. Chem. Soc.* **69,** 1402-5, (1947); *ibid.* **69,** 2245-6, (1947).

617. D. Hart, H. Eppig, and W. J. Powers, U.S. Patent 2,700,603 (1955).

618. Special Report USA CRDL, *Survey of Recent Investigations of Plastic-Bonded and Castable Smoke Compositions*, by Atlantic Research Corp., Alexandria, Va., AD 422745, (April 1963).

619. J. Whetstone, U.S. Patent 2,616,787 (1952).

620. A. Butchart, U.S. Patent 2,616,785 (1952).

621. P. O. Marti, Jr., U.S. Patent 2,901,317 (1959).

622. F. G. Nebel and T. L. Cramer, *Ind. & Eng. Chem.* **47,** 2393-6 (1955).

623. Wm. Ripley, NAD, Crane, Ind., *Investigation of the Burning Characteristics of the Lead Dioxide—Cupric Oxide—Silicon Starter Composition*, Rept. No. 41, AD 437978 (1964).

624. J. D. Edwards and R. I. Wray, *Aluminum Paint and Powder*, Reinhold Publishing Co., N. Y., 1955.

624a. J. A. Carrazza, D. E. Middlebrooks, and S. M. Kaye, Picatinny Arsenal, Dover, N. J., *Comparison of Mechanically Balled Magnesium with Atomized Magnesium for Use in Pyrotechnic Compositions*, PA-TR-3364, AD 638132, September 1966.

625. I. Hartmann et al., *Bureau of Mines Investigation* #4835,

"The Explosive Characteristics of Titanium, Zirconium, Thorium, Uranium and Their Hydrides," (1951).

626. Manufacturing Chemists Ass'n., Wash., D. C., *Chemical Safety Data Sheet SD*-92, "Zirconium and Hafnium Powder," (adopted 1966).

627. M. D. Banus and J. J. McSharry, U.S. Patent 2,688,575 (1954).

628. "Zirconium in Photoflash Bulbs," *Symposium on Zirconium and Zirconium Alloys*, ASM, Eighth Western Metal Congress & Exposition, 1953.

629. "Handling Zirconium," Sylvania Elect. Prod. Co., *Occupational Hazards*, (December 1954).

630. C. L. Roberts, "Junior's Chemistry Set," *National Safety News*, **72,** #5 (1955).

631. Th. Moeller and V. D. Aftandilian, *J. Am. Chem. Soc.* **76,** 5249-50 (1954); *C. A.* **49,** 2919-20 (1955).

632. H. C. Clauser and R. S. Long, U.S. Patent 2,651,567 (1953); *C. A.* **48,** 1004 (1954).

633. Burton Stevenson, *MacMillan Book of Proverbs, Maxims, and Famous Phrases*, MacMillan Co., N. Y., 1948 (6th Printing, 1965).

634. E. Crane and J. Kristal, Picatinny Arsenal, Dover, N. J., *New Formulas for Hand-Held Rocket Type Signals*, Tech. Note #50 (June 1960).

635. Winston W. Cavell, Frankford Arsenal, Phila., Pa., *A Survey of Some of the Recent Applications of Pyrotechnics to Small Arms Ammunition and Mild Detonating Fuse Systems*, Rept. A65-9 (4-5 Nov. 1965).

636. Wm. Ripley, NAD, Crane, Ind., *An Experimental Ashless Blue Flare Composition*, RDTR No. 34 (April 1963).

637. J. M. Dallavalle, *Micromeritics*, 2nd Ed., Pitman Publishing Corp., N. Y., 1948.

638. German Patent 1,158,884 (1963); *C. A.* **60,** 9094 (1964).

639. *The Shorter Bartlett's Familiar Quotations*, Permabooks, N. Y., 1959.

640. *Hackh's Chemical Dictionary*, Julius Grant (Editor), The Blakiston Co., Phila., (3rd ed., 1944); new edition in preparation, McGraw Hill Book Co., N. Y.

640a. *The Random House Dictionary of the English Language*, Unabridged Edition, Jess Stein—Editor-in-Chief, Laurence Urdang—Managing Editor, Random House, N. Y., 1966.

641. *Nouveau traité de chimie minérale* (Pascal), Masson & Cie, Paris, France (1965).

642. T. Urbanski, *Chemistry & Technology of Explosives* (Translation from Polish language), MacMillan Co., N. Y., 1964.
643. N. A. Toropov and V. P. Barzarhovskii, *High Temperature Chemistry of Silicates and Other Oxide Systems* (Translation from Russian language), Consultants Bureau, Plenum Press Data Div., N. Y., 1966.
644. *Angew. Chem. Intern. Ed. in Engl.* **5,** 53, (1966).
645. Frank H. Wilson III, Ventron Corp., Metal Hydrides Div., Beverly, Mass., *Personal Communication*, Feb. 1966.
646. W. Zimmermann, *Naturwiss.* **18,** 837, (1930).
647. *Information Sheet, Cast Infrared-Producing Flares*, Flare-Northern Div., Atlantic Research Corp.
648. R. H. Comyn et al., *Ind. Eng. Chem.* **52,** 995, (1961).
649. R. D. Sheeline, *Chem. Eng. Progr.* **61,** 77-82, (1965).
650. F. B. Cramer, U.S. Patent 3,022,149 (1962).
651. W. Shaafsma, U.S. Patent 3,092,527 (1963).
651a. E. Crane, B. Werbel, and G. Weingarten, Picatinny Arsenal, Dover, N. J., *Development of Burning-Type Colored Smokes*, PA-TR-3273, AD 637790, August 1966.
652. S. Dontas and P. Zis, *Wien. Klin. Wochschr.* **41,** 161-3, (1928).
652a. Col. Rex Applegate, Weapons for Riot Control, *Ordnance*, **51,** 604-609 (1967).
653. H. Ellern, U.S. Patent 3,120,184 (1964).
654. André Rollet, French Patent 1,001,058 (1953).
655. F. F. Farnsworth, et al., U.S. Patent 2,962,965 (1960).
656. Robert W. Bachi, U.S. Patent 2,866,862 (1958).
657. Picatinny Arsenal, Dover, N. J., *Cellulose Nitrate-Acetate Mixed Esters*, PA-TR-3105.
657a. J. N. Godfrey and M. G. DeVries, U.S. Patent 3,260,203 (1966); *C. A.* **65,** 10418, (1966).
658. E. F. Van Artsdalen, M. Levy, and J. B. Quinlan, Frankford Arsenal, Phila., Pa., *Combustible Ammunition for Small Arms*, Rept. R-1643.
659. S. Axelrod and V. Mirko, Picatinny Arsenal, Dover, N. J., *Surveillance and Cycling of Cast Combustible Cartridge Case Formulations*, Rept. No. 3066.
660. P. Karlowicz et al., *J. Electrochem. Soc.* **108,** 659-63, (1961); *C. A.* **55,** 25253, (1961).
661. P. Karlowicz, U.S. Patent 3,017,264 (1962).
662. M. F. Murphy and B. F. Larrick, U.S. Patent 3,110,638 (1963).
663. *C. A.* **55,** 11159 (1961).

Index

C

T

U

ERRATA

MILITARY AND CIVILIAN PYROTECHNICS

Herbert Ellern

page	*line*	*should read*
8	13 from bottom	(1968)
11	11 "	three
24	5 "	cacodyl
87	14	sporadically
95	col. 4 in table	kcal/g
104	2	oxidizer of choice
127	12 from bottom	*delete* case is
149	12 "	solid chlorine
156	2	smokes
156	13	*delete* in
181	5 "	*delete* 3
208	5 in table	quarrycord
261	3 and following	and purity display variations in pyrochemical behavior that are most baffling and disturbing to the user. Such powders may cause wide varieties of burning rate or, still worse, of initiability......
274	7 from bottom	(solid
293	13 "	*for* lowered *read* raised
"	12 "	*for* raised *read* lowered
302	8 "	filings
310	13	most
318	17	dipping wax
359	Formula 42	PA-PD-254, 24μ PA-PD-253, 147μ
363	10	70, 71
370	12	charcoal, fine
393	16	used of
399	9	Field expedient
426	bottom line	D.C.(reprinted in two parts, 1961)
429	Ref. 112	666, 146
423	Ref. 7	for Brooklyn read NY·

www.ingramcontent.com/pod-product-compliance
Lightning Source LLC
Chambersburg PA
CBHW021024210326
41598CB00016B/905

* 9 7 8 0 8 2 0 6 0 3 6 4 3 *